ANNALS OF THE NEW YORK ACADEMY OF SCIENCES
Volume 624

PULMONARY EMPHYSEMA

THE RATIONALE FOR
THERAPEUTIC INTERVENTION

Edited by George Weinbaum, Ralph E. Giles, and Robert D. Krell

The New York Academy of Sciences
New York, New York
1991

616.248 WEI

Library of Congress Cataloging-in-Publication Data

Pulmonary emphysema : the rationale for therapeutic intervention /
    edited by George Weinbaum, Ralph E. Giles, and Robert D. Krell.
        p. cm. — (Annals of the New York Academy of Sciences, ISSN
    0077–8923 : v. 624)
        Result of a conference held in Lake Buena Vista, Fla., May 16–18,
    1990, sponsored by the New York Academy of Sciences and the American
    Thoracic Society.
        Includes bibliographical references and index.
        ISBN 0-89766-659-3 (cloth : alk. paper). — ISBN 0-89766-660-7
    (pbk. : alk. paper)
        1. Emphysema, Pulmonary—Pathophysiology—Congresses.
    2. Proteinase—Inhibitors—Therapeutic use—Testing—Congresses.
    I. Weinbaum, George.   II. Giles, Ralph E.   III. Krell, Robert D.
    IV. New York Academy of Sciences.   V. American Thoracic Society.
    VI. Series.
        [DNLM: 1. Protease Inhibitors—metabolism—congresses.
    2. Pulmonary Emphysema—therapy—congresses.   W1 AN616YL v. 624 /
    WF 648 P9824 1990]
    Q11.N5 vol. 624
    [RC776.E5]
    500 s—dc20
    [616.2'48]
    DNLM/DLC
    for Library of Congress   91-15612
                                                                                          CIP

Bi-Comp/PCP
Printed in the United States of America
ISBN 0-89766-659-3 (cloth)
ISBN 0-89766-660-7 (paper)
ISSN 0077-8923

ANNALS OF THE NEW YORK ACADEMY OF SCIENCES

Volume 624
May 22, 1991

PULMONARY EMPHYSEMA

THE RATIONALE FOR THERAPEUTIC INTERVENTION[a]

Editors

GEORGE WEINBAUM, RALPH E. GILES, AND ROBERT D. KRELL

Conference Organizers

ROBERT D. KRELL, GEORGE WEINBAUM, AND RALPH E. GILES

CONTENTS

[a] This volume is the result of a conference entitled Pulmonary Emphysema: The Rationale for Therapeutic Intervention, which was sponsored by the New York Academy of Sciences and the American Thoracic Society and held on May 16–18, 1990 in Lake Buena Vista, Florida.

Financial assistance was received from:

Major funder
- ICI AMERICAS INC.

Contributors
- CIBA-GEIGY CORPORATION, PHARMACEUTICALS DIVISION
- E. I. DU PONT DE NEMOURS & COMPANY
- MERCK AND COMPANY
- MILES INC.
- SMITHKLINE BEECHAM

Preface

During the last 25 years, both basic and clinical research concerning pulmonary emphysema have matured. Considerable circumstantial evidence indicates that the pathogenetic mechanism leading to emphysema involves, in some way, a protease:protease inhibitor imbalance. A number of laboratories are now involved in developing protease inhibitors with an eye to therapeutic intervention for the arrest of pulmonary emphysema. This conference was organized to evaluate whether the database available to the scientific community is sufficiently solid to justify the initiation of presumably expensive clinical trials to test the efficacy of protease inhibitors in retarding the insidious progression of lung destruction associated with emphysema. The organizers asked each participant not only to review the most recent findings in the area under discussion, but also to be as critical as possible of the underlying rationale for considering protease inhibitor therapy a viable therapeutic path.

To design approaches for the future, we have to be fully cognizant of the history and current status of genetic predisposition and the risk factors associated with pulmonary emphysema. In addition, we must be informed as to how patients with emphysema are treated at the present time. The finding that a deficiency in the major serum protease inhibitor, alpha$_1$-proteinase inhibitor (α_1-PI; alpha$_1$-antitrypsin, AAT), is related to increased susceptibility to emphysema and the identification of cigarette smoking as a major risk factor in the development of the disease have been guideposts in directing research toward establishing how smoking relates to emphysema and which protease is the key to lung destruction.

Recent findings suggest that cigarette smoking causes retention of neutrophils in the lung and that these cells contain at least two enzymes that may overwhelm localized normal proteinase inhibitor defense mechanisms and destroy lung tissue. Superimposed on a potential excess proteolytic burden to which the lungs of smokers may be subjected is the response of the host. Destruction is balanced by repair which requires both gene activation and deposition of extracellular protein in an organized fashion, permitting normal function. Organization and functional interaction of the various components of the extracellular matrix may be compromised by increased proteolysis. Whether one or more proteolytic enzymes from neutrophils or alveolar macrophages work alone or in concert or with cell- or smoke-derived oxidants is unclear. This uncertainty makes the evaluation of clinical candidates all the more important, although indeed, a potentially challenging undertaking. The use of α_1-PI as augmentation therapy to replenish a key protein that is deficient in a small cohort of smokers who show rapid onset of emphysema represents the first clinically significant step. In addition, it is clear that new compounds that selectively inhibit a relevant proteolytic enzyme and that have worked in animal models of emphysema should also be considered for testing in humans. On the other hand, drugs that prevent enzyme secretion and have not been studied in models may need significant refinement before being evaluated in patients. In contrast, some interventions, such as lung transplantation, have had good clinical success, but are not commonplace treatment.

Probably the greatest hurdle to overcome is the establishment of a simple, reproducible, inexpensive method to monitor, at weekly or monthly intervals, the putative efficacy of the administered drug, to assure that a regimen utilizing a protease inhibitor is indeed inhibiting a destructive protease in the patient's lung.

As it also appears clear that proteinases may play crucial roles in other lung diseases, the same rational discourse that was extensively provided in this conference may be the seed for future conferences to examine potential therapies for other protease-related lung diseases.

Finally, the papers that follow are directed to the issues just summarized. The three of us wish to thank all the contributors to this volume. The poster presentations were cogent and added significantly to the scope of the conference. We also thank the New York Academy of Sciences and the American Thoracic Society for their sponsorship of the conference and the various companies listed in the table of contents for their financial support. The only sad note associated with the development of this *Annal* was the absence of two gifted researchers, one a basic scientist and the other a clinician. Because of their major contributions to the field of emphysema research, we dedicate this volume to the memories of Dr. Aaron Janoff and Dr. Philip Kimbel.

GEORGE WEINBAUM
RALPH E. GILES
ROBERT D. KRELL
Editors

Emphysema Before and After 1963

Historic Perspectives[a]

STEN ERIKSSON

*Department of Medicine
Malmö General Hospital
University of Lund
S-214 01, Malmö, Sweden*

By the end of the 1950s research on emphysema had indeed become very active, but it also appeared to have reached an impasse. The disease had been fairly well characterized in clinical, physiologic, and pathologic terms, but its etiology remained obscure. The atmosphere of resignation was apparent in the J. Burns Amberson lecture presented by Dr. George W. Wright at the annual meeting of the National Tuberculosis Association and the American Thoracic Society in 1963. This lecture was later published[1] with Dr. Jerome Kleinerman, a collaborator of Dr. Wright in the experimental induction of emphysema by inhalation of the oxidant gas nitrogen dioxide, as co-author.

In his lecture Wright systematically analyzed a variety of possible etiologic agents according to the revised Koch postulates. He suggested that the causal agent should: (1) act in *all* cases of emphysema in a logical fashion, (2) be recognizable and definable as a specific thing or condition, and (3) reproduce the disease in other susceptible animals. Wright was unable to suggest a putative candidate as agent. He was most inclined at that time to believe that the injury was of a vasculonecrotic nature, leading to necrosis and destruction of tissue with faulty repair, but he concluded his lecture on a pessimistic note, stating that "The current state of knowledge will not permit any summarization that is meaningful. Obviously, we do not believe the necessary causal agent of emphysema has been established."

Wright, in his lecture, emphasized the destructive element in emphysema. It is important to realize that this feature of the disease was first defined and accepted as late as 1958. The word emphysema is derived from the Greek word for inflation and is still commonly used for a state of inflation by gas or air of any tissue of the body. But emphysema is not only inflation. It is "a condition of the lung characterized by increase beyond the normal in the size of air spaces distal to the terminal bronchiole, either from dilatation or from destruction of their walls." This definition, presented at the well-known Ciba Guest Symposium in London in 1958, was further developed in 1961 by a committee of the World Health Organization. They deleted the criterion of dilatation of air spaces. One year later the American Thoracic Society[2] excluded altogether the concept of overinflation from the definition of emphysema. Emphysema was now defined as "an anatomic alteration of the lung characterized by an abnormal enlargement of the air spaces distal to the terminal, non-respiratory bronchiole, accompanied by destructive changes of the alveolar walls." This definition is retained today with the single

[a] This work was supported by grants from Anna and Edwin Bergers' Foundation.

1

qualification that obvious fibrosis should be absent, and it is applicable to both clinical and experimental emphysema.[3]

These anatomic concepts were largely based on new, improved, and simple techniques for preparing lung specimens in the inflated state thanks to work by Heard and by Gough and Wentworth. Their fixation techniques permitted the assessment of emphysema by the naked eye or with the dissection microscope and allowed selection of appropriate blocks for histologic study.[4,5] The panacinar and centrilobular types had been defined and were easily recognized by the use of these techniques.

Despite the lack of a comprehensive hypothesis to explain the etiology of emphysema, clinical diagnosis could be made with a reasonably high degree of accuracy. In classical correlative clinicopathologic studies based on the inflation techniques just discussed, Thurlbeck[6] could demonstrate an acceptable correlation between clinically diagnosed emphysema and the severity seen at necropsy. The ability to diagnose emphysema during life depended then, as today, on the combined use of history, physical examination, a correct interpretation of chest X-ray findings, and simple lung function tests. Laws and Heard[7] in England and others had defined the criteria for evaluation of the vascular tree in emphysema in 1962. They wrote: "A reduction in the calibre of the peripheral pulmonary arteries, often with an increased transradiancy of the background, due to reduction in the vascular bed, were the most reliable radiological signs of emphysema." The most significant functional abnormalities have been defined as: (a) expiratory air flow obstruction (FV_1), (b) increased residual volume (RV), (c) reduced negative intrapleural pressure, and (d) uneven distribution of inspired air. These abnormalities were all compatible with a reduction in elastic recoil of the emphysematous lung, identified as the most specific physiologic abnormality of emphysema.[8] It is understandable that interest in the connective tissue of the lung began to increase.

Laennec had described the clinicopathologic features of emphysema in 1838, and his concepts regarding etiology came to dominate the field for more than 150 years. One main feature of his hypothesis is that there is a constant association between chronic bronchitis (catarrh) and emphysema. The other essential assumption in Laennec's hypothesis is that partial bronchial obstruction raises the pressure in the lung distal to the obstructed bronchus and that this mechanical increase eventually leads to destruction of the tissue.

In the early 1960s several authors[9] voiced concern about the validity of this hypothesis. Several lines of evidence against Laennec's hypothesis were proposed. First, Lindskog's work on collateral ventilation made it seem unlikely that a marked increase in intraalveolar pressure would occur distal to a small bronchus. Secondly, bronchial asthma is generally not associated with emphysema as shown by the postmortem studies of Gough, and finally, experimental attempts to produce emphysema by obstruction of bronchi had been unsuccessful.

Considering the key role of reduced elastic recoil pressure in emphysema, investigators logically began to look for abnormalities in connective tissue, particularly in elastin. Briscoe and Loring[10] found an increasing amount of elastin in the lung with increasing age. Pierce et al.,[11] in their study of the collagen and elastin content in emphysematous lungs, could not find any reduction in total amounts. No quantitative abnormality was detected in elastin, and biochemical methods for detecting a putative qualitative abnormality at that time were unsatisfactory. In retrospect, it could be argued that an abnormality in elastin had to be present for loss of elasticity to have occurred. It was also evident from histologic studies that disruption and disintegration of elastic fibers are a frequent finding in emphysema.[12]

Rare conditions with obviously defective connective tissue had been observed to be associated with emphysema, but there was no general agreement that hereditary factors were important in the pathogenesis of emphysema. Some clinicians thought that observations of unusual familial cases or cases with atypical early onset of emphysema would perhaps give clues to the etiology of the disease. Among several reports I will just mention two that seemed important. Wimpfheimer and Schneider[13] in 1961 reviewed the literature on familial emphysema and presented two new families. Seebohm and Bedell[14] from Iowa City presented 10 cases of "primary pulmonary emphysema in young adults." These authors emphasized that shortness of breath was the first symptom, infections and bronchitis were absent, and breath sounds on the lung bases were weak or absent. Their findings did not fit the traditional Laennec hypothesis and again pointed towards a possible inherent defect in connective tissue, but these authors could not identify any biochemical correlate.

Against this background the discovery of alpha$_1$-antitrypsin (AAT) deficiency and its association with emphysema appeared at exactly the right time. The possibility of proleolytic mechanisms gave a completely new insight into the pathogenesis of emphysema. The early history of AAT deficiency has been reviewed in detail.[15] I will briefly return to some of the initial biochemical and clinical observations and the concepts concerning pathogenesis that these patients generated.

Laurell[15] had refined and improved the paper electrophoretic technique for separation of plasma proteins. Paper electrophoresis had gained increasing popularity among clinicians as a helpful diagnostic tool in many clinical settings. When Laurell first observed the missing α_1 band on the strips, he thought that the samples had been bacterially contaminated and that neuraminidase activity was responsible for the lack of a distinct band. Neuraminidase removes terminal sialic acid residues from glycoproteins and retards their electrophoretic mobility towards the anode. But no other glycoprotein was affected. The rest of the α_1 and all other zones seemed to be normal. We could deduce from data published in 1955 by Jacobsson[15] on the partition of antitryptic activity in serum after electrophoretic separation that the missing fraction corresponded to AAT.

This glycoprotein had been isolated 1 year earlier, in 1962, by Schultze et al.[15] at the Behring Werke in Marburg, Germany. The protein had the ability to inhibit trypsin activity and was therefore named AAT. Consequently we suggested the name AAT deficiency for the clinical deficiency state. In our first paper in 1963 we also noted that AAT in deficient sera had slightly decreased mobility consistent with a structural abnormality. We know today that this change of negative net charge is the result of the glutamic acid → lysine substitution in position 342 in the molecule.

In our report[16] in 1963 we described five patients with this deficiency with varied clinical presentations. Only three had obstructive lung disease with chronic bronchitis, bronchiectasis, and emphysema as the main diagnoses. One young and one elderly woman lacked obvious signs of obstructive lung disease. We concluded, however, that an association existed between degenerative pulmonary disease and AAT deficiency. We suggested that the primary cause was an inborn error of metabolism. The hereditary nature of the condition was rapidly confirmed and the type of inheritance was apparent. The first report of a large family was published in 1964.[17] Two homozygous siblings had severe emphysema. Their children had intermediate AAT levels and were thus heterozygotes and carried only one abnormal allele.

In 1965 I had collected a series of 33 deficient homozygous probands and performed extensive family studies.[18] The probands had been traced from a variety of files and hospitals and were in no way representative of an unbiased population sample. Most patients had come to diagnosis because of lung disease, resulting in a considerable ascertainment bias. The series, however, permitted a delineation of the characteristic clinical features of this new entity. It was evident that, as in other inherited conditions, the degree of clinical manifestation in AAT deficiency was highly variable, ranging from complete absence of clinical symptoms to early onset of obstructive lung disease with severe dyspnea. In my series many patients had early "primary" emphysema and resembled in this respect the cases described by Seebohm and Bedell[14] discussed earlier. Tissue destruction was prominent and bullous transformation frequent. The basal predominance was striking. Vascular markings were lacking at the lung bases, and blood flow was redistributed towards the apices. However, patients with bronchitis and even bronchiectasis were also observed. Postmortem examination in single cases using mounted paper sections according to Gough and Wentworth's method confirmed the general clinical impression that these patients had panacinar emphysema. This assumption was first confirmed by Talamo and associates[19] in 1966. Other typical features were pronounced weight loss and lack of hypercapnia until late in the course of this particular obstructive disease.

The most important observation emerging from this series was the concept of emphysema as the *primary defect* in these patients who often had received clinical diagnoses reflecting secondary events such as bronchitis, asthmatic bronchitis, and even bronchial asthma. I focused on the obligate destructive element of emphysema when suggesting the so-called proteolytic theory for its development, at this time a concept that seemed somewhat naive and oversimplified. Two main sources of proteases, namely, neutrophils, which were known to be sequestered in the lungs, and macrophages, were suggested. The true biologic significance of AAT at that time, however, was unknown, so that any attempt to explain the exact causal mechanism of emphysema in these patients remained speculative. The proteolytic concept received unexpected support from the independent observation in 1964 by Gross et al.,[20] demonstrating the experimental production of emphysema after intratracheal instillation of papain, a plant protease with broad spectrum proteolytic activity. A new era of experimental emphysema had begun.

Gradually, but not unexpectedly, considering the well-established abnormal properties of elastic tissue in emphysema just discussed, it became apparent that neutrophil elastase was the probable key enzyme involved in the pathogenetic process. Although several groups of investigators made significant contributions toward focusing interest in this direction, only a few can be mentioned here. Kueppers and Bearn[21] connected clinical and experimental medicine when they demonstrated the ability of AAT to inhibit proteolytic enzymes from neutrophils. They emphasized a possible role for elastase in the development of emphysema. In agreement with earlier work by Heimburger and Haupt,[22] they showed that AAT was an effective inhibitor of pancreatic elastase. Janoff and Scherer[23] first demonstrated elastase activity in the human neutrophil granulae.

Turino and coworkers[24] found serum antielastase deficiency in patients with AAT deficiency. Ohlsson[25] demonstrated the high affinity of AAT for purified neutrophil elastase, and Lieberman[26] induced lung digestion *in vitro* by leukoproteases that could be inhibited by AAT. We and others emphasized the early loss of elastic recoil force in deficient patients, and Senior et al.[27] used purified human neutrophil elastase to produce emphysema in animals. However, the most convincing evidence for the key role of elastase came from the kinetic studies on the association rate between AAT and various serine proteinases. In a series of publications in the late 1970s Travis[28] and coworkers could demonstrate a 1,000-

fold more rapid rate of association between neutrophil elastase and AAT than between AAT and trypsin. The most important singular function of AAT, therefore, seemed to be to control the activity of neutrophil elastase.

A second important contribution from the biochemists was the demonstration of the oxidative inactivation of AAT. The presence of a methionine residue in the P_1' position of the inhibitory site of AAT led to a number of studies on the oxidation of the reactive site of the protein by use of chemical oxidants, myeloperoxidase, or components in cigarette smoke. Oxidation results in a 1,000-fold loss of antielastase activity and a very prolonged half-time of inhibition. These data led to the development of the "oxidation hypothesis," which attempts to explain the garden variety of emphysema, that is, that seen in smokers with normal plasma levels of AAT.

The protease-antiprotease hypothesis has certainly provided many new directions in emphysema research. It is nonetheless important to realize that most of the evidence that supports this concept is still indirect. No matter how compelling the AAT deficiency state is as a clinical model, it has been difficult to demonstrate an increased rate of elastin breakdown, elastolysis, in these or other emphysematous patients. In the production of experimental emphysema, the best results are obtained with the use of elastases, but a limitation in interpreting the results is the condensed time course of these models as compared to the human situation. Returning to Wright's[1] revised Koch postulates, elastase appears to be a good candidate for a causative agent because:

1. Elastase may well be the causal agent in *all* cases of emphysema.
2. Elastase is well defined and recognizable. Neutrophil elastase has been demonstrated anatomically by electron microscopic study of elastin fibers in the lung interstitium.[29]
3. Elastase reproduces the disease in susceptible animals.

Hopefully, the current availability of replacement therapy[30] for AAT deficiency will allow demonstration of retardation in the development of emphysema. The result of augmentation therapy would thus be a final proof of the proteolytic hypothesis.

REFERENCES

1. WRIGHT G. W. & J. KLEINERMAN. 1963. A consideration of the etiology of emphysema in terms of contemporary knowledge. Am. Rev. Respir. Dis. **88:** 605–620.
2. AMERICAN THORACIC SOCIETY. 1962. Chronic bronchitis, asthma and pulmonary emphysema. A statement by the Committee on Diagnostic Standards for Nontuberculous Respiratory Diseases. Am. Rev. Respir. Dis. **85:** 762–768.
3. SNIDER, G. L., J. KLEINERMAN, W. M. THURLBECK & Z. H. BENGALI. 1985. The definition of emphysema. Report of a National Heart, Lung, and Blood Institute, Division of Lung Diseases Workshop. Am. Rev. Respir. Dis. **132:** 182–185.
4. GOUGH, J. & J. WENTWORTH. 1960. *In* Recent Advances in Pathology, 7th ed. J & A. Churchill Ltd. London.
5. HEARD, B. E. 1960. Pathology of human emphysema. Methods of study. Am. Rev. Respir. Dis. **82:** 792–799.
6. THURLBECK, W. M. 1963. A clinico-pathological study of emphysema in an American hospital. Thorax **18:** 59–67.
7. LAWS, J. W. & B. E. HEARD. 1962. Emphysema and the chest film: A retrospective radiological and pathological study. Br. J. Radiol. **35:** 750–761.
8. STEAD, W. W., D. L. FRY & R. V. EBERT. 1952. The elastic properties of the lung in normal men and in patients with chronic pulmonary emphysema. J. Lab. Clin. Med. **40:** 674–681.

9. EBERT, R. V. & J. A. PIERCE. 1963. Pathogenesis of pulmonary emphysema. Arch. Intern. Med. **111:** 34–43.
10. BRISCOE, A. M. & W. E. LORING. 1958. Elastin content of the human lung. Proc. Soc. Exp. Biol. Med. **99:** 162–164.
11. PIERCE, J. A., J. B. HOCOTT & R. V. EBERT. 1961. The collagen and elastin content of the lung in emphysema. Ann. Intern. Med. **55:** 210–222.
12. WRIGHT, R. R. 1961. Elastic tissue of normal and emphysematous lungs. Am. J. Pathol. **39:** 355–367.
13. WIMPFHEIMER, F. & L. SCHNEIDER. 1961. Familial emphysema. Am. Rev. Respir. Dis. **83:** 697–703.
14. SEEBOHM, P. M. & G. N. BEDELL. 1963. Primary pulmonary emphysema in young adults. Am. Rev. Respir. Dis. **87:** 41–46.
15. ERIKSSON, S. 1989. Alpha₁-antitrypsin deficiency: Lessons learned from the bedside to the gene and back again. Historic perspectives. Chest **95:** 181–189.
16. LAURELL, C.-B. & S. ERIKSSON. 1963. The electrophoretic alpha₁-globulin pattern of serum in alpha₁-antitrypsin deficiency. Scand. J. Clin. Lab. Invest. **15:** 132–140.
17. ERIKSSON, S. 1964. Pulmonary emphysema and alpha₁-antitrypsin deficiency. Acta Med. Scand. **175:** 197–205.
18. ERIKSSON, S. 1965. Studies in alpha₁-antitrypsin deficiency. Acta Med. Scand. **432:** 1–85.
19. TALAMO, R. C., J. B. BLENNERHASSETT & K. F. AUSTEN. 1966. Familial emphysema and alpha₁-antitrypsin deficiency. N. Engl. J. Med. **275:** 1301–1304.
20. GROSS, P., M. BUBYAK, E. TOLKEN & M. KASHAK. 1964. Enzymatically produced pulmonary emphysema. A preliminary report. J. Occup. Med. **6:** 481–484.
21. KUEPPERS, F. & A. BEARN. 1966. A possible experimental approach of the association of hereditary alpha₁-antitrypsin deficiency and pulmonary emphysema. Proc. Soc. Exp. Biol. Med. **121:** 1207–1209.
22. HEIMBURGER, N. & H. HAUPT. 1966. Zur Spezifität der Antiproteinasen des Humanplasmas für Elastase. Klin. Wchnschr. **44:** 1196–1199.
23. JANOFF, A. & J. SCHERER. 1968. Mediators of inflammation in leukocyte lysosomes. IX. Elastinolytic activity in granules of human polymorphonuclear leukocytes. J. Exp. Med. **128:** 1137–1151.
24. TURINO, G. M., R. M. SENIOR, B. B. GARG, S. KELLER, M. M. LEVI & I. MANDL. 1969. Serum elastase inhibitor deficiency and alpha₁-antitrypsin deficiency in patients with obstructive emphysema. Science **165:** 709–711.
25. OHLSSON, K. 1971. Neutral leukocyte proteases and elastase inhibited by plasma alpha₁-antitrypsin. Scand. J. Clin. Lab. Invest. **28:** 251–253.
26. LIEBERMAN, J. 1972. Digestion of antitrypsin-deficient lung by leukoproteases. *In* Pulmonary Emphysema and Proteolysis. Mittman C, ed.: 189–203. Academic Press, London.
27. SENIOR, R. M., H. TEGNER, C. KUHN, K. OHLSSON, B. C. STARCHER & J. A. PIERCE. 1977. The induction of pulmonary emphysema with human leukocyte elastase. Am. Rev. Respir. Dis. **116:** 469–475.
28. TRAVIS, J. 1989. Alpha₁-proteinase inhibitor deficiency. *In* Lung Biology in Health and Disease. Donald Massaro, ed. **41:** 1227–1246.
29. DAMIANO, V. V., A. TSANG, V. KUCICH, W. R. ABRAMS, J. ROSENBLOOM, P. KIMBEL, M. FALLAHNEJAD & G. WEINBAUM. 1986. Immunolocalization of elastase in human emphysematous lungs. J. Clin. Invest. **78:** 482–493.
30. HUBBARD, R. C. & R. G. CRYSTAL. 1988. Alpha₁-antitrypsin augmentation therapy for alpha₁-antitrypsin deficiency. Am. J. Med. **84**(suppl 6A): 52–62.

Risk Factors Associated with Chronic Obstructive Lung Disease

MILLICENT HIGGINS

National Heart Lung and Blood Institute
Bethesda, Maryland 20892

The term "risk factor" is now a household word usually thought of in connection with coronary heart disease, the condition for which it was first used. Risk factors can also identify men and women whose probability of developing chronic obstructive lung disease (COLD) is increased. Some risk factor-disease relationships are causal, other risk factors are manifestations of early stages in the disease process, and some are descriptors of subgroups of the population among whom the incidence of disease is increased for a variety of known and unknown reasons. Thus, information about risk factors may provide insights into the etiology and pathogenesis of COLD, and it may suggest opportunities for and approaches to prevention or treatment. Before considering some of the demographic, behavioral, biologic, and environmental characteristics that have been linked to the incidence of COLD, it is necessary to consider how COLD is defined for the purpose of this report.

COLD is not defined precisely or diagnosed according to standardized criteria despite recognition by epidemiologists and clinicians of the need for such a consistent approach. Measurements of the frequency and distribution of COLD are based on information that is recorded in the course of clinical practice on medical records and death certificates or that is collected in epidemiologic or clinical studies in which questionnaires are completed and limited physical examinations and simple tests of pulmonary function are performed.

FREQUENCY AND DISTRIBUTION OF COLD

Approximately 3–4% of the 2.1 million deaths per year in the United States are currently attributed to COPD and Allied Conditions, ICD9 codes 490–496, which include asthma. A larger number of death certificates have one or more of these conditions entered as a contributory cause, so that COPD and Allied Conditions are recognized as causing or contributing to 8% of all deaths.[1,2] Although there are no national data on how often these conditions are omitted from death certificates of affected patients, this number is probably substantial.[3] The most frequent COPD diagnosis is chronic airways obstruction (ICD9 496) which accounts for 70% of the total; in decreasing order of frequency entries are emphysema (18%), asthma (6%), chronic bronchitis (5%), and other conditions (1%) (FIG. 1). The numbers of deaths, hospitalizations, and physicians' office visits for COLD (excluding asthma) are shown in TABLE 1, which also shows that 5.3 million men and 7.5 million women are estimated to have chronic bronchitis and 1.3 million men and 0.8 million women are estimated to have emphysema, based on reports to the National Health Interview Survey.[4-6] Although deficiencies in these numbers are recognized, it is clear that COLD is an important problem in the United States.

Death rates by age, race, and sex are shown in FIGURE 2. Rates rise with

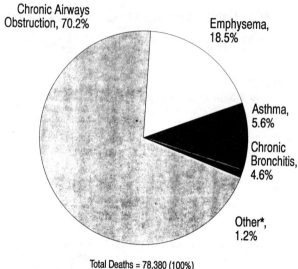

Total Deaths = 78,380 (100%)

*Bronchiectasis and Extrinsic Allergic Alveolitis

FIGURE 1. COPD and allied conditions deaths, percentage by subgroup, United States, 1987.

increasing age in all groups and rank from highest to lowest for white males, black males, white females, and black females. Old age, white race, and male sex are associated with higher death rates.

The differentials by sex and race have been apparent at least since 1960, but the sex differences have become smaller as death rates have risen more steeply in women than in men during the last 20 years (FIG. 3). In fact, death rates for COLD have leveled off or are declining in men below the age of 75 years.[7]

The objective of the National Longitudinal Mortality Study is to investigate socioeconomic, demographic, and occupational differentials in mortality. The population of approximately 1 million people of all ages was defined in the Census Bureau's survey of households; it is a representative sample of the noninstitutionalized population of the United States.[8] Deaths have been ascertained through

TABLE 1. COLD[a] Mortality and Morbidity: United States, 1987

	Men	Women
Deaths	45,000	29,000
Hospitalizations	183,000	171,000
Physicians' office visits	5.5 million	5.7 million
Prevalence: Chronic bronchitis	5.3 million	7.5 million
Emphysema	1.3 million	0.8 million
Population	118 million	125 million

[a] ICD/9 490–492, 494–496.

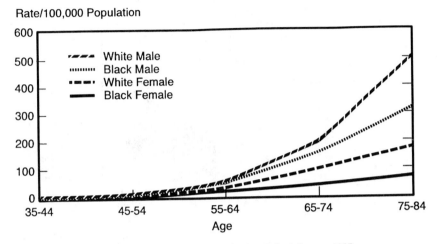

FIGURE 2. Death rates for COLD, United States, 1987.

the National Death Index for an average 3-year followup. Standardized mortality ratios (SMR) are shown in FIGURES 4 and 5 for 45–64-year-old men and women classified according to employment, income, education, marital status, and household size at enrollment. In both sexes SMRs for COPD were higher among those not in the labor force, those with family incomes less than $20,000, those with less than 12 years of education, and those who were not married or who lived alone. There are many possible explanations for these strong associations, including prevalent disease at entry to the study. Nevertheless, these data show that

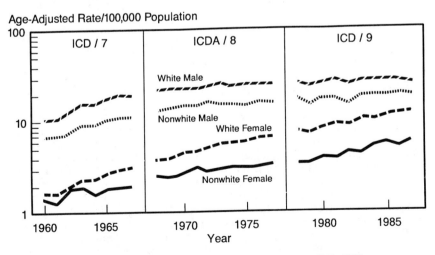

FIGURE 3. Death rates for COLD, United States, 1960–1987.

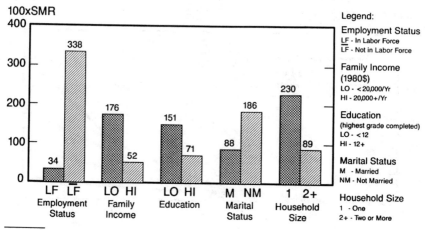

Source: NLMS, 1979-83 follow-up

FIGURE 4. COPD mortality by selected social and economic factors, white men aged 45–64.

social and economic factors identify men and women at high risk of dying of COPD. Although several of these social and economic factors are correlated, SMRs increased as the number of adverse factors increased.[8,9]

RISK FACTORS FOR COLD

In addition to the demographic and socioeconomic risk factors just described, behavioral, environmental, biologic, and familial risk factors are listed in TABLE 2.

CIGARETTE SMOKING

For over 25 years cigarette smoking has been established as the most important cause of chronic bronchitis, emphysema, and COLD. Overwhelming evidence links cigarette smoking to mortality and to measures of morbidity including diagnosed disease, respiratory symptoms, reduced pulmonary function, disability, absence because of sickness, impaired quality of life, and reduced life expectancy.[10] Mortality ratios for smokers compared with nonsmokers illustrate the strength of the association for the most serious adverse effect. Relative risks for death from emphysema and/or chronic bronchitis range from 3–15 for men in prospective epidemiologic studies in the United States and up to 25 for British physicians.[11] Relative risks and attributable risks for male and female smokers in the American Cancer Society's 25 and 50 States Studies (Cancer Prevention Studies I and II) are shown in TABLE 3. During the 6-year followup from 1959–1965, male and female smokers' death rates from COPD were 8.8 and 5.9 times higher than those of men and women who never smoked regularly. In the later time period (1982–1986) relative risks were 9.6 for male smokers and 10.5 for female smokers.

Estimates of the attributable risk of cigarette smoking depend on the relative risks of exposure and on the prevalence of exposure in the population.[11] Calculations indicate that 84% of COPD mortality in males is attributable to smoking, and this proportion did not change from 1965 to 1985; however, the number of excess deaths increased from 16,000 to 37,000 in part because of population growth and aging (TABLE 3). Among women, COPD deaths attributable to smoking made up 67% of all COPD deaths in 1965, but 79% in 1985, and this increase was due to changing smoking habits as well as to population growth and aging among women. Excess deaths among female smokers increased from 2,300 in 1965 to 20,000 in 1985 (TABLE 3 and ref. 11).

Numerous studies have shown that mortality and incidence rates increase with increasing exposure to cigarette smoke measured in terms of numbers of cigarettes smoked per day or duration of the smoking habit or the combination of the two usually expressed as pack years of exposure. Similarly, numerous studies show improvement in mortality and morbidity rates in smokers who quit compared with smokers who continued to smoke. An example of the reduction over time in COPD mortality among exsmokers is shown in TABLE 4. In the U.S. Veterans Study, exsmokers' death rates for COPD were higher than those of nonsmokers, and their mortality ratios ranged from 1.7 for veterans who formerly smoked less than 10 cigarettes per day to 6.7 for veterans who formerly smoked 40 or more cigarettes per day. Death rates were even higher among current smokers who had rates from 4–18 times those of nonsmokers depending on the number of cigarettes they smoked; on average, current smokers' death rates were over 11 times higher than those of nonsmokers. Mortality ratios of exsmokers varied according to the time since quitting. They remained elevated for 30 years but gradually declined after 10 years of cessation to levels 2–3 times those of nonsmokers. The peak in mortality ratios for COPD which occurred 5–10 years

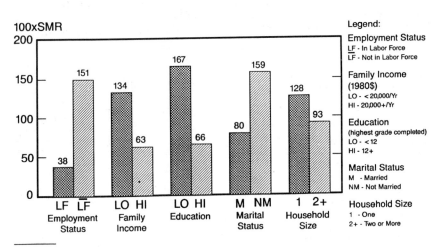

Source: NLMS, 1979-83 follow-up

FIGURE 5. COPD mortality by selected social and economic factors, white women aged 45–64.

TABLE 2. Risk Factors for COLD

Demographic factors
  Age
  Male sex
  Low socioeconomic status
Environmental and behavioral factors
  Cigarette smoke
  Occupational exposures
  Air pollution
  Infections of the respiratory tract
Genetic and constitutional factors
  Reduced pulmonary function
  Protease inhibitor phenotypes
  Bronchial hyperreactivity
  Low BMI
  Other respiratory conditions
Other familial factors
  Shared genes
  Shared environments or behaviors

after cessation may reflect smokers quitting for health reasons and experiencing even higher death rates than did continuing smokers during this period. Quitting smoking does not undo the permanent damage that occurs during the course of COPD.[12] Smoking habits were not ascertained during the followup period in the U.S. Veterans Study, and undoubtedly some exsmokers started smoking again, thus contributing to the increased mortality ratios among men who were ex-smokers at the start of the study. The effects of stopping smoking were reviewed in the Surgeon General's report (1990). Reductions in risk from the use of filter cigarettes and low tar, low nicotine cigarettes have been disappointing or absent especially in chronic airways obstruction; however, there appears to be a reduction in the prevalence of cough and the risk of lung cancer.[11]

The effects of exposure to environmental tobacco smoke on COLD are contro-

TABLE 3. COPD Mortality Rates and Risks: American Cancer Society

	Males		Females	
	CPS-I (1959–1965)	CCPS-II (1982–1986)	CPS-I (1959–1965)	CPS-II (1982–1986)
Death rate for COPD[a] in nonsmokers	9.5	8.7	4.0	4.0
Relative risks[b]				
Bronchitis + emphysema	8.8	—	5.9	—
COPD	—	9.6	—	10.5
Attributable risks				
1965		84%		67%
1985		84%		79%

[a] Age-adjusted annual death rate per 100,000 persons aged ≥35 years.
[b] Ratio of death rates for current smokers/nonsmokers.
Source: U.S. DHHS. Reducing the Health Consequences of Smoking. Publication No. (CDC) 89–8411.

versial, with some studies reporting slightly lower mean levels of pulmonary function in exposed workers or housewives and one study reporting higher death rates for chronic bronchitis, emphysema, or COPD among nonsmoking wives married to smokers. However, none of these differences was large, and some investigators did not find any differences between nonsmokers exposed and those not exposed to environmental tobacco smoke.[13]

REDUCED PULMONARY FUNCTION

Level of FEV_1 at the start of the observation period is a very important predictor of later frank chronic obstructive lung disease defined as an FEV_1 <65% of predicted and an FEV_1/FVC ratio of <80%.[12-16] In the Tecumseh Study incidence rates for males with initial levels of FEV_1 of 70%–84%, 85%–99%, and ≥100% of predicted were 14.5%, 3.1%, and 0.2%, respectively. The corresponding rates in women were: 12.6%, 2.4%, and 0.2%. Responsiveness to isopro-

TABLE 4. Mortality Ratios for COPD: U.S. Veterans Study, 1954–1980

	Smoking Status	Cigarettes/Day				
		0	<10	10–20	21–39	40+
	Nonsmoker	1.0	—	—	—	—
	Exsmoker[a]	—	1.7	3.8	5.5	6.7
	Current smoker	—	4.2	10.6	17.1	18.3

Nonsmoker	Current Smoker	Years of Cessation						
		<5	5–9	10–14	15–19	20–24	25–29	30+
1.0	11.3	9.8	13.0	10.8	5.8	4.7	2.7	2.2

[a] Ex-cigarette smokers who stopped for reasons other than doctor's orders.
Source: Rogot, E. Unpublished data.

terenol was also related to the incidence of COPD.[16] In general, those who responded most had lower FEVs initially, and those with the greatest response to isoproterenol had higher rates of developing disease. Nevertheless, in multivariate analyses FVC response to isoproterenol was a predictor of COLD in men when FEV_1% predicted was controlled in the analysis.[16]

COMPOSITE RISK FACTOR PROFILES

When the combined effects of several risk factors were evaluated in multivariate analyses, those that made clinically useful and statistically significant contributions to identifying persons at increased risk of developing COLD were: age, sex, smoking habits, and forced expiratory volume, adjusted for height.[14-17] The range of risks for 45-year-old men is illustrated in FIGURE 6. It was lowest in nonsmokers with normal FEVs and highest in smokers who smoked most cigarettes and had low FEVs. Smokers who stopped smoking during the 10-year observation period had substantially lower incidence rates of COLD than did smokers who continued to smoke (FIG. 6).

ALPHA₁-ANTITRYPSIN DEFICIENCY

Protease inhibitor phenotypes Z and null are associated with severe alpha₁-antitrypsin deficiency and emphysema at an early age. Homozygotes are rare and account for a very small fraction of cases. Interaction between Pi type and cigarette smoking has been noted in Swedish PiZ men and women.[18] Although survival was reduced for all PiZ men and women compared with all Swedish men and women, the reduction was substantially greater among PiZZ smokers than among PiZZ nonsmokers.[19] Age of onset of dyspnea was 13–15 years younger in PiZZ men and women who smoked than in PiZZ nonsmokers. Average annual rates of decline in FEV_1 were increased to above 100 ml per year in U.S. and Swedish subjects with impaired pulmonary function (FEV 30–65% of predicted).[19,20] Heterozygotes are more frequent in the population, but a hypothesis that they would be more susceptible to environmental exposures including cigarette smoke has not been supported, and heterozygotes have not been found to be at increased risk in population-based studies of defined populations. Selection bias was probably responsible for some reports of increased risk based on heterozygotes seen in clinical settings.

NHLBI is sponsoring a registry of patients with alpha₁-antitrypsin deficiency to collect information about the clinical and laboratory course of disease and to increase understanding of its natural history and factors that influence it. (See chapter by Turino.)

A major goal of this conference is to consider the rationale for therapeutic intervention on pulmonary emphysema in the context of the protease/antipro-

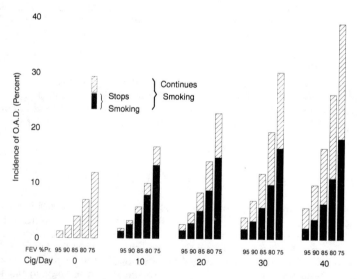

FIGURE 6. Predicted 10-year incidence of obstructive airways disease by cigarettes/day and FEV, % predicted, men aged 45.

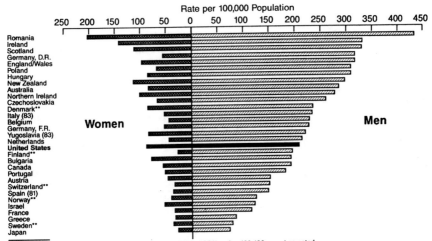

FIGURE 7. Death rates for COPD by sex, ages 65–74 for 31 countries, 1984 or latest data year.

tease hypothesis. Alpha$_1$-antitrypsin deficiency is genetically determined and therefore clusters in families, but it cannot account for the familial clustering of COLD or the familial resemblance in ventilatory lung function that has been found in several population-based epidemiologic studies. Similarly, the fact that members of families adopt similar smoking habits contributes to but does not fully account for familial aggregation because it is present among nonsmokers and persists when smoking habits are controlled. Additional research is needed to identify genetic markers and candidate genes and to evaluate interactions among genes, behaviors, and environmental exposures including infections, air pollution, and occupational hazards.

In summary, I have emphasized the major role of cigarette smoking as a cause of COLD and referred to evidence that smokers at greatest risk are those with alpha$_1$-antitrypsin-deficient phenotypes, with reduced pulmonary function, respiratory symptoms and illnesses, or positive family histories of COLD. But we should also recognize that heart disease mortality is 10 times higher than COPD mortality and that cigarette smoking and low pulmonary function are associated with increased mortality from coronary heart disease, lung cancer, and death from all causes combined.[21–23] Therapeutic interventions on smoking should be especially vigorous in persons with other major risk factors including elevated serum cholesterol levels, elevated blood pressure, diabetes, and obesity. Declining mortality from cardiovascular disease and aging of the population will lead to increased prevalence and mortality from COPD in the United States unless preventive and therapeutic strategies are implemented. The appreciably lower death rates from COPD in Scandinavian and Southern European countries shown in FIGURE 7 are an incentive to further efforts in the United States. Recent downward trends in mortality at younger ages provide further encouragement for strengthening efforts to prevent and treat COLD in the United States.

SUMMARY

Chronic obstructive lung disease (COLD) includes emphysema, chronic airways obstruction, and chronic bronchitis but not asthma. COLD mortality and morbidity rates are higher in men than women, in whites than blacks, and they increase with increasing age and with decreasing socioeconomic status. Death rates for COLD are approximately 10 times higher in cigarette smokers than in nonsmokers, and about 80% of COLD deaths in the United States are attributable to cigarette smoking. Among smokers, the quantity and duration of cigarette use are strongly related to mortality and morbidity, but susceptibility varies among individuals. Incidence of COLD is related inversely to pulmonary function and possibly to increased bronchial reactivity. Alpha$_1$-antiprotease deficiency, respiratory infections and symptoms, lean body build, and exposures to occupational hazards or air pollution have been associated with increased risks of COLD in epidemiologic and clinical studies. Susceptible individuals can reduce their risks of clinical disease by stopping smoking. Clinical trials are underway to determine whether bronchodilators reduce the rates of decline in pulmonary function.

REFERENCES

1. National Center for Health Statistics, Vital Statistics of the United States, 1987, Vol. II, Mortality, Part A. Washington, Public Health Service, 1990.
2. National Center for Health Statistics. Public Use Data Tape Documentation. Multiple Cause of Death for ICD9, 1986 Data. U.S. Government Printing Office, No. 216–475/36370, 1988.
3. HIGGINS, M. W. & J. B. KELLER. 1989. Trends in COPD morbidity and mortality in Tecumseh, Michigan. Am. Rev. Respir. Dis. **140:** S42–8.
4. GRAVES, E. J. 1989. Detailed Diagnoses and Procedures, National Hospital Discharge Survey, 1987. National Center for Health Statistics. Vital and Health Statistics, Series 13, No. 100.
5. Unpublished data from the National Ambulatory Medical Care Survey, 1985. National Center for Health Statistics. Personal communication, 1990.
6. SHOENBORN, C. A. & M. MARINO. 1988. Current Estimates from the National Health Interview Survey; United States, 1987. National Center for Health Statistics. Vital and Health Statistics, Series 10, No. 166.
7. HIGGINS, M. W. & T. J. THOM. Incidence, prevalence, and mortality: Intra- and intercountry differences. 1989. In Clinical Epidemiology of Chronic Obstructive Pulmonary Disease. M. J. Hensley & N. A. Saunders, eds.: 23–43. Marcel Dekker, Inc. New York.
8. ROGOT, E., P. D. SORLIE, N. J. JOHNSON, C. S. GLOVER & D. W. TREASURE. 1988. A Mortality Study of One Million Persons by Demographic, Social, and Economic Factors: 1979–1981 Follow-up. U.S. Longitudinal Mortality Study. NIH Publication No. 88–2896.
9. ROGOT, E. Personal communication.
10. The Health Consequences of Smoking. Chronic Obstructive Lung Disease: A Report of the Surgeon General, US Rockville, MD, Department of Health and Human Services, Office on Smoking and Health, 1984.
11. U.S. Department of Health and Human Services. 1989. Reducing the Health Consequences of Smoking: 25 Years of Progress. A Report of the Surgeon General. U.S. Department of Health and Human Services, Public Health Services, Centers for Disease Control, Center for Chronic Disease Prevention and Health Promotion, Office on Smoking and Health. DHHS Publication No. (CDC) 89–8411.
12. FLETCHER, C., R. PETO, C. TINKER et al. 1976. The Natural History of Chronic Bronchitis and Emphysema. Oxford University Press. Oxford.

13. U.S. Department of Health and Human Services. 1986. The Health Consequences of Involuntary Smoking. A Report of the Surgeon General. U.S. Department of Health and Human Services, Public Health Services, Centers for Disease Control, Center for Chronic Disease Prevention and Health Promotion, Office on Smoking and Health. DHHS Publication.

14. HIGGINS, M. W., J. B. KELLER, M. BECKER, W. HOWATT, J. R. LANDIS, H. ROTMAN, J. G. WEG & I. HIGGINS. 1982. An index of risk for obstructive airways disease. Am. Rev. Respir. Dis. **125:** 144–151.

15. HIGGINS, M. W., J. B. KELLER, J. R. LANDIS, T. H. BEATY, B. BURROWS, D. DEMETS, J. E. DIEM, I. T. T. HIGGINS, E. LAKATOS, M. D. LEBOWITZ, H. MENKES, F. E. SPEIZER, I. A. TAGER & H. WEILL. 1984. Risk of chronic obstructive pulmonary disease. Collaborative assessment of the validity of the Tecumseh index of risk. Am. Rev. Respir. Dis. **130:** 380–385.

16. HIGGINS, M. 1984. Epidemiology of COPD. State of the art. Chest **85S:** 3S–8S.

17. HIGGINS, M. W. & J. B. KELLER. 1983. Estimating your patient's risk of COPD. J. Respir. Dis. **4:** 97–108.

18. LARSSON, C. 1978. Natural history and life expectancy in severe alpha$_1$-antitrypsin deficiency, PiZZ. Acta Med. Scand. **204:** 345.

19. WU, M. & S. ERIKKSON. 1988. Lung function, smoking and survival in severe alpha$_1$-antitrypsin deficiency, PiZZ. J. Clin. Epidemiol. **41:** 1157–1165.

20. BUIST, A. S., B. BURROWS, S. ERIKSSON, C. MITTMAN & M. WU. 1983. The natural history of air-flow obstruction in PiZ emphysema: Report of a NHLBI Workshop. Am. Rev. Respir. Dis. **237:** S43–S45.

21. HIGGINS, M. W. & J. B. KELLER. 1970. Predictors of mortality in the adult population of Tecumseh. Respiratory symptoms, chronic respiratory disease, and ventilatory lung function. Arch. Environ. Health **21:** 418–424.

22. BEATY, T. H., B. H. COHEN, C. A. NEVILL, H. A. MENKES, E. L. DIAMOND & C. J. CHEN. 1982. Impaired pulmonary function as a risk factor for mortality. Am. J. Epidemiol. **116:** 102–113.

23. PETO, R., F. E. SPEIZER, A. L. COCHRANE, F. MOORE, C. M. FLETCHER, C. M. TINKER, I. T. T. HIGGINS, R. G. GRAY, S. M. RICHARDS, J. GILLILAND & B. NORMAN-SMITH. 1983. The relevance in adults of air-flow obstruction, but not of mucus hypersecretion, to mortality from chronic lung disease: Results from 20 years of prospective observation. Am. Rev. Respir. Dis. **128:** 491–500.

Natural History and Clinical Management of Emphysema in Patients with and without Alpha₁-Antitrypsin Inhibitor Deficiency

GERARD M. TURINO

Department of Medicine
St. Luke's-Roosevelt Hospital Center
John H. Keating, Sr., Professor of Medicine
Columbia University College of Physicians & Surgeons
New York, New York 10025

DEFINITIONS

The entity of pulmonary emphysema as a destructive lesion of the lung parenchyma has been known for over 150 years; however, only in the last 25 years have new insights into the possible cellular and chemical mechanisms underlying this destructive process been achieved. The first definitive description of pulmonary emphysema in human lungs is attributed to Dr. Matthew Baillie, a physician and pathologist in both London and Edinburgh, who published his work in the latter part of the eighteenth century in a volume entitled, "Morbid Anatomy of Some of the Most Important Parts of the Human Body."[1] In a section entitled "Air Cells of the Lungs Enlarged," he stated that "the lungs are sometimes although I believe very rarely formed into pretty large cells so as to resemble somewhat the lungs of an amphibious animal. Of this I have now seen three instances. Enlargement of the cells cannot well be supposed to arise from any other cause than the air being not allowed to common free egress from the lungs and therefore accumulating in them. It is not improbable also that this accumulation may sometimes breakdown two or three contiguous cells into one and thereby form a cell of a very large size." This was a remarkably accurate description of destruction of alveoli, septal tissue, and enlargement of alveolar airspaces in pulmonary emphysema.

Conceptually, destruction of the lungs is the major feature of pulmonary emphysema and is found in two forms of chronic obstructive airway disease: (1) the rather pure emphysema that may begin with dyspnea only as the major symptom without evidence of chronic bronchitis; and (2) chronic bronchitis itself which was defined by the World Health Organization in 1959 and whose definition still stands: "A condition of subjects with chronic or recurrent excess mucus secretion into the bronchial tree occurring on most days for at least three months of the year for at least two successive years."[2] In contrast, pulmonary emphysema has been defined as enlargement and destruction of the alveolar containing portions of the lung without significant tissue fibrosis.[3] Chronic obstructive pulmonary disease (COPD) may be defined as a process characterized by the presence of chronic bronchitis or emphysema that may lead to the development of airway obstruction. On the basis of National Health Interview Surveys[4] it is estimated that approximately 10 million Americans suffer from COPD, of which 7.5 million have chronic bronchitis and 2.5 million have emphysema. COPD is responsible for over 60,000 deaths per year and comprises the fifth largest cause of death in the United States. In countries outside the United States and especially the Far East the prevalence of COPD may actually be higher.

The two other major forms of airway obstructive disease that are seen clinically are bronchial asthma and cystic fibrosis of the pancreas or mucoviscidosis. Both of these latter conditions involve primarily bronchial tissue and do not demonstrate generalized emphysema throughout the lung. Although we recognize that there are pure forms of chronic bronchitis leading to marked airway obstruction and pure forms of pulmonary emphysema, most patients present with some features of both and clinically have been designated as having COPD. Patients therefore may complain of dyspnea and some degree of cough and sputum. As the disease advances, in addition to airway obstruction, there will be hypoxemia, hypercapnia, and in sustained hypoxemia cor pulmonale. As a result of the destruction of alveoli throughout the lung in pulmonary emphysema, a loss of tissue support of airways occurs with severe collapse under forced expiration and in advanced disease even under normal expiration.

Much effort has been devoted to describing the types of alveolar destruction with respect to which portion of the pulmonary lobule is affected, and central lobular emphysema has been distinguished from the panacinar form in which the entire acinus undergoes destruction. However, both forms have been described in the same lungs.[5] Lungs of smokers who develop emphysema seem more prone to the central lobular form of emphysema, whereas patients with alpha₁-antitrypsin deficiency usually demonstrate the panacinar form.

NATURAL HISTORY OF COPD

Throughout the 1960s and 1970s a number of studies described the clinical manifestations of patients with COPD, and long-term followup studies were performed to ascertain the natural history of the disease with respect to airway obstruction and survival as well as some of the parameters of lung function. Most of these studies indicated a markedly accelerated death rate in COPD once the FEV_1 began to diminish below 1.2 liters. Thus, as shown by Burrows and Earle,[6] the percentage of survival of individuals with FEV_1 below 0.75 liters was approximately 20% in 6 years. Those with a maximum voluntary ventilation below 30% of predicted also demonstrated approximately 20% survival at the end of 7 years. Similarly, if the arterial PCO_2 exceeded 52 mm Hg, survival was less than 10% at the end of 6 years, and if the diffusing capacity was less than 10 ml/mm Hg, survival was approximately 10% at 7 years (FIG. 1). Similar results by Boushy *et al.*[7] also demonstrated that the decrease in FEV_1 per year in COPD was in the range of 100–200 ml, whereas the normal individual has a decrease of approximately 20 ml. Boushy *et al.*[7] demonstrated that a high pulmonary vascular resistance also was associated with reduced survival, indicating that the onset of cor pulmonale was a bad prognostic sign.

One of the few clinical and morphologic correlations in COPD was made by Nagai *et al.*[8] who studied 48 patients 53–74 years of age who were followed on intermittent positive pressure breathing and on whom autopsy examinations were performed. Ninety-two percent of these individuals had chronic bronchitis and all were current or past smokers. This study demonstrated that the severity of emphysema correlated positively with a decreased FEV_1, decreased body weight, right ventricular hypertrophy, and increased residual volume. The severity of emphysema correlated negatively with the bronchodilator response, variability of expiratory air flow, and the amount of bronchial smooth muscle. Increases in bronchial smooth muscle were positively correlated with an increased FEV_1, an increased PaO_2, a decreased $PaCO_2$, and a decreased right ventricular size.

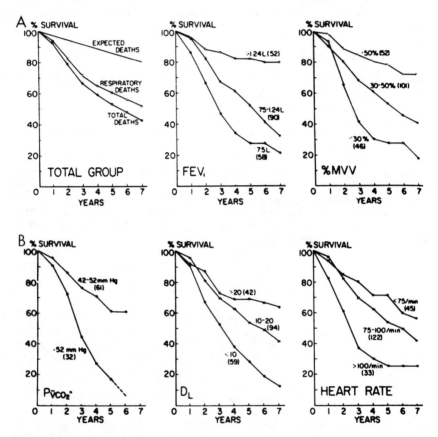

FIGURE 1A and B. Survival curves calculated by life-table method for various groups of patients and related to certain measurements of cardiac and pulmonary function. Characteristics defining the group are on the graphs. The numbers of patients in each group are in parentheses. (Reproduced, with permission, from Burrows and Earle.[6])

However, although these variations and their manifestations of COPD occurred in this population and all were smokers, it is not apparent as to what exactly created the difference in the severity of the disease or the various characteristics of the anatomic abnormalities in relation to the clinical abnormalities. Individual responses to smoking effects in this population were therefore quite variable and probably dependent on additional factors related to lung disease rather than smoking alone.

Thus far, pulmonary emphysema has been associated with alpha₁-antitrypsin deficiency in individuals with the severest levels of deficiency, usually below 50 mg% in the serum. The deficiency states in which associations with emphysema are clear-cut are the Z homozygote and null or null Z phenotypes, and even in these groups there is an association of the severity of emphysema with the intensity of cigarette smoking. The natural history of patients with alpha₁-antitrypsin deficiency of the Z phenotype shows a loss of FEV_1 of approximately 100 ml per year with a much reduced mean survival when compared with the normal population.[9,10]

DEVELOPMENT OF THE PROTEASE-ANTIPROTEASE CONCEPT IN EMPHYSEMA

By the early 1960s the approach to understanding the destruction of pulmonary parenchyma in emphysema began to focus on biochemical and cellular mechanisms, and this rapid change in approach occurred because of experimental observations in the United States and clinical observations in Sweden.

Investigation of proteases such as elastases, collagenases, and hyaluronidase began as a means of understanding the role of matrix constituents in pulmonary mechanics.[11,12] It could be demonstrated that intratracheal or intravenous administration of pancreatic elastase resulted in the mechanical changes of increased distensibility and the apparent destruction of alveolar tissue. This was demonstrated quite clearly for papain by Gross et al.[13] in 1964.

At the same time in Sweden certain families with alpha₁-antitrypsin deficiency were recognized to have an increased susceptibility to pulmonary emphysema of the panacinar type.[14,15] Emerging from these two different types of observations, one in the laboratory and one in the clinic, came the concept of an imbalance between proteases capable of degrading pulmonary parenchymal connective tissue and the inhibition of proteases to prevent this destruction. As work proceeded, there also was a focus on elastases because the administration of collagenase to the lungs in vivo did not result in alveolar destruction, and in fact in some instances there was a decrease rather than an increase in lung compliance.[16] Along with this focus on elastin came the demonstration that in experimental animals exposed to elastases or papain, parenchymal elastin was reduced as anatomic emphysema developed. In a series from our own laboratory,[17] eight patients were studied physiologically during life and their lungs studied for biochemical measurements of elastin at autopsy or following pulmonary surgery. At least five of these individuals had a statistically significant decrease in elastin content in the pulmonary parenchyma measured by analysis of elastin content as a proportion of crude connective tissue by identifying desmosines and isodesmosines[18] (FIG. 2). Also, in experimental animals a decrease in pulmonary elastin was demonstrated after elastase-induced or papain-induced emphysema.[19]

It also could be demonstrated that in experimental animals elastin is capable of resynthesis after degradation, but despite resynthesis there was no reconstitution of alveolar architecture.[20] These studies served to reinforce the concept that elastin degradation was an important component of the alveolar destruction in emphysema, and that once destruction of elastin had occurred, the anatomic and physiologic abnormalities might remain because of an inability to restore elastin continuity.

Clinical and anatomic studies have clearly indicated that exposure to tobacco

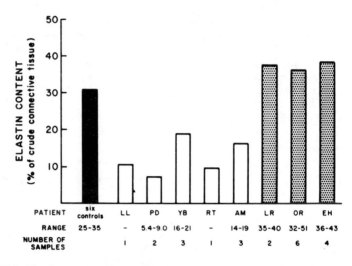

FIGURE 2. Elastin content of lung parenchyma expressed as a percentage of crude connective tissue in eight patients with anatomically diagnosed pulmonary emphysema compared with the lungs of six normal control subjects. The height of the bar for each patient is the mean value from analysis of the number of samples indicated. (Reproduced, with permission, from Chrzanowski et al.[17])

smoke is an important etiologic element in pulmonary emphysema in individuals who are not deficient in alpha$_1$-antitrypsin concentration in the serum.[21] Moreover, individuals who had inherited the ZZ type deficiency had an acceleration in the development of emphysema when they smoked.[22]

The role of exposure to tobacco smoke in defining the mechanisms leading to pulmonary emphysema has been elucidated in two directions. Studies by Johnson and Travis[23] as well as Janoff et al.[24] demonstrated that oxidation of the methionine-active center in the alpha$_1$-antitrypsin molecule (position 358) results in a loss of inhibitory function of the protein for elastase. Thus, individuals exposed to tobacco smoke could inactivate the alpha$_1$-antitrypsin in the alveolar lining and thereby increase the elastase burden on alveolar septal tissue. Also, in our own laboratory Osman et al.[25] showed that hamsters given intratracheal elastase and then exposed to tobacco smoke for 1 week developed emphysema of greater severity and were unable to resynthesize elastin to the same degree as that of animals given elastase but not exposed to tobacco smoke. It was also demonstrated in these studies that the lysyl oxidase activity in the lungs of these animals was significantly reduced, thus lowering elastin resynthesis. Lysyl oxidase is required for both elastin and collagen cross-linking. In the usual clinical presentation of emphysema, patients who are exposed to tobacco smoke may have a functional deficiency of alpha$_1$-antitrypsin in the alveolar lining fluid as a result of oxidative mechanisms acting on the methionine active center. A smaller group of patients with clinical emphysema, perhaps only 1–2%, have an inherited deficiency of alpha$_1$-antitrypsin in blood and alveolar lining fluid (FIG. 3).

ELASTASES THAT AFFECT THE LUNG

Although much has been learned of the inherited variations in alpha₁-antitrypsin structure and function, much more needs to be learned of the elastases that reach the lung through neutrophils and macrophages. Experimental studies over the past two decades have identified the elastases that may play a role in lung elastin degradation *in vivo*. Neutrophil elastase is a serine protease with a molecular weight of 33,000 and contains 18–20% carbohydrate. It is stored in azurophilic granules in amounts of approximately 3 pg per cell.[26] Neutrophil elastase is discharged into tissues when the cell encounters objects to be phagocytized and after cells die.[27] It is noteworthy that although neutrophil elastase degrades the amorphous component of elastin,[28] it also degrades core proteins of proteoglycan molecules in connective tissue ground substance,[29] collagen types 3 and 4,[30] as well as fibronectin.[31]

Thus far in understanding the physiologic role of neutrophil elastase it seems that its predominant function is bactericidal[32,33] and bacterial digestion.[34]

Human monocytes also contain an elastase[35] that is biochemically and antigenically closely related to neutrophil elastase. Monocyte elastase is found predominantly on the surface of a monocyte in the plasma membrane[36,37] and is present to the extent of about 3% of that in neutrophils. Monocyte elastase can degrade amorphous elastin, fibronectin, and serum amyloid A.[36] In culture, monocytes replace the serine protease with a metalloenzyme that contains a zinc atom at the catalytic site.[38]

Much of the work on macrophage elastase has been done in murine macrophages. Murine peritoneal[39] and alveolar[40] macrophages synthesize a metalloelastase that has a molecular weight of 22,000. Mouse macrophage elastase is not inhibited by alpha₁-antitrypsin and is capable of digesting that inhibitor.[41] Al-

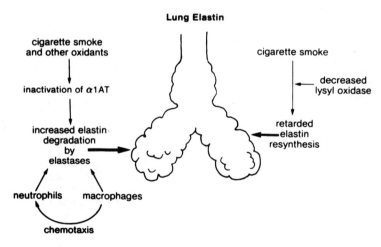

FIGURE 3. A schema demonstrating the combined effects of exposure to tobacco smoke to reduce alpha₁-antitrypsin inhibitory function by oxidants (*left side of figure*) and decreased resynthesis of elastin by decreasing lysyl oxidase activity (*right side of figure*). (See text.)

though alveolar macrophages contain approximately $1/200$th of the amount of elastase present in human neutrophils, it could play a significant role in the degradation of pulmonary elastin because of the large numbers of macrophages in pulmonary tissue. Although the metalloproteinase of alveolar macrophages is not inhibited by serum alpha$_1$-antitrypsin, other inhibitors of this proteinase may be present within the macrophage.

We need to better understand the individual variations among the normal population as well as patients with emphysema with respect to the concentration and composition of elastases in the cells. In this regard some studies have indicated increased concentrations of neutrophil elastase in patients with COPD.[42] We also have much to learn about the behavior of elastases elaborated *in situ* in lung tissue with respect to those critical factors that determine the protease inhibitor interactions that result in alveolar septal breakdown. This conference will focus on some of these factors related to the mechanisms of emphysema.

QUESTIONS ON THE NATURAL HISTORY OF COPD AND THERAPY

With respect to patients with phenotypic abnormalities of alpha$_1$-antitrypsin, thus far only the Z homozygous form has been definitively associated with pulmonary emphysema. Within that group, not all individuals with Z type alpha$_1$-antitrypsin deficiency have emphysema because approximately 60% of individuals with the Z phenotype have an association with emphysema if they are nonsmokers.[43]

One of the intriguing aspects about the natural history of emphysema is the steady decline in pulmonary function that occurs in patients with COPD in terms of loss of FEV_1, reductions in diffusing capacity, and abnormalities in gas exchange. Once the process has begun, certain factors are brought to bear that result in a steady decline in lung function that far exceeds that in the normal aging population.

In experimental animals also, some progression of emphysema has been demonstrated. Snider and Sherter[44] showed progressive increase in lung volume measured as total lung capacity at 25 cm of water inflation pressure as well as an increase in lung compliance observed in up to 52 weeks of followup of these animals, with most of the increase occurring by the 26th week after a single injection of elastase intratracheally.

Thus, both the natural history of emphysema in human subjects and that of elastase-induced emphysema indicate that once the process of destruction is initiated, there is a progression and perpetuation of the emphysematous process through a series of mechanisms that as yet are not well understood.

In this regard, some recent observations of Oldmixon *et al.*[45] have shown very well that elastin is deployed in heavy cables at the junction points of alveolar septae and that these areas of heavy elastin deposition may be tension-bearing structures for the alveolar septal walls that in themselves have relatively reduced amounts of elastin and therefore depend on surface forces to maintain their configuration. It is therefore conceivable that the lung, by virtue of the deployment of elastin in concentrated zones, may be especially vulnerable to losses of alveolar architecture once these reinforced cable areas undergo degradation and disruption (FIG. 4).

Lastly, I would like to comment on two areas of study in pulmonary emphysema and COPD that may provide significant information on the effects of therapy. Two of the more important therapeutic measures in COPD have been cessation of smoking and the use of bronchodilator agents for reducing airway obstruction. A multicentered clinical trial is underway with three arms of study and involving 6,000 patients who have early manifestations of airway obstruction.[46] The first arm, called special intervention, involves smoking intervention, but it uses a placebo inhaler. There is a followup for maintenance of the program every 4 months and an annual evaluation for 5 years. The second arm involves smoking intervention but active bronchodilatation with the agent Atrovent. The third arm is usual care with no special efforts toward smoking intervention or

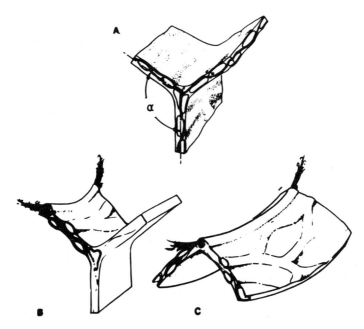

FIGURE 4. Sketches of various types of septal borders and junctions between septal walls. A connective tissue cable runs within borders reinforcing it and by its tension and curvature exerting tensions counterbalancing septal tensions. These septal cables are high in elastin content. (Reproduced, with permission, from Oldmixon *et al.*[45])

active use of a bronchodilator. There are 10 participating clinical centers in this study as well as a data collection center. The 6,000 participants are between 35 and 59 years of age, are cigarette smokers, but are free of serious disease. This study has the prospect of evaluating the effects on the disease of effective smoking cessation as well as the role of bronchodilatation over a followup period of 5 years. Until this point there has not been a well-controlled evaluation of the long-term effects of bronchodilatation in determining the rate of progression of COPD.

A second study on a national scale concerns a national registry for patients with ZZ phenotype alpha₁-antitrypsin deficiency.[47] Approximately 23 centers throughout the country have been established under the auspices of the Lung

Division of the National Heart, Lung and Blood Institute. Patients are evaluated in a standardized protocol of pulmonary function testing, historical characteristics, and chest X-rays and are being followed in the registry whether or not they wish to have alpha$_1$-antitrypsin replacement. Medical providers are accepting the expense of such therapy. Approximately 400 patients are already registered. Some interesting observations are already being demonstrated with respect to the frequency of reversible airway obstruction as a presenting manifestation in approximately 30–40% of such individuals. This observation raises the prospect that a significant proportion of patients with alpha$_1$-antitrypsin deficiency are presenting with asthmatic manifestations.

The primary objective of the registry is to characterize the clinical course of severe alpha$_1$-antitrypsin deficiency whether or not the patient is receiving augmentation therapy. The registry, because of its inherent study design, is not a clinical trial to evaluate the efficacy of alpha$_1$-antitrypsin augmentation therapy, but rather a mechanism to collect and analyze clinical data useful for understanding the natural history of this deficiency in individual patients. Some primary outcomes will be the rate of decline in FEV$_1$ as well as the rate of mortality. Some secondary outcomes are the clinical status of the patients and the functional status with respect to modification of symptoms such as dyspnea.

Thus, in reviewing the status of our understanding of the development of COPD and of emphysema it seems most likely that the variations in individual clinical presentation, pulmonary morphology, and clinical course suggest a role for multiple risk factors in the natural history of the disease. In the past, smoking has clearly been associated with the severity and progression of COPD, but interest in the role of airway hyperreactivity is now increased in both COPD associated with smoking as well as pulmonary emphysema associated with alpha$_1$-antitrypsin deficiency. The level of alpha$_1$-antitrypsin itself may be a factor particularly when smoking is involved, and it still remains a possibility that individuals who have intermediate levels of alpha$_1$-antitrypsin in serum but have a history of heavy smoking may be at a higher risk for developing pulmonary emphysema. It is also becoming evident that other genetic factors may play a role in the severity of COPD in certain families that have a history of COPD. It is noteworthy that approximately 15–18% of smokers develop COPD, suggesting that certain individuals have a predisposition to develop COPD. A number of factors may come together to create an increased risk for the development of COPD in any given individual once the risk factors are present (FIG. 5). In this regard, childhood

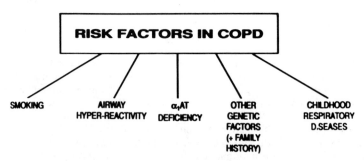

FIGURE 5. The course and severity of COPD in the individual patient may be the consequences of multiple risk factors rather than single mechanisms. (See text.)

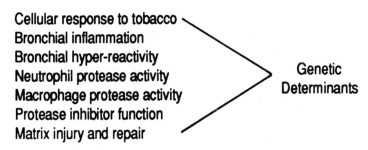

FIGURE 6. The various forms of tissue injury in chronic obstructive lung disease may be genetically determined and ultimately determine the natural history of the individual patient. (See text.)

respiratory diseases, which in many instances may be viral infections such as influenza and respiratory syncytial virus in infancy, may alter the respiratory tract structurally to increase its susceptibility to hyperreactivity or poor growth.

In the final analysis, when one looks at the several pathogenic variables present in the development of COPD, it is conceivable that each of these pathogenic mechanisms may have a genetic determinant in terms of its severity and the role it plays in the ultimate course of the patient with COPD (Fig. 6). Some of these factors are the cellular response to tobacco, the tendency toward bronchial inflammation, bronchial hyperreactivity, neutrophil protease activity, macrophage protease activity, protease inhibitor function, and factors of matrix injury and repair. Ultimately the rate of tissue repair in the first phases of tissue injury may be an important factor in maintaining anatomic integrity of the pulmonary parenchyma. By understanding these factors and some of the genetic determinants we may begin to understand why significant proportions of the population can smoke and yet not develop COPD and a significant proportion of individuals who have homozygous deficiency of alpha₁-antitrypsin have not yet developed significant evidence of pulmonary emphysema or airway obstruction. As methods to increase our insights into cellular and chemical mechanisms related to tissue destruction and bronchial hyperreactivity develop, we may begin to understand more fully the development of COPD and pulmonary emphysema in human beings.

REFERENCES

1. BAILLIE, M. 1807. The Morbid Anatomy of Some of the Most Important Parts of the Human Body, 3rd Ed. W. Bulmer & Co. London.
2. CIBA FOUNDATION GUEST SYMPOSIUM. 1959. Terminology, definitions and classification of chronic pulmonary emphysema and related conditions. Thorax **14:** 286–299.
3. SNIDER, G. L., J. KLEINERMAN, W. M. THURLBECK & Z. BENGALI. 1985. The definition of emphysema: Report of a National Heart, Lung and Blood Institute, Division of Lung Diseases, Workshop. Am. Rev. Respir. Dis. **132:** 182–185.
4. NATIONAL CENTER FOR HEALTH STATISTICS. 1985. Current estimates from the national health review survey. United States. Series 10, No. 160.
5. THURLBECK, W. M. 1983. Overview of the pathology of pulmonary emphysema in the human. Clin. Chest Med. **4:** 337–350.

6. .Burrows, B. & R. H. Earle. 1969. Course and prognosis of chronic obstructive lung disease. N. Engl. J. Med. **280:** 397–404.
7. Boushy, S. F., H. K. Thompson, L. B. North, A. R. Beale & T. R. Snow. 1973. Prognosis in chronic obstructive pulmonary disease. Am. Rev. Respir. Dis. **108:** 1373–1383.
8. Nagai, A., W. W. West, J. L. Paul & W. M. Thurlbeck. 1985. The National Institutes of Health intermittent positive pressure breathing trial: Pathology studies. II. Correlation between morphological findings, clinical findings and evidence of expiratory air-flow obstruction. Am. Rev. Respir. Dis. **132:** 946–953.
9. Larsson, C. 1978. Natural history and life expectancy in severe alpha$_1$-antitrypsin deficiency PiZ. Acta Med. Scand. **204:** 345–351.
10. Buist, A. S., B. Burrows, S. Eriksson, C. Mittman & M. Wu. 1983. The natural history of air-flow obstruction in PiZ emphysema. Am. Rev. Respir. Dis. **127**(suppl): S43–S45.
11. Turino, G. M., R. V. Lourenco & G. H. McCracken. 1964. The role of connective tissues in pulmonary mechanics. Presented at National Meeting of American Society for Clinical Investigation. J. Clin. Invest. **43:** 1297.
12. Turino, G. M., R. V. Lourenco & G. H. McCracken. 1968. The role of connective tissues in large pulmonary airways. J. Appl. Physiol. **25:** 645–653.
13. Gross, P., M. Babjak, E. Tolker & M. Kaschak. 1964. Enzymatically produced pulmonary emphysema: A preliminary report. J. Occup. Med. **6:** 481–484.
14. Laurell, C. B. & S. Eriksson. 1963. The electrophoretic alpha-1-globulin pattern of serum in alpha-1 antitrypsin deficiency. Scand. J. Clin. Lab. Invest. **15:** 132–140.
15. Erickson, S. 1964. Pulmonary emphysema and alpha$_1$-antitrypsin deficiency. Acta Med. Scand. **175**(FASCL): 197–205.
16. Johanson, W. G., Jr. & A. K. Pierce. 1972. Effects of elastase, collagenase and papain on structure and function of rat lungs in vitro. J. Clin. Invest. **51:** 288–293.
17. Chrzanowski, P., S. Keller, J. Cerreta, I. Mandl & G. M. Turino. 1980. Elastin content of normal emphysematous lung parenchyma. Am. J. Med. **69:** 351–359.
18. Keller, S., I. Mandl & G. M. Turino. 1981. Determination of the relative amounts of elastin in lung tissue. Biochem. Med. **25:** 74–80.
19. Turino, G M., S. Keller, P. Chrzanowski, M. Osman, J. Cerreta & I. Mandl. 1980. Lung elastin content in normal and emphysematous lungs. Bull. Eur. de Physiopath. Respir. **16**(Suppl.): 43–56.
20. Osman, M., S. Keller, J. M. Cerreta, P. Leuenberger, I. Mandl & G. M. Turino. 1980. Effect of papain induced emphysema on canine pulmonary elastin. Proc. Soc. Exp. Biol. Med. **164:** 471–477.
21. Auerbach, O., E. C. Hammond, L. Garfinkel & C. Benante. 1972. Relation of smoking and age to emphysema: Whole lung section study. N. Engl. J. Med. **286:** 853–857.
22. Tobin, M. J., P. J. L. Cook & D. C. S. Hutchinson. 1983. Alpha$_1$-antitrypsin deficiency. The clinical features and physiological features of pulmonary emphysema in subjects homozygous for Pi type Z. A survey of the British Thoracic Association. Br. J. Dis. Chest **77:** 14–27.
23. Johnson, D. & J. Travis. 1979. The oxidative inactivation of human alpha-1-proteinase inhibitor: Further evidence for methionine at the reactive center. J. Biol. Chem. **254:** 4022–4026.
24. Janoff, A., H. Carp, D. K. Lee & R. T. Drew. 1979. Cigarette smoke inhalation decreases alpha-1 antitrypsin activity in rat lung. Science **206:** 1313–1314.
25. Osman, M., J. O. Cantor, S. Roffman, S. Keller, G. M. Turino & I. Mandl. 1985. Cigarette smoke inhalation impairs elastin resynthesis in hamsters with elastase-induced emphysema. Am. Rev. Respir. Dis. **132:** 640–643.
26. Ohlsson, K. & I. Olsson. 1974. The neutral proteases of human granulocytes: Isolation and partial characterization of granulocyte elastase. Eur. J. Biochem. **42:** 519–527.
27. Weissman, G., J. E. Smolen & H. M. Korchak. 1980. Release of inflammatory mediators from stimulated neutrophils. N. Engl. J. Med. **330:** 27–34.

28. JANOFF, A. & J. SCHERER. 1968. Mediators of inflammation in leukocyte lysosomes. IX. Elastinolytic activity in granules of human polymorphonuclear leukocytes. J. Exp. Med. **128:** 1137–1155.
29. KAISER, H., R. A. GREENWALD, G. FEINSTEIN & A. JANOFF. 1976. Degradation of cartilage proteoglycan by human leukocyte granule neutral proteases; a model of joint injury. II. Degradation of isolated bovine nasal cartilage proteoglycan. J. Clin. Invest. **57:** 625–632.
30. MAINARDI, C. L., D. L. HASTY, J. M. SAYER & A. H. KANG. 1980. Specific cleavage of human type III collagen by human polymorphonuclear leukocyte elastase. J. Biol. Chem. **255:** 12006–12010.
31. MCDONALD, J. A. & D. G. KELLEY. 1980. Degradation of fibronectin by human leukocyte elastase. Release of biologically active fragments. J. Biol. Chem. **255:** 8848–8858.
32. ODEBERG, H. & I. OLSSON. 1976. Microbicidal mechanisms of human granulocytes: Synergistic effects of granulocyte elastase and myeloperoxidase or chymotrypsin-like cationic protein. Infect. Immunol. **14:** 1276–1283.
33. REST, R. F., S. H. FISCHER, Z. Z. INGHAM & J. F. JONES. 1982. Interactions of Neisseria gonorrhoeae with human neutrophils: Effects of serum and gonococcal opacity on phagocyte killing chemiluminescence. Infect. Immunol. **36:** 737–744.
34. BLONDIN, J., A. JANOFF & J. C. POWERS. 1978. Digestion of E. coli proteins by human neutrophil elastase and chymotrypsin-like enzyme (cathepsin G): Experiments with a cell-free system and living leukocytes. *In* Neutral Proteases of Human Polymorphonuclear Leukocytes. K. Havemann & A. Janoff, eds.: 39–55. Urban & Schwartzenberg. Baltimore/Munich.
35. SENIOR, R. M., E. J. CAMPBELL, J. A. LANDIS, R. R. COX, C. KUHN & H. S. KOREN. 1982. Elastase of U-937 monocyte-like cells. Comparisons with elastases derived from human monocytes and neutrophils and murine macrophage-like cells. J. Clin. Invest. **69:** 384–393.
36. LAVI, G., D. ZUCKER-FRANKLIN & E. C. FRANKLIN. 1980. Elastase-type proteases on the surface of human blood monocytes: Possible role in amyloid formation. J. Immunol. **125:** 175–180.
37. ZUCKER-FRANKLIN, D., G. LAVIE & E. C. FRANKLIN. 1981. Demonstration of membrane-bound proteolytic activity on the surface of mononuclear leukocytes. J. Histochem. Cytochem. **29:** 451–456.
38. SANDHAUS, R. A., K. M. MCCARTHY, R. A. MUSSON & P. M. HENSON. 1983. Elastolytic proteinases of the human macrophage. Chest **83**(suppl): S60–S62.
39. WERB, Z. & S. GORDON. 1975. Elastase secretion by stimulated macrophages. J. Exp. Med. **142:** 361–377.
40. WHITE, R. R., H. S. LIN & C. KUHN. 1977. Elastase secretion by peritoneal exudative and alveolar macrophages. J. Exp. Med. **146:** 802–808.
41. BANDA, M. J., E. J. CLARK & Z. WERB. 1980. Limited proteolysis by macrophage elastase inactivates human alpha 1-proteinase inhibitor. J. Exp. Med. **152:** 1563–1570.
42. RODRIGUEZ, J. R., J. E. SEALS, A. RADIN, J. S. LIN, I. MANDL & G. M. TURINO. 1979. Neutrophil lysosomal elastase activity in normal subjects and in patients with chronic obstructive lung disease. Am. Rev. Respir. Dis. **119:** 409–417.
43. HUTCHINSON, J. H. 1988. Symposium on Alpha₁-Antitrypsin deficiency. Am. J. Med. **84**(suppl 6A): 3–12.
44. SNIDER, G. L. & C. B. SHERTER. 1977. A one year study of the evolution of elastase induced emphysema in hamsters. J. Appl. Physiol. **43:** 721–729.
45. OLDMIXON, E. H., J. P. BUTLER & F. G. HOPPIN, JR. 1989. Lengths and topology of alveolar septal borders. J. Appl. Physiol. **67:** 1930–1940.
46. ANTHONISEN, N. R. 1989. Lung health study. Am. Rev. Respir. Dis. **140:** 871–873.
47. BUIST, A. S., B. BURROWS, A. COHEN, R. CRYSTAL, R. FALLAT, J. GADEK & G. M. TURINO. 1989. Guidelines for the approach to the patient with severe hereditary Alpha₁-Antitrypsin deficiency. Am. Rev. Respir. Dis. **140:** 1494–1497.

Neutrophil Kinetics in Normal and Emphysematous Regions of Human Lungs[a]

M. L. BRUMWELL,[b] W. MacNEE,[c] C. M. DOERSCHUK,[d]
B. WIGGS,[d] AND J. C. HOGG[d,e]

[b]Department of Thoracic Surgery
University of British Columbia and St. Paul's Hospital
Vancouver, British Columbia

[d]University of British Columbia Pulmonary Research Laboratory
St. Paul's Hospital
Vancouver, British Columbia

Epidemiologic studies[1,2] have established that centrilobular emphysema (CLE) is associated with the cigarette smoking habit, and a popular hypothesis to explain this lung destruction is that smoking causes a functional protease imbalance in the peripheral lung.[3,4] The central role of the neutrophil (PMN) in this imbalance is supported by several lines of evidence. Postmortem studies of sudden death from nonpulmonary causes have shown that PMNs accumulate in the airspace and tissue of the respiratory bronchioles of cigarette smokers.[5] The products of cigarette smoke activate PMNs *in vitro*,[4] causing them to release proteolytic enzymes from their lysosomal granules and generating oxygen-derived free radicals via the PMN membrane NADPH oxidase system. The free radicals generated by the PMN act with others present in the cigarette smoke to oxidize the active site of the major plasma protease inhibitor (α_1-PI). This reduces the inhibitor's ability to bind and inactivate elastase by a factor of 2000,[6] shifting the normal proteolysis-antiproteolysis balance in a direction that favors lung destruction.

Until recently, experiments designed to test the hypothesis that a functional protease imbalance causes CLE were based on bronchoalveolar lavage. This technique provides cells from the airway and alveolar surface, but it does not sample the lung microvascular space. This means that the large pool of marginated PMNs present in the lung microvessels[7] is not considered. Recent studies from our laboratory[8] have shown that the presence of smoke in the airspaces delays and possibly activates the PMNs within the lung microvascular marginated pool. The present study further examines neutrophil kinetics in human lungs

[a] This work was supported by MRC MT-4219 and the British Columbia Lung Association (C.M.D. and B. W.) and the George Saxton Fellowship from the British Columbia Lung Association (to M.L.B.). W. MacNee is a recipient of the Dorothy Temple Cross British MRC Travelling Fellowship. C. M. Doerschuk is a recipient of a Parker B. Francis Fellowship in Pulmonary Research.

[c] Presently Senior Lecturer in Respiratory Medicine, University of Edinburgh, Edinburgh, Scotland.

[e] Address for correspondence: James C. Hogg, M.D., UBC Pulmonary Research Laboratory, St. Paul's Hospital, 1081 Burrard Street, Vancouver, B.C., Canada V6Z 1Y6.

removed at surgery to determine if there were differences in neutrophil kinetics between normal and emphysematous regions of the same lung. The techniques employed were adapted from those previously described in animal experiments[9-12] that have been modified for use in the operating theater on patients undergoing lung resection for cancer.[13]

METHODS

The study was based on seven patients who required lung resection for bronchogenic carcinoma and was approved by the Human Experimentation Committees of the University of British Columbia and St. Paul's Hospital. After the diagnosis of lung cancer had been made and surgical treatment selected, the protocol was explained to the patients by a physician who was not involved in the study.

Three hours before surgery, a total of 100 ml of blood was withdrawn from an antecubital vein. Four milliliters of blood was used to determine the white blood cell count and differential, and the remainder was used for neutrophil preparation. All solutions and materials used to harvest, isolate, and label neutrophils were sterile and pyrogen free, and the procedures were performed in a laminar flow hood (Envirco model ESC, Albuquerque, New Mexico).

CELL PREPARATION

Erythrocytes (RBCs)

Twenty-five milliliters of whole blood in 10 ml of ACD (Frosst, Kirkland, Quebec) was incubated with 400–600 μCi of ^{51}Cr for 30–40 minutes. Thereafter, 100 mg of ascorbic acid (Roche, Etobicoke, Ontario) was added for 5 minutes, and the blood was centrifuged at 600 g for 10 minutes. The plasma was removed and the RBCs were resuspended to the original blood volume with 0.9% saline solution. Labeling efficiency was 92–96%. The hematocrit of the labeled RBCs (^{51}Cr-RBC) was determined, and a sample was retained for gamma counting. The final RBC injectate volume was about 18 ml in each case.

Neutrophils (PMNs)

Forty milliliters of whole blood anticoagulated with 5 ml of ACD was centrifuged at 600 g for 10 minutes. The plasma layer was aspirated and centrifuged at 1,300 g for a further 10 minutes, and the resulting platelet-poor plasma (PPP) layer was saved.

A percoll density gradient was prepared by adding PPP to 90% percoll (Pharmacia, Uppsala, Sweden) in 0.9% saline solution to obtain percoll concentrations of 50% and 42%. Two milliliters of the 50% percoll were overlaid with 2 ml of the 42% solution.

Thirty milliliters of whole blood in 4 ml of ACD was gently combined with 12–15 ml of Gentran 70 (Dextran 70, MW 75,900, Travenol, Mississauga, Ontario) in 0.9% saline solution and allowed to settle for 40–60 minutes. The resulting leukocyte-rich plasma (approximately 25 ml) was removed, washed with 10 ml of

PMN buffer, and centrifuged at 290 g for 10 minutes. The cell pellet was resuspended in 4 ml of PPP, overlaid above the percoll gradient, and then spun for 10 minutes at 290 g. The mononuclear cells and platelets remained at the top of the gradient and the PMNs descended to a wide layer between the 42% and 50% percoll. The PMN layer was removed, washed with 10 ml of PMN buffer, and centrifuged at 290 g for 10 minutes. Contaminating RBCs were lysed by adding 20–22 ml of 0.2% hypotonic saline solution for 30 seconds and restoring the tonicity with a similar volume of 1.6% saline solution. After centrifugation at 290 g for 10 minutes, the cell pellet was incubated with 300–800 μCi of ^{111}In in 1–1.5 ml PMN buffer for 10 minutes. PPP was added to take up any free indium, and after centrifugation the cell pellet was resuspended in 3–4 ml PPP. The indium-labeled (^{111}In-PMN) PMN cell count was $1.16–5.27 \times 10^8$, cell purity was $98 \pm 2\%$, and labeling efficiency was 62.7–85.3%. The number of ^{111}In cpm/ml injectate was determined by weight.

INTRAOPERATIVE STUDIES

After the patient was anesthetized, an arterial line was placed in the radial artery and a thermodilution Swan-Ganz catheter was positioned in the main pulmonary artery (PA) by threading it up from a peripheral vein. Repeated measurements (average 8, range 6–12) of heart rate, systemic and pulmonary arterial blood pressure, and cardiac output were made during the course of the operation. Surgery proceeded until the blood supply to the lobe(s) to be resected was isolated and ready for ligation. The lung was then reexpanded, and both lungs mechanically ventilated at a tidal volume of 100 ml/kg and a frequency of 10 breaths per minute.

A bolus of ^{99m}Tc-labeled macroaggregated albumin (MAA) was slowly injected into the right atrial port of the Swan-Ganz catheter as a marker of pulmonary blood flow. A withdrawal pump (Harvard Laboratories) was used to obtain a reference flow of blood from the main pulmonary artery at a constant rate over a period of 1 minute during this injection. The ^{51}Cr-RBCs were injected into the right atrium as a marker of blood volume followed by the ^{111}In-PMNs. Ten minutes after the MAA injection, a radial artery blood sample (end-blood sample) was obtained and the pulmonary vessels were ligated.

STUDIES OF THE RESECTED LUNG

The lung specimen was immediately inflated with 7% glutaraldehyde in order to fix the blood into a solid and prevent it from leaking out of the vessels when the lung was cut. After approximately 24 hours of fixation, the specimen was sliced sagittally (usually into three slices) so that the lateral slice was the uppermost at the time of resection. The middle slice was saved, and samples of the tumor, resection margins, and lymph nodes were taken to allow preparation of an accurate surgical pathology report. The remaining lung (usually one lateral and one medial slice) was divided into pieces, and the presence or absence of gross emphysema was recorded for each sample before it was placed into preweighed scintillation vials. These lung samples, the end-blood sample, the reference blood flow, and samples of the labeled neutrophil and erythrocyte injectates were placed in a gamma counter (Beckman Model 7000). The counts for ^{99m}Tc, ^{51}Cr, and ^{111}In were obtained, corrected for decay and overlap, and used in the following calculations.

Calculations

Transit Time

The transit time (seconds) in each lung sample was determined using a method that has been fully described elsewhere.[10–13] Briefly, the blood volume in each lung sample was calculated by dividing the ^{51}Cr counts in each sample by the ^{51}Cr counts per milliliter in the end-blood sample, and the blood flow was determined by dividing the lodged ^{99m}Tc-MAA counts in the sample by an appropriate reference flow. Both the lung sample and reference flow values were corrected for MAAs that did not lodge in the pulmonary vasculature by assuming that the ratio of "free" ^{99m}Tc-MAA : ^{51}Cr-RBC counts per milliliter of end-blood was constant throughout the vascular bed. The "free" ^{99m}Tc-MAA counts in each lung sample were calculated as the product of this ratio and the ^{51}Cr-RBC counts in the lung sample. These were subtracted from the total ^{99m}Tc-MAA counts in each sample to obtain the "lodged" ^{99m}Tc-MAA counts. The ^{99m}Tc-MAA counts per milliliter per second in the reference flow were multiplied by the cardiac output to obtain the total ^{99m}Tc-MAA counts delivered to the lungs. The ^{99m}Tc-MAA counts that lodged in the lung were calculated by subtracting the product of the ^{99m}Tc-MAA counts per milliliter in the end-blood, and the total blood volume[13] from the delivered counts. The MAA counts that failed to lodge ranged from 1.5–4.9% of the MAA counts in the reference flow in the seven patients studied.

Neutrophil Retention

The marginated ^{111}In-PMN counts were calculated by subtracting the circulating counts from the total ^{111}In-PMN counts in the sample. The circulating intravascular ^{111}In-PMN counts in each lung sample were determined by dividing the ratio of ^{111}In-PMN : ^{51}Cr-RBC counts per milliliter in the end-blood by ^{51}Cr-RBC counts in the sample. The PMN counts delivered to each lung piece were the total ^{111}In counts injected multiplied by the fraction of the total pulmonary blood flow that went to the piece. The PMN retention (%) was the marginated PMN counts divided by the delivered PMN counts in each piece.

RESULTS

Table 1 shows the data concerning sex, age, smoking history, preoperative pulmonary function studies, intraoperative hemodynamics, surgical procedure, and estimate of emphysema in the resected lung for each patient. Because a frozen section taken during the operation showed tumor at the resection margin of the right lower lobe of patient 7, the right upper and middle lobes were also resected. This second specimen became available 30 minutes after the injection of labeled PMNs and erythrocytes.

TABLE 1. Data on Seven Patients Who Underwent Lung Resection

Patient No.	Sex & Age (yr)	Smoking (cigs/yr)	Smoking (yr)	% Predicted			Correct FEV$_1$	FEV$_1$/FVC	DCO	HR	BP	PAP	CO (ml/sec)	Lung Resected	Score
				TLC	FRC	RV									
1	F, 67	840	0	111	128	142	80	63	85	76.5	121/64	26/22	117	Rlung	20
2	F, 71	588	2	118	133	155	100	71	—	71.3	131/69	30/15	58	LLL	0
3	M, 68	960	3	99	109	123	93	78	57	58.5	120/73	37/18	97	LUL	2
4	M, 64	1,440	7	105	123	100	110	70	49	75	119/75	21/14	104	RLL	40
5	M, 58	1,360	10	111	113	99	115	71	81	70	114/56	27/12	125	RUL	37
6	M, 67	2,520	9	101	128	138	89	80	59	71.3	95/63	28/17	57	RUL	15
7	M, 67	2,200	0	103	139	131	71	57	85	75.5	120/70	30/15	113	RLL,RML +RUL	25
Mean	66.0	1,415.4	4.4	106.9	124.7	126.9	94	70	69.3	71.2	117±11	30±6	96±28		
SD	4.1	714.4	4.2	6.7	10.7	21.1	15.7	8	15.9	6.1	67±7	16±3			

ABBREVIATIONS: TLC = total lung capacity; FRC = functional residual capacity; RV = residual volume; FEV$_1$ = forced expiratory volume in one second; FEV$_1$/FVC = forced expiratory volume in one second/forced vital capacity %; DCO = diffusing capacity % predicted; HR = heart rate; PAP = pulmonary artery pressure; CO = cardiac output; RUL = right upper lobe; LLL = left lower lobe, etc.

Erythrocyte Transit Time

The frequency distribution of total lung transit times in the lung samples without gross emphysema is compared to those with gross emphysema in FIG-URE 1. The mean transit time in nonemphysematous lung regions was 4.6 ± 0.2 seconds when all lung samples (*n* = 380) were considered. The transit times were similar (4.9 ± 0.5 seconds) in the 62 samples with gross emphysema.

The mean transit time was 5.6 ± 0.4 seconds through the upper region samples (*n* = 156) and 4.1 ± 0.3 seconds through the lower lung region samples (*n* = 224) (*p* <0.05). Using the mean value for all the samples for each of seven patients, the overall transit time was 5.5 ± 1.4 seconds, the upper lung regions had an average transit time of 6.6 ± 1.8 seconds, and the lower lung regions had a transit time of 4.7 ± 1.3 seconds. All but one patient showed longer transit times in the upper region.

In patient 7 the mean transit time in the right lower lobe, which was removed first, was 3.9 ± 0.4 seconds, and this value was closely similar to that observed in the right upper and middle lobes (4.1 ± 0.42 seconds) which were removed after the surgeon was told tumor was present at the initial resection line.

PMN Retention

In nonemphysematous lung, the mean PMN retention in the lung at 10 minutes after the labeled cells were injected was 26.0 ± 1.0% when all the lung samples (*n* = 380) were pooled and 27.8 ± 5.4% when the mean values for each of the seven patients were considered. The patients' mean PMN retention was 32.0 ±

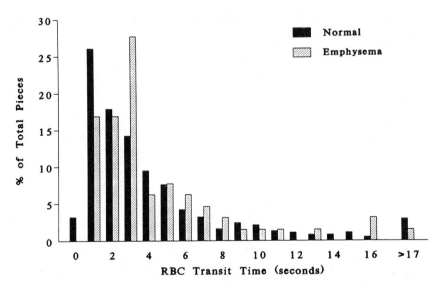

FIGURE 1. Frequency distribution of measured regional transit times of 378 samples from seven subjects without emphysema and 64 samples from four subjects with emphysema. There is no difference between the two groups.

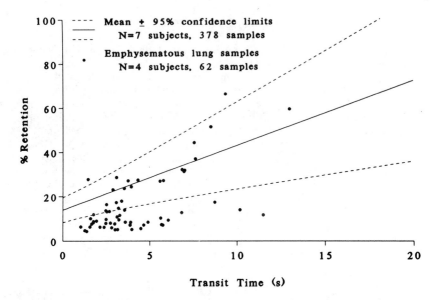

FIGURE 2. Mean ± 95% confidence limits for the relationship between transit times and neutrophil retention in 378 samples of nonemphysematous lung from seven subjects. The 62 emphysematous samples from four subjects are shown as individual data points. Analysis of group data and those from individual experiments showed no differences between emphysematous and nonemphysematous regions.

6% in upper lung regions and 25 ± 6% in lower lung regions, and a clear difference between upper and lower regions was present in five of seven patients. The mean PMN retention was 29.9 ± 1.5% in all upper region samples ($n = 156$) and 23.2 ± 1.3% in all lower lung region samples ($n = 224$) ($p < 0.05$).

The mean ± 95% confidence limits of the regression of PMN retention against transit time for the nonemphysematous regions of all seven lungs is compared to the individual points for the samples containing macroscopic emphysema in FIGURE 2. This shows no evidence of excess PMN retention in the emphysematous as compared to the normal regions when all lung samples are considered. A separate analysis comparing normal and emphysematous regions of individual cases showed no difference between normal and diseased regions of the same lung.

DISCUSSION

The pulmonary microvasculature is well known to be the major site of PMN margination within the vascular system.[7] Previous studies in humans[8] and in animals[11,12] have suggested that PMNs are delayed with respect to RBCs in the capillary bed because the PMNs require more time than do RBCs to deform to the diameter of capillary segments that are narrower than they are. The multisegmental nature of the lung capillary bed allows RBCs to stream around segments filled with slower moving PMNs which concentrates them with respect to RBCs and accounts for a large marginated pool of PMNs in the lung.[7,11]

Previous work from our laboratory has shown a mean pulmonary RBC transit time of 1.4 seconds in the rabbit,[12] 2.9 seconds in the dog,[10] 2.8 seconds in swine,[14] and 4.9 seconds in man,[8] with all four studies showing shorter transit times in the more gravity-dependent regions of the lungs. The results from the present study compare favorably with those from a previous study in humans[13] with mean transit times of 4.6 vs 4.9 seconds, and median transit times of 3.1 vs 3.5 seconds, respectively. Similarly, comparison of the regional transit times shows upper lung transit time was 5.6 vs 5.5 seconds and lower lung transit times was 4.0 vs 3.8 seconds, respectively. This regional difference in transit time is attributed to dilatation of capillaries in the dependent regions of the lung[15] which has a greater effect on blood flow than on blood volume.

The percentage of delivered neutrophils remaining in the pulmonary vasculature 10 minutes after they are injected has been shown to correlate with RBC transit time in the animal studies[12,16] and in awake human volunteers.[13] The present study confirms this relationship and shows that the data obtained in these resected lungs are closely similar to those in previous reports (FIG. 3). Patient 7 is of interest in this regard because two lobes were resected 20 minutes apart. The mean transit time remained constant over the time (3.9 at 10 minutes vs 4.0 at 30 minutes), whereas PMN retention showed a 43% decrease in the time period between resections. This decrease is consistent with the washout rate previously calculated in the studies of human volunteers.[8]

FIGURE 2 shows that PMN retention was not enhanced in lung samples when gross emphysema was present, suggesting that the lung remaining around regions of destruction has normal neutrophil traffic under the conditions of this study. This result is of interest because of our previous work on awake human volun-

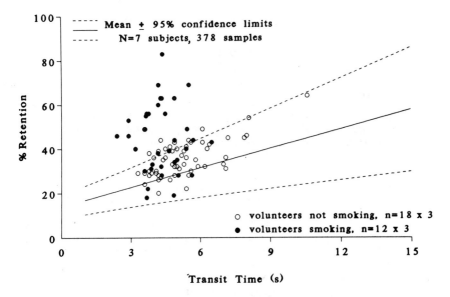

FIGURE 3. The present work is compared to that previously done on awake human volunteers in reference 8. Note the good agreement between this study and the study of volunteers who were not smoking (*open circles*) at the time they were studied. However, some of the volunteers who smoked during the experiment (*closed circles*) showed excess neutrophil retention.

teers[13] (FIG. 3) and more recent studies in rabbits[17] in which active cigarette smoking delayed PMNs in the lung microvascular bed. This delay would be expected to occur in the upper regions of the lung where work in several species[11,12,14,16] has shown greater PMN retention and longer RBC transit times because the upper lung has the greatest discrepancy between PMN and capillary size.[15] Therefore, the present and previous studies[8,17] are consistent with the hypothesis that cigarette smoke may activate the PMN and narrow the pulmonary capillary bed to increase the marginated PMN pool.

In a recent review of the mechanism of tissue destruction by neutrophils, Weiss[18] has argued that the membrane NADPH oxidase and the myeloperoxidase

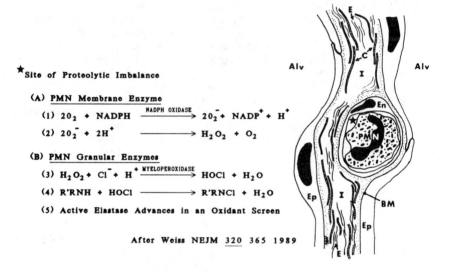

★Site of Proteolytic Imbalance

(A) PMN Membrane Enzyme

$$(1)\ 2O_2 + NADPH \xrightarrow{NADPH\ OXIDASE} 2O_2^- + NADP^+ + H^+$$

$$(2)\ 2O_2^- + 2H^+ \longrightarrow H_2O_2 + O_2$$

(B) PMN Granular Enzymes

$$(3)\ H_2O_2 + Cl^- + H^+ \xrightarrow{MYELOPEROXIDASE} HOCl + H_2O$$

$$(4)\ R'RNH + HOCl \longrightarrow R'RNCl + H_2O$$

(5) Active Elastase Advances in an Oxidant Screen

After Weiss NEJM 320 365 1989

FIGURE 4. We postulate that the proteolytic imbalance responsible for emphysema is based on neutrophils delayed in the microvessels by cigarette smoke. We speculate that these neutrophils are activated and generate an oxidant screen by a combination of the NADPH membrane oxygenase and granular myeloperoxidase. This oxidant screen inactivates α_1-PI and allows active proteolytic enzymes to diffuse the short distance from the PMN to the elastin and collagen in the alveolar wall.

released from the granules of activated PMN act in common to provide an oxidant screen for proteolytic enzymes such as elastase to advance into the tissue. We speculate (FIG. 4) that this mechanism could be critically important in explaining how neutrophils delayed in the marginating pool by cigarette smoke might destroy the alveolar wall. Whether or not this hypothesis is correct must be determined by future experiments.

ACKNOWLEDGMENTS

The authors wish to acknowledge B. A. Martin, Dr. W. Doll from the Department of Anaesthesia, and Dr. H. Ling and Dr. A. Gerein from the Department of Surgery for assistance with these experiments.

REFERENCES

1. FLETCHER, C., R. PETO, C. TINKER & F. E. SPEIZER. 1976. An 8-year study of early chronic obstructive lung disease in working men in London. *In* The Natural History of Chronic Bronchitis and Emphysema. Oxford University Press.
2. U.S. Department of Health and Human Services. 1984. Chronic Obstructive Lung Disease. Health Consequences of Smoking. A Report of the Surgeon General. Rockville, MD: US Government Printing Offices. Public Health Service Publication No. 84–50205.
3. GADEK, J. E., G. A. FELS & R. G. CRYSTAL. 1979. Cigarette smoking induces a functional anti-protease deficiency in the lower respiratory tract of humans. Science **206:** 1315–1316.
4. JANOFF, A. 1983. Biochemical links between cigarette smoking and pulmonary emphysema. J. Appl. Physiol. **55:** 285–293.
5. NIEWOEHNER, D. E., J. KLEINERMAN & D. B. RICE. 1974. Pathologic changes in the peripheral airways of young cigarette smokers. N. Engl. J. Med. **291:** 755–758.
6. JOHNSON, D. & J. TRAVIS. 1979. The oxidative inactivation of human alpha-1 antiprotease inhibitor. J. Biol. Chem. **254:** 4022–4026.
7. HOGG, J. C. 1987. Neutrophil kinetics and lung injury. Physiol. Rev. **67:** 1249–1295.
8. MACNEE, W., B. WIGGS, A. S. BELZBERG & J. C. HOGG. 1989. Effect of cigarette smoking on neutrophil kinetics in human lung. N. Engl. J. Med. **321:** 924–928.
9. MARTIN, B. A., J. L. WRIGHT, H. V. THOMMASEN & J. C. HOGG. 1982. The effect of pulmonary blood flow on the exchange between circulating and marginating pools of polymorphonuclear leukocytes in dog lungs. J. Clin. Invest. **69:** 1277–1288.
10. HOGG, J. C., B. A. MARTIN, S. LEE & T. MCLEAN. 1985. Regional differences in erythrocyte transit in normal lungs. J. Appl. Physiol. **59:** 1266–1271.
11. HOGG, J. C., T. MACLEAN, B. A. MARTIN & B. WIGGS. 1988. Erythrocyte transit and neutrophil concentration in the lung. J. Appl. Physiol. **65:** 1217–1225.
12. DOERSCHUK, C. M., M. F. ALLARD, B. A. MARTIN, A. MACKENZIE, A. P. AUTOR & J. C. HOGG. 1987. Marginated pool of neutrophils in rabbit lung. J. Appl. Physiol. **63:** 1806–1815.
13. MACNEE, W., B. A. MARTIN, B. R. WIGGS, A. S. BELZBERG & J. C. HOGG. 1989. Regional pulmonary transit times in humans. J. Appl. Physiol. **66:** 844–850.
14. OHGAMI, M., C. M. DOERSCHUK, D. ENGLISH, P. M. DODEK & J. C. HOGG. 1989. Kinetics of radiolabelled neutrophils in swine. J. Appl. Physiol. **66:** 1881–1885.
15. GLAZIER, J. B., J. M. B. HUGHES, J. E. MALONEY & J. B. WEST. 1969. Measurements of capillary dimensions and blood volume in rapidly frozen lungs. J. Appl. Physiol. **26:** 65–76.
16. MARTIN, B. A., B. R. WIGGS, S. LEE & J. C. HOGG. 1987. Regional differences in neutrophil margination in dog lungs. J. Appl. Physiol. **63:** 1253–1261.
17. BOSKEN, C. H., C. M. DOERSCHUK, D. ENGLISH & J. C. HOGG. 1989. The effect of acute cigarette smoke exposure on neutrophil transit through the rabbit lung (Abstr. #R-529). Clin. Invest. Med. **12:** B87.
18. WEISS, S. J. 1989. Tissue destruction by neutrophils. N. Engl. J. Med. **320:** 365–376.

Nonsteroidal Antiinflammatory Drugs

STEVEN B. ABRAMSON

Department of Rheumatology
Hospital for Joint Diseases Orthopaedic Institute and
Department of Medicine
N.Y.U. School of Medicine
New York, New York 10003

Nonsteroidal antiinflammatory drugs (NSAIDs) have antiinflammatory, analgesic, and antipyretic properties. They are among the most highly prescribed groups of therapeutic agents in the United States. This review outlines the mechanisms of action of the nonsteroidal drugs.

NSAIDs derive much of their antiinflammatory properties from their capacity to inhibit the biosynthesis of prostaglandins. Prostaglandins are released in response to injury and have the capacity to provoke vasodilatation, erythema, and hyperalgesia. However, it is well established that NSAIDs have multiple biologic effects that may provide alternative mechanisms by which they reduce the inflammatory response. Several recent articles review the variety of effects that NSAIDs exert on the inflammatory process.[1–3]

EFFECTS OF NSAIDs ON ARACHIDONIC ACID METABOLISM

The ability of NSAIDs to inhibit the cyclooxygenase enzyme, combined with the known proinflammatory properties of prostaglandins such as PGE_2 and PGI_2, has led to the conclusion that the antiinflammatory activity of NSAIDs could be attributed to their capacity to inhibit cyclooxygenase. A schematic model of the prostaglandin H synthase, which catalyzes the first step in the cascade that produces prostaglandins, was recently developed[4] (FIG. 1). The enzyme is localized primarily in the endoplasmic reticulum and has been purified as a dimer of 60-kD subunits. The amino acid sequence of the enzyme has been deduced and the gene cloned. Prostaglandin synthase catalyzes two reactions: the biodeoxygenation of arachidonic acid to form prostaglandin G_2 (cyclooxygenase activity) and the reduction of hydroperoxide to the corresponding alcohols (peroxidase activity). The cyclooxygenase activity is inhibited by most nonsteroidal antiinflammatory drugs. A prominent protease-sensitive region is located near Arg^{253}. NSAIDs not only inhibit the cyclooxygenase activity but also render the cyclooxygenase resistant to attack by proteases. Interestingly, this property is not shared by aspirin which, in contrast to other NSAIDs, actually makes the prostaglandin H synthase less resistant to trypsin. The model postulates that NSAIDs partially occlude the arachidonic acid binding site as well as deny access of trypsin to Arg^{253}. In contrast, aspirin, which acetylates Ser^{506}, does not prevent access to the protease-sensitive Arg^{253}.

40

ANTIINFLAMMATORY EFFECTS OF NSAIDs INDEPENDENT OF ARACHIDONIC ACID METABOLISM

One problem with the hypothesis that NSAIDs derive all of their effects from the capacity to inhibit cyclooxygenase is the fact that nonacytelated salicylates and phenylbutazone are weak inhibitors of prostaglandin synthesis and yet effective antiinflammatory agents. Sodium salicylates does reduce prostaglandins in carageenan-induced inflammation;[5] however, it is unclear whether this effect is

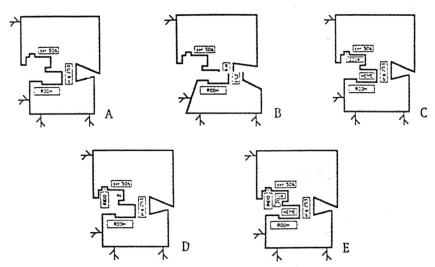

FIG. 6. Schematic model for the arrangement of functional areas on the synthase surface. *N*-Linked polysaccharides are shown as the *protruding branched objects*; Ser[506] is the residue acetylated by aspirin; Arg[253] is the site of cleavage by trypsin; and the *box* with *ROOH* indicates the peroxidase substrate-binding site. *Panel A*, one of the two subunits in the native purified synthase. *Panel B*, a subunit after proteolytic cleavage into 38 (*upper*)- and 33 (*lower*)-kDa fragments. *Panel C*, a subunit with bound arachidonate (20:4) and heme (*HEME*). *Panel D*, a subunit with bound indomethacin (*INDO*) that has been acetylated (*Ac*) by aspirin. *Panel E*, a subunit of indomethacin-impaired synthase, poised for cyclooxygenase catalysis, with bound indomethacin (*INDO*), arachidonate (20:4), and heme (*HEME*).

FIGURE 1. Schematic model of prostaglandin H synthase. Figure and legend reproduced from Kulmacz[4] with permission.

due to an inhibition of the cyclooxygenase or to a more global reduction in the inflammatory response. Some NSAIDs, such as indomethacin, increase the generation of lipoxygenase products. This phenomenon has been suggested to result from the blocking of the cyclooxygenase pathway that permits excess arachidonic acid to be metabolized via the lipoxygenase pathway.[2,6,7] This mechanism has been used to account for aspirin-induced bronchospasm, because in the presence of aspirin, increased products of lipoxygenase pathways such as leukotrienes C4, D4, and E4 could induce bronchospasm. Interestingly, patients with aspirin-induced bronchospasm can tolerate nonacetylated salicylates.[7] It has been speculated that this results from the fact that nonacetylated salicylates do not effectively inhibit the cyclooxygenase enzyme and, therefore, do not promote the formation of bronchoconstricting leukotrienes. There are reports that NSAIDs inhibit the 5-lipoxygenase enzyme in a variety of systems, although the concentrations required are higher than those likely to be achieved therapeutically.[8]

Significant evidence exists that many of the antiinflammatory effects of NSAIDs are unrelated to the inhibition of arachidonic acid metabolism. This is not surprising given the observation that nonacetylated salicylates are poor inhibitors of prostaglandin synthesis. It is also consistent with the observation that the antiinflammatory effects of NSAIDs require doses of drug that are in excess of those required to inhibit prostaglandin synthesis. Forrest and Brooks[2] demonstrated that indomethacin and piroxicam inhibit endotoxin-induced plasma leakage and polymorphonuclear leukocyte accumulation in the rat subcutaneous air pouch, an effect that could not be explained by the inhibition of prostaglandin synthesis. NSAIDs have been shown, both *in vitro* and *in vivo,* to inhibit a variety of functions of neutrophils including adhesiveness, aggregation, chemotaxis, degranulation, and superoxide anion generation. These effects are independent of the capacity to inhibit prostaglandin because sodium salicylate is equally potent to aspirin with regard to its effect on neutrophil functions.[1] In addition, although the NSAIDs studied, other than salicylate, are all effective inhibitors of prostaglandin synthesis, they demonstrate different profiles of inhibition of neutrophil functions. For example, whereas all NSAIDs effectively inhibited neutrophil aggregation, only some (piroxicam and indomethacin) consistently inhibited superoxide anion generation.[8] This may be due to previously observed independent effects of these two NSAIDs on the cell free NADPH-oxidase superoxide anion generating system, a property not shared by salicylate.[6] Interestingly, the capacity of NSAIDs to inhibit neutrophil functions was not shared by the analgesic agent acetaminophen (FIG. 2).[9] The mechanisms by which NSAIDs interfere with neutrophil activation are unknown. The studies by Minta and Williams[6] suggest that they

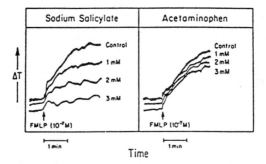

Fig. 1. Effect of sodium salicylate and acetaminophen on FMLP-induced aggregation. Neutrophils were preincubated with buffer or drug (5 min, 37°C) before exposure to FMLP (10^{-7} M). Change in light transmission was recorded over time. Aggregation curves are representative of three experiments.

FIGURE 2. Effects of sodium salicylate and acetaminophen on neutrophil aggregation. Figure and legend reproduced from Abramson *et al.*[9] with permission.

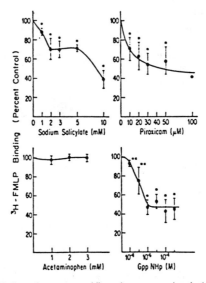

Fig. 3. FML[³H]P binding to human neutrophil membrane preparations in the presence of varying concentrations of sodium salicylate, piroxicam, acetaminophen, and GppNHp. Membranes were preincubated with the indicated concentration of agent for 15 min, then incubated with 1 nM FML[³H]P for 30 min at 25°C. Results represent the mean ± SEM of three experiments performed in duplicate (*$P < 0.01$, **$P < 0.05$).

FIGURE 3. Effects of NSAIDs on FMLP binding. Figure and legend reproduced from Abramson et al.[9] with permission.

may act at more than a single site (e.g., ligand binding, NADPH-oxidase). A number of studies have demonstrated that NSAIDs, but not acetaminophen, interfere with the binding of chemoattractant ligands to specific cell surface receptors[9] (FIG. 3). In addition, it has been demonstrated that NSAIDs, which are anionic, lipophilic compounds, insert into the lipid bilayer of neutrophils where they interrupt the function of the GTP-binding regulatory protein (G protein).[1,9] The G protein in the neutrophil, as in other cell types, transduces signals from stimuli outside the cell across the cell membrane. The interruption of G protein dependent of events by NSAIDs may explain why these drugs interfere with ligand-induced cell activation, but have little effect on agents that activate the cell via different mechanisms (such as phorbol esters and ionophores). The stimulus dependence of the NSAID effect helps explain divergent reports in the literature regarding the capacity of these agents to inhibit neutrophil activation. For example, Forrest found that indomethacin inhibits peptone-induced neutrophil accumulation in a model using a rat subcutaneous air patch, yet it has only a marginal effect on monosodium urate crystal-induced neutrophil accumulation.

EFFECTS OF NSAIDs ON THE IMMUNE SYSTEM

It is now appreciated that prostaglandins have complex effects on the immune response. Not only are they capable of inducing signs of inflammation, but they also have potential antiinflammatory effects. Prostaglandins of the E series have the capacity to inhibit the activation of a variety of cells such as macrophages, lymphocytes, and neutrophils. The effect of prostaglandins on the humoral immune response apparently is achieved through effects on T-cell regulation: the activity of suppressor T cells, in particular, is inhibited by prostaglandin E_2.[10] *In vivo,* it has been demonstrated that the rheumatoid factor and C-reactive protein production can be reduced in a subset of patients taking nonsteroidal antiinflammatory drugs.[11,12]

REFERENCES

1. ABRAMSON, S. & G. WEISSMAN. 1989. The mechanisms of action of nonsteroidal antiinflammatory drugs. Arthritis & Rheum. **31:** 1–9.
2. FORREST, N. & P. M. BROOKS. 1988. Mechanism of action of nonsteroidal antirheumatic drugs. Bailliere's Clin. Rheum. **2:** 275–293.
3. HOCHBERG, M. C. 1989. NSAIDs: Mechanisms and pathways of action. Hosp. Pract. **24:** 185–209.
4. KULMACZ, R. J. 1989. Topography of prostaglandin H synthase. J. Biol. Chem. **264:** 14136–14144.
5. HIGGS, G. A., J. A. SALMON, B. HENDERSON & J. R. VANE. 1987. Pharmacokinetics of aspirin and salicylate in relation to inhibition of arachidonate cyclooxygenase and antiinflammatory activity. Proc. Natl. Acad. Sci. USA **84:** 1417–1420.
6. MINTA, J. O. & M. D. WILLIAMS. 1985. Some nonsteroidal antiinflammatory drugs inhibit the generation of superoxide anions by activated polymorphs by blocking ligan-receptor interactions. J. Rheumatol. **12:** 751–757.
7. ROBINSON, D. R., M. SKOSKIEWICZ, K. J. BLOCH *et al.* 1986. Cyclooxygenase blockade elevates leukotrienne E_4 production during acute anaphylaxis in sheep. J. Exp. Med. **163:** 1509–1517.
8. BOCTOR, A. M., M. EICKHOLD & T. A. PUGSLEY. 1986. Meclofenamate sodium is an inhibitor of both the 5-lipoxygenase and cyclooxygenase pathways of the arachidonic acid cascade in vitro. Prost. Leukotriennes & Med. **23:** 229–238.
9. ABRAMSON, S. B., B. CHERKSEY, D. GOODE, J. LESZCCYNSKA-PIZAK, M. R. PHILLIPS, L. BLAU & G. WEISSMANN. 1990. Nonsteroidal antiinflammatory drugs exert differential effects on neutrophil function and plasma membrane viscosity. Inflammation **14:** 11–30.
10. GOODWIN, J. S., J. L. CEUPPENS & N. GUALDE. 1984. Control of the immune response in humans by prostaglandins. **7:** 79–92.
11. CEUPPENS, J. L., M. A. RODRIGUEZ & J. S. GOODWIN. 1982. Non-steroidal antiinflammatory agents inhibit the synthesis of IgM rheumatoid factor in vitro. Lancet **i:** 528–530.
12. KIRSCH, J. J., P. E. LIPSKY, A. E. POSTLETHWAITE, R. E. SCHROHENLOHER, A. SAWAY & W. J. KOOPMAN. 1990. Correlation of serologic indicators of inflammation with effectiveness of nonsteroidal antiinflammatory drug therapy in rheumatoid arthritis. Arthritis Rheum. **33:** 19–29.
13. SETTIPANE, G. A. 1988. Aspirin sensitivity and allergy. Biol. Med. & Pharmacother. **42:** 493–498.

Putative Role of Neutrophil Elastase in the Pathogenesis of Emphysema[a]

GORDON L. SNIDER,[b] DAVID E. CICCOLELLA,
SHIRLEY M. MORRIS, PHILLIP J. STONE, AND
EDGAR C. LUCEY

Pulmonary Center and Biochemistry Department
Boston University School of Medicine, and
Boston Veterans Affairs Medical Center
Boston, Massachusetts 02130

It has been known for almost a century that fraying and rupture of the elastic fibers occur in areas of alveolar wall destruction, which characterizes emphysema.[1] The elastase-antielastase hypothesis of emphysema posits that emphysema occurs in smokers because the antielastase defenses of the lungs cannot adequately protect the alveolar walls from damage by their own proteolytic enzymes; neutrophil elastase has been proposed as the most likely culprit among these enzymes.[2,3] There is powerful direct evidence for this hypothesis in persons with homozygous alpha$_1$-protease inhibitor (α_1-PI) deficiency. However, the evidence is largely indirect in smokers with adequate, protective levels of α_1-PI.

This essay will first address the question of whether elastin destruction is the final common pathway in the development of all emphysema. It will then review the evidence for and against the primacy of neutrophil elastase in the pathogenesis of emphysema associated with the destruction of elastic tissue. Our position is that elastin destruction is not the *sine qua non* of all forms of emphysema; rather we believe that emphysema is a stereotyped response of the lungs to certain types of injury, and elastin destruction may or may not be a part of the process. In emphysema associated with elastin destruction, the case for the role of neutrophil elastase is not proven. However, neither can it be excluded as a major factor in the train of events leading to emphysema. What is clear is that we have very little understanding of the precise way in which neutrophil elastase might lead to human emphysema.

PATTERNS OF HUMAN EMPHYSEMA

Emphysema in humans is usually treated by authors as a single entity characterized by respiratory airspace enlargement with evidence of destruction of airspace walls. In actuality, however, different patterns of emphysema have been recognized: centriacinar emphysema, which includes as a subclass the centrilobu-

[a] This work was supported by NIH grant HL-19717 and The Department of Veterans Affairs Medical Research Service.
[b] Address for correspondence: Gordon L. Snider, M.D., Chief, Medical Service, Boston VA Medical Center, 150 South Huntington Ave., Boston, MA 02130.

45

lar emphysema of smokers, begins in the respiratory bronchioles; panacinar emphysema, the most frequent type of emphysema seen in severe α_1-PI deficiency, initially involves all the alveolar ducts in a secondary lung lobule; subpleural or paraseptal emphysema begins in alveoli adjacent to fibrous planes; and airspace enlargement with fibrosis, formerly known as paracicatricial emphysema, is associated with scars of various sorts. The varying morphologic characteristics of these forms of emphysema suggest that they may differ in pathogenesis. The background factors, age and sexual predilection, and the preferential localization within the lungs these various types of emphysema (centrilobular emphysema predominantly in the upper half of the lungs, panacinar mostly at the lung bases, and subpleural emphysema mostly peripheral and adjacent to septa) further suggest different pathogeneses.

ANIMAL MODELS OF EMPHYSEMA NOT INDUCED BY ELASTASE TREATMENT

A number of animal models of emphysema have been developed using methods other than the instillation of crude or purified preparations of proteolytic enzymes. Treatment with cadmium salts, hyperoxia, or other oxidants, and severe starvation fall into this category. These models of airspace enlargement will be discussed in terms of the possible role of elastolysis in their genesis. Experience with tobacco smoke as an emphysema-inducing agent will also be reviewed in this section.

Cadmium

Some patients who survive severe lung injury with vaporized cadmium salts develop airspace enlargement with fibrosis.[4] Subsequently, models of airspace enlargement with fibrosis were developed by several investigators[5,6] using cadmium chloride ($CdCl_2$). The addition of the lathyritic agent, beta-aminoproprionitrile (BAPN), to the $CdCl_2$ treatment resulted in marked enhancement of airspace enlargement.[7] A recent study[8] in which lung elastin was radiolabeled in the neonatal period demonstrated that neonatally formed lung elastin is not destroyed in $CdCl_2$ airspace enlargement with fibrosis in hamsters. Hoidal *et al.*[9] also recently showed that rendering hamsters neutropenic with antineutrophil globulin did not inhibit the airspace enlargement produced by combined $CdCl_2$-BAPN treatment. Therefore, it seems likely that $CdCl_2$-induced airspace enlargement with fibrosis is not caused by neutrophil elastase.

A number of observations suggest that this finding may be important in relation to centrilobular emphysema in smokers. Fibrosis is a part of the microbullae of centrilobular emphysema;[10] cigarette smoke is an important source of cadmium;[11] cadmium is found in emphysematous lungs in direct proportion to the severity of emphysema;[12] finally, compliance of the microbullae of centrilobular emphysema is less than that of normal lungs and much less than that of the lungs that contain them.[13] The compliance of whole lung is a net measurement, reflecting the algebraic sum of areas with varying degrees of alteration in lung compliance. Saetta *et al.*[14] recently described discontinuities in the walls of normal-sized respiratory airspaces of smokers that are more frequent than those in nonsmokers. These may be the lesions that account for the increased compliance of the lungs of smokers with mild emphysema, despite the presence in the lungs of centrilobular lesions with decreased compliance. Elastase-antielastase imbalance may account for the widespread discontinuities in alveolar walls, whereas the centrilobular lesions may represent focal airspace enlargement with fibrosis.

Hyperoxia

Exposure of rats to hyperoxia has been shown to give rise to airspace enlargement and lung volumes without alteration of compliance or alveolar number. The increase in collagen content of the lungs was almost 50%. Treatment of the oxygen-exposed rats with an analog of proline that prevents synthesis of cross-linked collagen prevented the development of these changes.[15] Exposure of lung tissue slices to hyperoxia with and without added protease inhibitors suggests that collagen degradation is not mediated by proteases.[16]

The mechanism of airspace enlargement in these hyperoxic rats is not known. However, because rats grow throughout their life spans and because the number of alveoli did not change in this model, abnormal growth of alveoli may be the cause of the enlargement. Like the model of airspace enlargement with fibrosis due to cadmium, this model raises the question of whether elastin degradation is a necessary prelude to airspace enlargement.

Other Oxidants

A number of laboratories have observed airspace enlargement with exposure of rats,[17–19] hamsters,[20–22] mice,[23] rabbits,[24] and dogs[25] to NO_2. Hamsters exposed to 30 ppm of NO_2 showed decreased lung elastin and collagen content 4 and 10 days (respectively) after starting exposure. The collagen content returned to control levels by day 14 of exposure. Total elastin did not return to control levels until after NO_2 exposure had stopped.[26] During exposure, neutrophils accumulated rapidly in the lungs. There is suggestive but not definitive evidence that the neutrophils released elastase during their migration through the lung interstitium.[18] Airspace enlargement has also been shown with ozone,[27] but less is known of the mechanism of airspace enlargement than in the NO_2 model.

Starvation

Severe starvation induces airspace enlargement in rats[28–31] and hamsters.[32] The mean linear intercept is increased, and the internal surface area of the lungs and the alveolar number are decreased. Saline-filled volume pressure relations corrected for lung volume were found to be shifted up and to the left only in young rats by Sahebjami and MacGee[33] but were found to be unaltered by Karlinsky *et al.*[32] and by Kerr *et al.*[31] Lung DNA content was unchanged,[34] suggesting no increase in the number of lung cells, such as neutrophils, in this model. Lung collagen and elastin were decreased compared to those in fed controls but not compared to baseline values;[31,32] this observation does not exclude increased turnover of connective tissues. The mechanism of this form of emphysema is obscure, but it may be related to disordered growth of the rodent lung. It does not appear to be related to neutrophil elastase.

Tobacco Smoke

Prolonged exposure of rats and hamsters to cigarette smoke has failed to consistently produce emphysema, perhaps because of the highly efficient filtering ability of the rodent's upper respiratory tract. Airspace enlargement and alveolar wall fibrosis were induced in tracheotomized beagle dogs by exposure to smoke of 2–7 cigarettes per day for 2–4 months; a dose response relation was not observed.[35] Cigarette smoke exposure did not induce emphysema in hamsters, but it did potentiate the emphysema induced by pancreatic elastase.[36] Increased alveolar wall permeability seems the most likely explanation for this observation.

ANIMAL MODELS OF EMPHYSEMA ASSOCIATED WITH ELASTIN DAMAGE

Elastase-Induced Emphysema

Extensive published data have established the concept that when enzymes are instilled into the lungs of experimental animals, only elastolytic enzymes, human neutrophil elastase (HNE) among them, can induce emphysema. Proteolytic but nonelastolytic enzymes do not do so. Homogenates of leukocytes,[37,38] as well as highly purified preparations of neutrophil elastase, produce emphysema in experimental animals.[39–41] Treatment with HNE causes decreased lung elastin content,[40] increased airspace size, and increased lung volumes,[39–41] but all of the changes are much less severe than are those induced by comparable or even smaller doses of pancreatic elastase. The emphysema persists indefinitely but without progression.[42] Increasing the dose of neutrophil elastase in order to produce more severe emphysema results in death of the animals from pulmonary hemorrhage.[40,41]

The reasons for the lesser severity of the emphysema induced by human neutrophil than pancreatic elastase are incompletely known. Destruction of alveolar epithelium is much greater in hamster lung after human neutrophil elastase than after pancreatic elastase administration.[43] Increased bleeding into the lung resulting in inactivation of human neutrophil elastase by α_1-PI has been a suggested mechanism.[40] However, the half-life in the lung of instilled enzymatically active human neutrophil elastase is similar to that of pancreatic elastase.[44] Various elastases have well-described differences in substrate preference.[39,45] Such differences along with differing accessibility of the enzymes to elastic fibers in the alveolar walls likely explain the differing severity of emphysema induced by HNE and pancreatic elastase. The emphysema induced by HNE may be potentiated by cathepsin G and by neutrophil proteinase 3.[46,47]

Emphysema and Impaired Cross-Linking of Elastin

In the blotchy mouse (a model of genetically occurring emphysema) and in young animals fed lathyrogens, emphysema results from impaired cross-linking of elastin.[3] Elastases do not appear to play a role in these models.

EMPHYSEMA, A STEREOTYPED RESPONSE OF THE LUNGS TO INJURY

The production of experimental emphysema by the diverse agents reviewed in this essay, which is only a partial list,[3] suggests that airspace enlargement is a stereotyped response of the lungs to a wide variety of injuries (FIG. 1). It seems clear from the experimental data that not all airspace enlargement occurs on a background of elastin destruction, although elastin damage results in emphysema. Injury and repair of connective tissue play a central role in all disease processes leading to airspace enlargement. The airspace enlargement may be considered to result from a repair process that attempts to preserve the basic structure of the lungs. That structure consists of airspaces communicating with the external environment, which are created by vascular tissue specialized for molecular gas exchange.

NEUTROPHILS

Neutrophil Enzymes

Neutrophil elastase is a neutral serine protease that, along with cathepsin G, occurs in the azurophilic granules of the neutrophil. The enzyme is, in a sense,

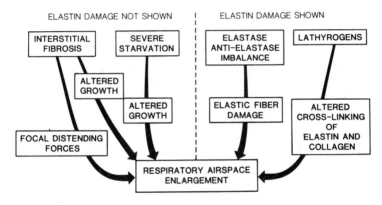

FIGURE 1. Schema of relation between elastin damage and respiratory airspace enlargement. Elastin damage is not shown in airspace enlargement associated with fibrosis or starvation. Focal distending forces are a key factor in the former state; altered alveolar growth likely plays a role in both of these lesions. Elastin damage does occur in elastase-antielastase imbalance and after lathyrogen treatment in young animals. Respiratory airspace enlargement is a stereotyped response of the lungs to injury, and elastin damage is not a final common pathway of its pathogenesis.

misnamed because it is a protease of broad specificity which, in addition to its ability to solubilize elastin, has the potential for cleaving a variety of other proteins and for attacking intact cells. As discussed elsewhere in this conference, human neutrophil granules also contain another elastolytic serine protease, designated proteinase 3.[48] Two metallo-proteases, collagenase and gelatinase, are found in the specific granules and tertiary granules, respectively.

Specific granules contain receptors for the *Escherichia coli* chemoattractant, f-met-leu-phe;[49] they readily fuse with the plasma membrane as they approach the inflammatory site and can release their contents into the extracellular milieu. Azurophilic granules fuse with phagosomes and do not generally release their contents extracellularly,[50,51] although evidence exists of "leakage" of granule contents into the extracellular space, especially if the phagocytosed particle is large.[52] If elastase or cathepsin G does enter the extracellular milieu, as certainly happens after cell death, an elaborate antiprotease defense exists to protect against tissue injury.

Neutrophil Margination

A portion of the circulating neutrophils passing through the lungs during each circulation adhere transiently to the capillary endothelium, forming the marginated pool of cells in the pulmonary vascular bed.[53] Normally sparse in the alveolar space, neutrophils can migrate from the capillaries into the alveolar space in great numbers in response to concentration gradients of chemotactic substances. These phagocytic cells have the potential for releasing their granule proteases while marginated in the capillary, while migrating through the thin or thick portion of the alveolar walls, or after gaining access to the alveolar space. Neutrophils can exclude α_1-PI from penetrating into zones of contact with opsonized surfaces,[54] and stimulated adherent neutrophils can degrade surface-bound substrates *in vitro* even in the presence of high concentrations of protease inhibitors.[55,56]

Emphysema Induced by Eliciting Neutrophils into the Lungs

Intravenous administration of endotoxin causes intravascular pulmonary leukocyte sequestration. Mild emphysema arose after 10 weekly intravenous injections of endotoxin in Rhesus monkeys[57] or 50 intravenous injections over 17 weeks in dogs.[58] Combined treatment of rats with D-galactosamine and repeated injections of endotoxin resulted after 10 weeks in emphysema.[59] The D-galactosamine produces liver disease, which decreases plasma α_1-PI levels; treatment with endotoxin alone produced milder changes. It is postulated that the emphysema in these experiments is induced by release of enzymes from sequestered leukocytes; however, it is not yet proven that elastin destruction has indeed occurred.

Unpublished observations from our laboratory have shown that the intratracheal instillation of 10–100 μg of *E. coli* endotoxin into hamsters elicits marked pulmonary neutrophilia, with from 50 to 100 $\times$ 10^6 PMNs lavageable from the lungs. A small amount of elastolytic activity can be detected in BAL fluid using SDS-treated tritiated elastin as substrate (Fig. 2). Mild but definite respiratory airspace enlargement is noted following BAL 24 hours after endotoxin treatment. Instillation of the degranulating agent, f-met-leu-phe, at the height of pulmonary neutrophilia resulted in increased elastolytic activity in the BAL fluid but no increase in airspace enlargement. This observation suggests that the emphysema is caused by release of elastase from neutrophils into the interstitium of the lungs rather than into the alveolar spaces. Elastin destruction has not yet been proven in this model.

ALTERNATE SOURCES OF ELASTASE

Macrophages

As will be discussed, macrophages are increased four- to fivefold in the lungs of smokers as compared with nonsmokers. These cells are concentrated in the centriacinar zones of the lungs, the region where the centrilobular emphysema of smokers occurs. Thus, there has been continuing interest in these cells as alternate sources of elastolytic enzyme systems. Human macrophages do not elaborate a measurable amount of metalloelastase into culture medium, but they are capable of degrading elastin when cultured on elastin-rich ^{3}H-lysine-labeled extracellular matrices deposited by rat smooth muscle cells[60] or when cultured in direct contact with radiolabeled elastin.[61] The elastin degradation was inhibited by the tissue inhibitor of metalloproteases, but it was inhibited minimally or not at all by inhibitors of cysteine proteases or by the serine protease inhibitor eglin-c. Cooperative interaction has also been shown between a neutral protease, plasminogen activator, and an acid protease.[60] The plasminogen activator may enhance elastin degradation by removing glycoproteins covering the elastic fibers.

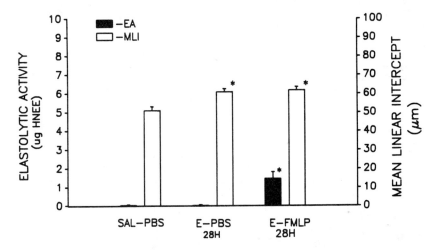

FIGURE 2. Bar diagram showing elastolytic activity in bronchoalveolar lavage (BAL) fluid (*solid bar*) and alveolar mean linear intercept (MLI) (*clear bar*). Two groups of six hamsters were treated with 10 µg of *E. coli* endotoxin (E) at 0 hours and either phosphate-buffered saline solution (E-PBS) or f-met-leu-phe (E-FMLP) at 24 hours. A control group received only normal saline solution at 0 hours and phosphate-buffered saline solution at 24 hours (SAL-PBS). All animals were killed at 28 hours; elastolytic activity was measured in BAL fluid using SDS-treated ^{3}H-elastin as substrate; the MLI was measured on histologic sections. Note that elastolytic activity is measurable only in the E-FMLP group. However, the MLI, which is significantly increased in both treatment groups, is not different in the E-PBS and E-FMLP groups.

EVIDENCE FROM HUMANS THAT HNE CAUSES EMPHYSEMA

Investigators have sought evidence for direct involvement of HNE in human emphysema in five ways. They have measured elastase activity in circulating neutrophils or alveolar macrophages; used immunologic methods to measure small amounts of elastase bound to antiproteases in blood; measured neutrophils and elastase activity in BAL fluid; measured degradation products of elastin in blood and urine; and used immunohistologic methods to demonstrate the presence of elastases in lung tissue in relation to their substrates. The last mentioned method is the only one that provides direct evidence of a role for proteases in lung diseases. The other methods are all indirect, and there is no way of knowing how well they reflect the microenvironments of the lungs. Attempts have been made to give relevance to these studies by relating these measurements to an index of disease severity, such as lung function measurements, or by comparing a diseased population with a suitable control population.

Blood Leukocyte Elastase Levels

Although it has not yet been proven by replacement therapy, there is little question that the emphysema in severe alpha$_1$-protease inhibitor deficiency is caused by insufficient protease inhibitor to balance the effects of elastase(s). Some investigators have reported an association between blood leukocyte elastase concentration and airflow obstruction, in homozygous or heterozygous α_1-PI deficiency,[62–65] whereas others have not;[66,67] the same controversy surrounds persons with normal levels of α_1-PI, with some reports supporting the concept[66,67] and others not.[68]

Proteases in Blood

Serum immunoreactive leukocyte elastase concentration in chronic obstructive bronchitis has been found to be double that in age-matched controls; however, patients with other active lung diseases have similarly increased leukocyte elastase concentrations in serum.[69]

Studies of Bronchoalveolar Lavage Fluid

The percentage of neutrophils in BAL fluid of smokers is unchanged, but their absolute number is increased because of a four- to fivefold increase in the total number of cells in the BAL fluid of smokers. Macrophages secrete chemotactic factors and are likely responsible for the increased number of neutrophils; they also secrete a neutrophil degranulating agent and can ingest and later release neutrophil elastase. Small amounts of elastase-like esterase activity have been shown in BAL fluid of smokers compared with nonsmokers.[70,71] Much of this activity appears to be due to a metalloprotease.[71]

Elastin Degradation Products in Blood and Urine

Investigators have not found elevated levels of desmosine in the urine of smokers with chronic obstructive pulmonary disease as compared with normal smokers or nonsmokers.[72,73] Urinary desmosine values were similar in normal subjects, PiZZ subjects with emphysema, and subjects with interstitial lung diseases; none of these markers were elevated in PiZZ children.[74]

Ongoing studies in Stone's laboratory, as yet unpublished, using an HPLC technique and isotope dilution, have found ratios of urinary desmosine to creatinine that are 10–20% of previously reported values. The reason for this may be the measurement by previous methods of interfering substances in the urine. Pyridinoline, a collagen cross-link derived from hydroxylysine, with a similar structure to that of desmosine, has been identified as one such possible substance; there may be others.[75,76] If up to 90% of previously recorded desmosine values were impurities, a doubling of desmosine output would have increased the measured value by only 10–20%, an amount well within the experimental error of these methods. Therefore, it is possible that the reported absence of increased desmosine in the urine of smokers and persons with emphysema is a spurious result due to a methodologic error.

Studies with an ELISA[77] have shown that patients with COPD have elevated plasma elastin peptide levels compared with levels in normal nonsmokers; normal smokers had intermediate values, but the three groups overlapped.

Immunocytochemical Localization of Elastase in Lung

Controversy surrounds a report that provides direct evidence of the protease-antiprotease hypothesis of emphysema in smokers. One group[78] has shown immunoultrastructurally that neutrophil elastase is bound to elastic fibers in the lungs of smokers and that the amount of elastase is proportional to the amount of emphysema that is present. Another group reports inability to reproduce these findings.[79]

Morris, in unpublished observations from our laboratories, has made interesting observations comparing the *in vitro* effects of porcine pancreatic elastase (PPE) and HNE on human lung parenchyma in short-term organ culture. Using a postembedding method and preimmune goat serum to block nonspecific binding, she found that both PPE and HNE bound to elastic fibers in the lungs (FIG. 3). Antibody to elastase was found only on elastic fibers that exhibited the granular texture indicative of digestion by elastase. Even in HNE-treated specimens, the anti-HNE antibody did not bind to normal appearing, intact, electron-lucent elastic fibers. Neither anti-PPE nor anti-HNE stained elastic fibers in lung samples that had not been incubated in the appropriate elastase.

In this study, Morris also confirmed the animal studies that lung tissue is more susceptible to attack by PPE than by HNE. When equal amounts of the two enzymes are used, PPE attacks a greater number of elastic fibers than does HNE and it does so more rapidly and completely; the immunoultrastructural data were consonant with biochemical measurements of proteins in the supernatants of the treated lung specimens. The PPE and HNE solubilized protein at similar rates, but the elastin content of the PPE-solubilized protein was 8% as compared with 1% for the HNE-solubilized protein. The alveolar epithelium was always intact in areas where damaged elastic fibers were seen in PPE-treated tissues. On the contrary, in HNE-treated tissues, the damaged elastic fibers were always seen in areas where the basal lamina was denuded of epithelial cells or in a few places where the epithelial cells appeared to be in the process of separating from the basal lamina. There is evidence that PPE gains rapid access to the interstitium of the lungs by transport in pinocytotic vesicles;[80] although HNE was not studied as rigorously as was PPE, evidence for transport of HNE by pinocytotic vesicles was not obtained.

PUTATIVE ROLE OF NEUTROPHIL ELASTASE

As summarized partially herein and in more detail elsewhere,[2,3] there is much evidence, mostly indirect, that elastases play an important role in the pathogenesis of emphysema. Although neutrophil elastase appears at first blush to be the

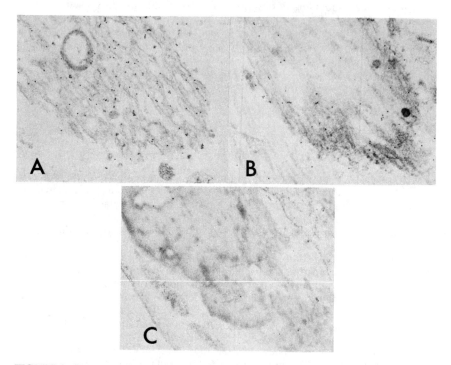

FIGURE 3. Demonstration by immunogold (*circular black particles*) of elastase on elastic fibers in human lung after incubation in either porcine pancreatic elastase (PPE) or human neutrophil elastase (HNE). All tissue was incubated in 200 μg/ml of enzyme for 45 minutes at 37°C. Magnification × 23,000. **(A)** Severely damaged elastic fiber after exposure to PPE. This section was incubated in anti-PPE followed by a secondary antibody conjugated to colloidal gold. Colloidal gold particles are seen over both peripheral and interior regions of the fiber. **(B)** An elastic fiber that shows evidence of damage only at the periphery after exposure to HNE. Colloidal gold particles are seen primarily over the damaged peripheral areas. The section was incubated in anti-HNE followed by a secondary antibody conjugated to colloidal gold. **(C)** An elastic fiber that shows no damage after incubation in HNE and very few colloidal gold particles. The section was also incubated in anti-HNE and colloidal gold-conjugated secondary antibody.

most plentiful intracellular elastase, when the potential capability of the macrophage to solubilize elastin is considered, the primacy of neutrophil elastase is no longer so certain. Furthermore, we have little understanding of the accessibility to lung tissues of intracellular neutrophil elastase.

All circulating neutrophils pass through the pulmonary capillaries at regular intervals. Neutrophils may enter the lungs in large numbers in response to chemotactic gradients. The *potential* elastase burden of the lungs may be estimated by the product of the total number of neutrophils in the lungs at any point in time and the concentration of elastase in the neutrophils. However, the elastase in the neutrophils is packaged in the azurophilic granules and cannot readily reach the external milieu. Consequently, the *actual* elastase burden of the lungs, that is, the amount of extracellular elastase reaching the interstitium or alveolar spaces of the lungs, is a small portion of the potential elastase burden of the lungs.

In summary, neutrophil elastase likely is important in the genesis of emphysema in humans. But the pathogenesis of emphysema is multifactorial, and little is as yet known of the events in the alveolar walls that lead to the development of airspace enlargement. Neutrophil elastase is surely not the only elastase of importance in the pathogenesis of emphysema. Much work still needs to be done to determine just where neutrophil elastase fits into the complex picture of pathogenesis of emphysema that is now emerging.

SUMMARY

Emphysema in humans takes several different forms: centrilobular, panacinar, paraseptal, and airspace enlargement with fibrosis. The varying morphologic and background features of these forms of emphysema suggest that they differ in pathogenesis. Elastic fiber rupture and fraying are a feature of emphysema. Experimental emphysema may be induced by human neutrophil elastase and other elastolytic enzymes but not by nonelastolytic proteases. Disruption of elastic fibers also appears to be the underlying feature of lathyrogen-induced airspace enlargement and of the emphysema in the blotchy mouse. However, there is no evidence of elastic fiber destruction in cadmium-induced airspace enlargement with fibrosis or in emphysema associated with hyperoxia or severe starvation. Thus, elastic fiber disruption is not common to all forms of experimental emphysema. We posit that airspace enlargement may be a stereotyped response of the lungs to different injuries. Emphysema can be induced in experimental animals by repeated induction of pulmonary neutrophilia. However, the evidence for involvement of neutrophil elastase in human emphysema is not clear: there are studies using a variety of approaches that weigh on both sides of the question. There is also *in vitro* evidence that alveolar macrophages can degrade elastin or elastic fibers with which they are in contact by means of a metalloelastase or the cooperative action of plasminogen activator and an acid cysteine protease. We conclude that the pathogenesis of emphysema is complex. Neutrophil elastase likely plays a major role in the development of some forms of emphysema, but our understanding of the interactions between the alveolar walls and neutrophils is still fragmentary.

REFERENCES

1. Orsos, F. 1907. Uber das elastiche Gerust der normalen und der emphysematosen Lungen. Beitr. Pathol. **41:** 95–121.
2. Janoff, A. 1985. Elastase and emphysema. Current assessment of the protease-antiprotease hypothesis. Am. Rev. Respir. Dis. **132:** 417–411.
3. Snider, G. L., E. C. Lucey & P. J. Stone. 1986. Animal models of emphysema. Am. Rev. Respir. Dis. **133:** 149–169.

4. LANE, R. E. & A. C. P. CAMPBELL. 1954. Fatal emphysema in two men making a copper cadmium alloy. Br. J. Ind. Med. **11:** 118–122.
5. THURLBECK, W. M. & F. D. FOLEY. 1963. Experimental pulmonary emphysema: The effect of intratracheal injection of cadmium chloride solution in the guinea pig. Am. J. Pathol. **42:** 431–441.
6. SNIDER, G. L., J. A. HAYES, A. L. KORTHY & G. P. LEWIS. 1973. Centrilobular emphysema experimentally induced by cadmium chloride aerosol. Am. Rev. Respir. Dis. **108:** 40–48.
7. NIEWOEHNER, D. E. & J. R. HOIDAL. 1982. Lung fibrosis and emphysema: Divergent responses to a common injury?. Science **217:** 359–360.
8. SNIDER, G. L., E. C. LUCEY, B. FARIS, Y. JUNG-LEGG, P. J. STONE & C. FRANZBLAU. 1988. Cadmium chloride-induced airspace enlargement with interstitial pulmonary fibrosis is not associated with destruction of lung elastin. Implications for the pathogenesis of human emphysema. Am. Rev. Respir. Dis. **137:** 918–923.
9. HOIDAL, J. R., D. E. NIEWOEHNER, N. V. RAO & M. S. HIBBS. 1985. The role of neutrophils in the development of cadmium-chloride-induced emphysema in lathyrogen-fed hamsters. Am. J. Pathol. **120:** 22–29.
10. LEOPOLD, J. G. & J. GOUGH. 1957. The centrilobular form of hypertrophic emphysema and its relation to chronic bronchitis. Thorax **12:** 219–235.
11. LEWIS, G. P., W. J. JUSKO, L. L. COUGHLIN & S. HARTZ. 1972. Contribution of cigarette smoking to cadmium accumulation in man. Lancet **1:** 291–292.
12. HIRST, R. N., H. M. PERRY, M. G. CRUZ & J. A. PIERCE. 1973. Elevated cadmium concentration in emphysematous lungs. Am. Rev. Respir. Dis. **108:** 30–39.
13. HOGG, J. C., S. N. NEPSZY, P. T. MACKLEM & W. M. THURLBECK. 1969. Elastic properties of the centrilobular emphysematous space. J. Clin. Invest. **48:** 1306–1312.
14. SAETTA, M., H. GHEZZO, W. D. KIM, M. KING, G. E. ANGUS, N. S. WANG & M. G. COSIO. 1985. Loss of alveolar attachments in smokers: A morphometric correlate of lung function impairment. Am. Rev. Respir. Dis. **132:** 894–900.
15. RILEY, D. J., R. A. BERG, N. H. EDELMAN & D. J. PROCKOP. 1980. Prevention of collagen deposition following pulmonary oxygen toxicity in the rat by cis-4-hydroxy-l-proline. J. Clin. Invest. **65:** 643–651.
16. CURRAN, C. F., M. A. AMORUSO, B. D. GOLDSTEIN, D. J. RILEY, N. H. EDELMAN & R. A. BERG. 1984. Direct cleavage of soluble collagen by ozone or hydroxyl radicals; possible mechanism of emphysema produced by exposure to oxidant gases. Chest **85:** 43S–45S.
17. FREEMAN, G., S. C. CRANE, R. J. STEPHENS & N. J. FURIOSI. 1968. Pathogenesis of the nitrogen dioxide-induced lesion in the rat lung: A review and presentation of new observations. Am. Rev. Respir. Dis. **98:** 429–443.
18. GLASGOW, J. E., G. G. PIETRA, W. R. ABRAMS, J. BLANK, D. M. OPPENHEIM & G. WEINBAUM. 1987. Neutrophil recruitment and degranulation during induction of emphysema in the rat by nitrogen dioxide. Am. Rev. Respir. Dis. **135:** 1129–1136.
19. BLANK, J., J. E. GLASGOW, G. G. PIETRA, L. BURDETTE & G. WEINBAUM. 1988. Nitrogen-dioxide-induced emphysema in rats: Lack of worsening by beta-aminoproprionitrile treatment. Am. Rev. Respir. Dis. **137:** 376–379.
20. LAM, C., M. KATTAN, A. COLLINS & J. KLEINERMAN. 1983. Long-term sequelae of bronchiolitis induced by nitrogen dioxide in hamsters. Am. Rev. Respir. Dis. **128:** 1020–1023.
21. KLEINERMAN, J., M. P. C. IP & R. E. GORDON. 1985. The reaction of the respiratory tract to chronic NO_2 exposure. In The Pathologist and the Environment. Internat. Acad. Pathol. Monograph. Scarpelli, D. G., J. E. Craighead & N. Kaufman, eds.: 200–210. Williams & Wilkins. Baltimore, MD.
22. LAFUMA, C., A. HARF, F. LANGE, L. BOZZI, J. L. PONCY & J. BIGNON. 1987. Effect of low-level NO_2 chronic exposure on elastase-induced emphysema. Environ. Res. **43:** 75–84.
23. BLAIR, W. H., M. C. HENRY & R. EHRLICH. 1969. Chronic toxicity of nitrogen dioxide: Effect on histopathology of lung tissue. Arch. Environ. Health **18:** 186–192.
24. HAYDON, G. B., J. T. DAVIDSON & G. A. LILLINGTON. 1967. Nitrogen dioxide-induced emphysema in rabbits. Am. Rev. Respir. Dis. **95:** 797–805.

25. HYDE, D., J. ORTHOEFFER, D. DUNGWORTH, W. TYLER, R. CARTER & H. LUM. 1978. Morphometric and morphologic evaluation of pulmonary lesions in beagle dogs chronically exposed to high ambient levels of air pollutants. Lab. Invest. **38:** 455–469.
26. KLEINERMAN, J. & M. P. C. IP. 1979. Effect of nitrogen dioxide on elastin and collagen contents of lung. Arch. Environ. Health **34:** 228–232.
27. LUCEY, E. C., P. J. STONE, T. G. CHRISTENSEN, J. L. TURNER & G. L. SNIDER. 1990. Human neutrophil elastase-induced and ozone-induced emphysema in hamsters are additive (abstr.) Am. Rev. Respir. Dis. **141:** A736.
28. SAHEBJAMI, H. & C. L. VASSALLO. 1979. Effects of starvation and refeeding on lung mechanics and morphometry. Am. Rev. Respir. Dis. **119:** 443–451.
29. SAHEBJAMI, H. & J. A. WIRMAN. 1981. Emphysema-like changes in the lungs of starved rats. Am. Rev. Respir. Dis. **124:** 619–624.
30. HARKEMA, J. R., J. L. MAUDERLY, R. E. GREGORY & J. A. PICKRELL. 1984. A comparison of starvation and elastase models of emphysema in the rat. Am. Rev. Respir. Dis. **129:** 584–591.
31. KERR, J. S., D. J. RILEY, S. LANZA-JACOBY, R. A. BERG, H. C. SPILKER, Y. Y. SHIU & N. H. EDELMAN. 1985. Nutrition emphysema in the rat: Influence of protein depletion and impaired lung growth. Am. Rev. Respir. Dis. **131:** 644–650.
32. KARLINSKY, J. B., R. H. GOLDSTEIN, B. OJSERKIS & G. L. SNIDER. 1986. Lung mechanics and connective tissue levels in starvation-induced emphysema in hamsters. Am. J. Physiol. **251:** R282–R288.
33. SAHEBJAMI, H. & J. MACGEE. 1985. Effects of starvation on lung mechanics and biochemistry in young and old rats. J. Appl. Physiol. **58:** 778–784.
34. SAHEBJAMI, H. & J. MACGEE. 1982. Effects of starvation and refeeding on lung biochemistry in rats. Am. Rev. Respir. Dis. **126:** 483–487.
35. FRASCA, J. M., O. AUERBACH, H. W. CARTER & V. R. PARKS. 1983. Morphologic alterations induced by short-term cigarette smoking. Am. J. Pathol. **111:** 11–20.
36. HOIDAL, J. R. & D. E. NIEWOEHNER. 1983. Cigarette smoke inhalation potentiates elastase-induced emphysema in hamsters. Am. Rev. Respir. Dis. **127:** 478–481.
37. MASS, B., T. IKEDA, D. R. MERANZE, G. WEINBAUM & P. KIMBEL. 1972. Induction of experimental emphysema. Cellular and species specificity. Am. Rev. Respir. Dis. **106:** 384–391.
38. FONZI, L. & G. LUNGARELLA. 1979. Elastolytic activity in rabbit leukocyte extracts. Effects of the whole leukocyte homogenate on the rabbit lung. Exp. Molec. Pathol. **31:** 486–491.
39. JANOFF, A., B. SLOAN, G. WEINBAUM, V. DAMIANO, R. A. SANDHAUS, J. ELIAS & P. KIMBEL. 1977. Experimental emphysema induced with purified human neutrophil elastase: Tissue localization of the instilled protease. Am. Rev. Respir. Dis. **115:** 461–478.
40. SENIOR, R. M., H. TEGNER, C. KUHN, K. OHLSSON, B. C. STARCHER & J. A. PIERCE. 1977. The induction of pulmonary emphysema with human leukocyte elastase. Am. Rev. Respir. Dis. **116:** 469–475.
41. SNIDER, G. L., E. C. LUCEY, T. G. CHRISTENSEN, P. J. STONE, J. D. CALORE, A. CATANESE & C. FRANZBLAU. 1984. Emphysema and bronchial secretory cell metaplasia induced in hamsters by human neutrophil products. Am. Rev. Respir. Dis. **129:** 155–160.
42. LUCEY, E. C., P. J. STONE, T. G. CHRISTENSEN, R. BREUER & G. L. SNIDER. 1988. An 18-month study of the effects on hamster lungs of intratracheally administered human neutrophil elastase. Exp. Lung Res. **14:** 671–686.
43. KUHN, C., J. SLODKOWSKA, T. SMITH & B. STARCHER. 1980. The tissue response to exogenous elastase. Bull. Eur. Physiolpathol. Respir. (Suppl) **16:** 127–139.
44. STONE, P. J., E. C. LUCEY, J. D. CALORE, M. P. MCMAHON, G. L. SNIDER & C. FRANZBLAU. 1988. Defenses of the hamster lung against human neutrophil and porcine pancreatic elastase. Respiration **54:** 1–15.
45. SENIOR, R. M., D. R. BIELEFELD & B. C. STARCHER. 1976. Comparison of the elastolytic effects of human leukocyte elastase and porcine pancreatic elastase. Biochem. Biophys. Res. Commun. **72:** 1327–1334.

46. LUCEY, E. C., P. J. STONE, R. BREUER, T. G. CHRISTENSEN, J. D. CALORE, A. CATANESE, C. FRANZBLAU & G. L. SNIDER. 1985. Effect of combined human neutrophil cathepsin G and elastase on induction of secretory cell metaplasia and emphysema in hamsters, with *in vitro* observations on elastolysis by these enzymes. Am. Rev. Respir. Dis. **132:** 362–366.

47. KAO, R. C., N. G. WEHNER, K. M. SKUBITZ, B. H. GRAY & J. R. HOIDAL. 1988. Proteinase 3: A distinct human polymorphonuclear leukocyte proteinase that produces emphysema in hamsters. J. Clin. Invest. **82:** 1963–1973.

48. HOIDAL, J. 1990. Structural and functional characterization of proteinase 3: A potential participant in lung destruction. Ann. N.Y. Acad. Sci. This volume.

49. FLETCHER, M. P. & J. I. GALLIN. 1983. Human neutrophils contain an intracellular pool of putative receptors for the chemoattractant N-formyl-methionyl-leucylphenylalanine. Blood **62:** 792–799.

50. MALECH, H. L. & J. I. GALLIN. 1987. Neutrophils in human diseases. N. Engl. J. Med. **317:** 687–694.

51. GALLIN, J. I. 1984. Neutrophil specific granules in health and disease. Clin. Res. **32:** 320–328.

52. WEISSMAN, G., R. B. ZURIER, P. J. SPIELER & I. M. GOLDSTEIN. 1971. Mechanisms of lysosomal enzyme release from leukocytes exposed to immune complexes and other particles. J. Exp. Med. **134:** 149s–165s.

53. HOGG, J. 1990. Neutrophil kinetics in the lungs of smokers and non-smokers. Ann. N.Y. Acad. Sci. In press.

54. CAMPBELL, E. J. & M. A. CAMPBELL. 1988. Pericellular proteolysis by neutrophils in the presence of proteinase inhibitors: Effects of substrate opsonization. J. Cell. Biol. **106:** 667–676.

55. CAMPBELL, E. J., R. M. SENIOR, J. A. MCDONALD & D. L. COX. 1982. Proteolysis by neutrophils: Relative importance of cell-substrate contact and oxidative inactivation of proteinase inhibitors *in vitro*. J. Lab. Invest. **70:** 845–852.

56. WEISS, S. J. & S. REGIANI. 1984. Neutrophils degrade subendothelial matrices in the presence of alpha-1-proteinase inhibitor: Cooperative use of lysosomal proteinases and oxygen metabolites. J. Clin. Invest. **73:** 1297–1303.

57. WITTELS, E. H., J. J. COALSON, M. H. WELCH & C. A. GUENTER. 1974. Pulmonary intravascular leukocyte sequestration: A potential mechanism of lung injury. Am. Rev. Respir. Dis. **109:** 502–509.

58. GUENTER, C. A., J. J. COALSON & J. JACQUES. 1981. Emphysema associated with intravascular leukocyte sequestration: Comparison with papain-induced emphysema. Am. Rev. Respir. Dis. **123:** 79–84.

59. BLACKWOOD, R. A., J. MORET, I. MANDL & G. M. TURINO. 1984. Emphysema induced by intravenously administered endotoxin in an alpha$_1$-antitrypsin-deficient rat model. Am. Rev. Respir. Dis. **130:** 231–236.

60. CHAPMAN, H. A., J. J. REILLY & L. KOBZIK. 1988. Role of plasminogen activator in degradation of extracellular matrix protein by live human alveolar macrophages. Am. Rev. Respir. Dis. **137:** 412–419.

61. SENIOR, R. M., N. L. CONNOLLY, J. D. CURY, H. G. WELGUS & E. J. CAMPBELL. 1989. Elastin degradation by human alveolar macrophages: A prominent role of metalloproteinase activity. Am. Rev. Respir. Dis. **139:** 1251–1256.

62. GALDSTON, M., E. L. MELNICK, R. M. GOLDRING, V. LEVYTSKA, C. A. CURASI & A. L. DAVIS. 1977. Interactions of neutrophil elastase, serum trypsin inhibitory activity, and smoking history as risk factors for chronic obstructive pulmonary disease in patients with MM, MZ, and ZZ phenotypes for alpha$_1$-antitrypsin. Am. Rev. Respir. Dis. **116:** 837–846.

63. KIDOKORO, Y., T. C. KRAVIS, K. M. MOSER, J. C. TAYLOR & I. P. CRAWFORD. 1977. Relationship of leukocyte elastase concentration to severity of emphysema in homozygous alpha$_1$-antitrypsin-deficient persons. Am. Rev. Respir. Dis. **115:** 793–803.

64. ABBOUD, R. T., J. M. RUSHTON & S. GRZYBOWSKI. 1979. Interrelationships between neutrophil elastase, serum alpha$_1$-antitrypsin, lung function and chest radiography in patients with chronic airflow obstruction. Am. Rev. Respir. Dis. **120:** 31–40.

65. LAM, S., R. T. ABBOUD, M. CHAN-YEUNG & J. M. RUSHTON. 1979. Neutrophil elastase and pulmonary function in subjects with intermediate alpha₁-antitrypsin deficiency (MZ phenotype). Am. Rev. Respir. Dis. **119:** 941–951.

66. RODRIGUEZ, J. R., J. E. SEAL, A. RADIN, J. S. LIN, I. MANDL & G. M. TURINO. 1979. Neutrophil lysosomal elastase activity in normal subjects and in patients with chronic obstructive pulmonary disease. Am. Rev. Respir. Dis. **119:** 409–417.

67. KRAMPS, J. A., W. BAKKER & J. H. DIJKMAN. 1980. A matched-pair study of the leukocyte elastase-like activity in normal persons and in emphysematous patients with and without alpha-1-antitrypsin deficiency. Am. Rev. Respir. Dis. **121:** 253.

68. BINDER, R., P. J. STONE, J. D. CALORE, D. M. DUNN, G. L. SNIDER, C. FRANZBLAU & C. R. VALERI. 1985. Serum antielastase and neutrophil elastase levels in PiM phenotype cigarette smokers with airflow obstruction. Respiration **47:** 267–277.

69. STOCKLEY, R. A. & K. OHLSSON. 1982. Serum studies of leucocyte elastase in acute and chronic lung diseases. Thorax **37:** 114–117.

70. JANOFF, A., L. RAJU & R. DEARING. 1983. Levels of elastase activity in bronchoalveolar lavage fluids of healthy smokers and nonsmokers. Am. Rev. Respir. Dis. **127:** 540–544.

71. NIEDERMAN, M. S., L. L. FRITTS, W. W. MERRILL, R. B. FICK, R. A. MATTHAY, H. Y. REYNOLDS & J. B. L. GEE. 1984. Demonstration of a free elastolytic metalloenzyme in human lung lavage fluid and its relationship to alpha₁-antiprotease. Am. Rev. Respir. Dis. **129:** 943–947.

72. DAVIES, S. F., K. P. OFFORD, M. G. BROWN, H. CAMPE & D. NIEWOEHNER. 1983. Urine desmosine is unrelated to cigarette smoking or to spirometric function. Am. Rev. Respir. Dis. **128:** 473–475.

73. HAREL, S., A. JANOFF, S. Y. YU, A. HUREWITZ & E. H. BERGOFSKY. 1980. Desmosine radioimmunoassay for measuring elastin degradation *in vitro*. Am. Rev. Respir. Dis. **122:** 769–773.

74. PELHAM, F., M. WEWERS, R. CRYSTAL, A. S. BUIST & A. JANOFF. 1985. Urinary excretion of desmosine (elastase cross-links) in subjects with PiZZ alpha-1-antitrypsin deficiency, a phenotype associated with hereditary predisposition to pulmonary emphysema. Am. Rev. Respir. Dis. **132:** 821–823.

75. GUNJA-SMITH, Z. 1985. An enzyme-linked immunosorbent assay to quantitate the elastin crosslink desmosine in tissue and urine samples. Anal. Biochem. **147:** 258–264.

76. LAURENT, P., L. MAGNE, J. DEPALMAS, J. BIGNON & M-C. JAURAND. 1988. Quantitation of elastin in human urine and rat pleural mesothelial cell matrix by a sensitive avidin-biotin ELISA for desmosine. J. Immunol. Meth. **107:** 1–11.

77. KUCICH, U., P. CHRISTNER, M. LIPPMANN, P. KIMBEL, G. WILLIAMS, J. ROSENBLOOM & G. WEINBAUM. 1985. Utilization of a peroxidase antiperoxidase complex in an enzyme-linked immunosorbent assay of elastin-derived peptides in human plasma. Am. Rev. Respir. Dis. **131:** 709–713.

78. DAMIANO, V. V., A. TSANG, U. KUCICH, W. R. ABRAMS, J. ROSENBLOOM, P. KIMBEL, M. FALLAHNEJAD & G. WEINBAUM. 1986. Immunolocalization of elastase in human emphysematous lungs. J. Clin. Invest. **78:** 482–493.

79. FOX, B., T. B. BULL, A. GUZ, E. HARRIS & T. D. TETLEY. 1988. Is neutrophil elastase associated with elastic tissue in emphysema? J. Clin. Pathol. **41:** 435–440.

80. MORRIS, S. M., H. M. KAGAN, P. J. STONE, G. L. SNIDER & J. T. ALBRIGHT. 1986. Ultrastructural changes in hamster lung 15 minutes to 3 hours after exposure to elastase. Anat. Rec. **215:** 134–143.

Proteinase-3 (PR-3):
A Polymorphonuclear Leukocyte
Serine Proteinase

N. V. RAO,[a] NANCY G. WEHNER,[b]
BRUCE C. MARSHALL,[a] ANNE B. STURROCK,[a]
THOMAS P. HUECKSTEADT,[a] GOPNA V. RAO,[a]
BEULAH H. GRAY,[b] AND JOHN R. HOIDAL[a,c]

[a]Department of Pulmonary Medicine
University of Utah Medical Center
Salt Lake City, Utah 84132

[b]Department of Microbiology
University of Minnesota
Minneapolis, Minnesota 55455

The current explanation for the development of emphysema is the proteinase pathogenesis hypothesis, which holds that progressive destruction of the interstitium is due to an excess of proteolytic enzymes (particularly elastase) in relation to the availability of proteolytic inhibitors. Inasmuch as a major endogenous source of elastase in the human lung is the polymorphonuclear leukocyte (PMNL), emphasis has been placed on the effects of PMNL-derived elastase and its inhibitors in regulating the degradation of lung elastin. Experimental emphysema has been produced by a single intratracheal injection of purified human PMNL elastase (HLE) in laboratory animals. Potent synthetic inhibitors have been developed for HLE, and several are effective in preventing emphysema in animal models of HLE-induced disease.

To consider only HLE-mediated injury in examining the pathogenesis of emphysema in the human, however, is too narrow a focus. A direct relation of HLE to human disease remains to be demonstrated. Recent studies suggest that the increased risk of emphysema associated with cigarette smoking may be due partially to the effect of phagocyte- or smoke-derived oxidants in the lung. Another variable immediately tangent to the HLE pathogenesis concept is that other PMNL or alveolar macrophage (AM) proteinases may participate in matrix destruction. Cathepsin G and an AM cysteine proteinase(s) both have elastolytic capabilities in vitro.[1,2] Moreover, cathepsin G has been reported to act synergistically with HLE in the solubilization of elastin in vitro,[3] although results have been discrepant.[4,5] Studies to date that have failed to demonstrate that these proteinases can produce emphysema[4] do not exclude their participation in the induction of the disease.

[c] Address for correspondence: John R. Hoidal, Pulmonary Division, Rm. 4R240, 50 North Medical Drive, University of Utah Medical Center, Salt Lake City, UT 84132.

In preliminary studies, we marshalled evidence that another human PMNL serine proteinase, proteinase-3 (PR-3), also causes experimental emphysema.[6] The studies to be described are aimed at elucidating the biochemical and functional properties of PR-3, an essential first step toward determining its role in physiologic and pathophysiologic processes.

METHODS

Purification of Human PMNL Proteinases

PR-3, HLE, and cathepsin G were purified from an extract of PMNL granules using Matrex Gel Orange A chromatography followed by cation exchange chromatography on Bio-Rex 70 as previously described.[6] The purity and molecular mass of each proteinase were determined by SDS-PAGE followed by silver staining of the gels. Purity was also ascertained by discontinuous nondenaturing gel electrophoresis followed by staining of the gels for esterase activity with α-naphthyl acetate as substrate. Protein concentration of purified enzymes was determined by the method of Hartree.[7]

Amino Acid Composition and N-Terminal Amino Acid Sequence of PR-3

Purified PR-3 was desalted and hydrolyzed in constant boiling HCl at 110°C *in vacuo*. Hydrolysates were dried and derivatized with phenyl isothiocyanate. HPLC separations of derivatized samples were carried out at ambient temperature on Bio-Rad Biosil ODS-5S column using 30 mM of sodium phosphate buffer, pH 6.3, and acetonitrile as solvents.[8] Peaks were assigned by comparison with standard amino acid mixtures.

The N-terminal amino acid sequence of PR-3 was determined by sequential Edman degradation in a Beckman 890 D spinning cup sequenator, using 0.1 M of Quodral buffer and Polybrene carrier.[9] Phenylthiohydantoin derivatives were identified by HPLC on a Hewlett-Packard 1084B instrument using an Ultrasphere ODS column (0.46 × 15 cm, 5 μm particle size, endcapped) eluted with a gradient of acetonitrile in 0.05 M sodium acetate, pH 4.5.[10] Search of the Swiss Prot 13 database (release 3.0, February 1990) for protein sequences homologous to PR-3 was carried out using PC/Gene software (Intelligenetics, Mountain View, California).

Proteolytic Activity of PR-3

Elastin degradation was assayed by determining the ability of PR-3 to solubilize bovine ligament elastin over a pH range from 6.5 to 8.9 using the procedure of Stone *et al.*[11] Tritiated powdered elastin was washed and resuspended in 0.2 M sodium phosphate buffer, pH 6.5 or 7.4, or sodium bicarbonate buffer, pH 8.9. The proteinase in physiologic saline solution was added to a 5-mg aliquot of ^{3}H-elastin and the reaction mixture was incubated at 37°C for 6 hours. After incubation, the contents of each tube were filtered through medium porosity filter paper to remove the insoluble elastin. The rate of degradation was determined by measuring the radioactivity in the filtrate.

Hydrolysis of synthetic peptides was determined using peptide substrates coupled to chromogenic or fluorogenic groups. The method used for these substrates was that described by Barrett,[12] with some modification. Briefly, 50 μl of substrate (10 mg/1.1 ml DMSO) was added to 0.1 M Tris buffer, pH 7.0. The reaction was started by the addition of the proteinase for a final volume of 3.0 ml. The release of p-nitrophenol and p-nitroaniline was measured by the increase in optical density at 347.5 nm and 410 nm, respectively. The release of 7-amino-4-methylcoumarin (AMC) from MCA substrates was monitored flurometrically with excitation and emission wavelength of 380 nm and 440 nm, respectively.

Inhibition of PR-3

The ability of natural inhibitors to prevent PR-3 activity was determined for alpha$_1$-proteinase inhibitor (α_1-PI), alpha$_1$-antichymotrypsin (α_1-Achy), and alpha$_2$-macroglobulin (α_2-M). For α_1-PI and α_1-Achy, equimolar concentrations of PR-3 and inhibitor in a final volume of 975 μl of 0.05 M phosphate buffer, pH 7.5, were incubated for selected periods of time. Residual PR-3 activity was then determined by the addition of 25 μl of 20 mM Boc-ala-ONp in methanol.

The inhibition of α_2-M was measured according to Virca and Travis.[13] Briefly, PR-3 (25 pmol) was incubated with α_2-M in 0.5 ml of 0.05 M phosphate buffer, pH 7.5, at 25°C for 10 minutes. Subsequently, 25 μl of α_1-PI (0.5 nmol) were added to the reaction mixture. After a 5-minute incubation, the mixture was brought up to 975 μl with phosphate buffer, and then 25 μl of 20 mM Boc-Ala-ONp in methanol was added. The α_1-PI-resistant esterase activity of PR-3 (that bound to α_2-M) was monitored at 347.5 nm.

Immunologic Properties of PR-3

Monoclonal antibodies were prepared by initially immunizing 6-week-old female BALB/c J mice with purified PR-3 (in complete Freunds' adjuvant). After three boosts, spleen cells were harvested and fused with murine myeloma P3-X63-Ag 8.653 cells. After 10–14 days of growth in hypoxanthine-aminopterin-thymidine selection culture medium, supernatants from hybridomas were evaluated for anti-PR-3 activity.

Dot blot analysis[14] using anti-PR-3 monoclonal antibodies was performed on purified PMNL proteinases. Indicated amounts of purified enzymes were spotted on nitrocellulose paper. The blots were blocked with 5% nonfat dry milk, 0.05% Tween 20, and phosphate-buffered saline solution, pH 7.4, overnight at 4°C. Blots were incubated with antibodies diluted in the blocking solution for 1 hour at room temperature followed by 16 hours at 4°C. After washing with 0.05% Tween 20 in phosphate-buffered saline solution, the blots were reacted with affinity-purified rabbit anti-mouse IgG (heavy and light chains) conjugated to alkaline phosphatase.

RESULTS

SDS polyacrylamide gel electrophoresis of PR-3 revealed a major band of 26.8 kD and two minor bands of slightly larger molecular mass, the largest estimated at 28.6 kD. Values calculated after SDS polyacrylamide gel electrophoresis with nonreducing conditions were identical to those determined with reducing conditions.

The amino acid composition of PR-3 is shown in TABLE 1 and compared to that of HLE and cathepsin G as reported by Travis *et al.*[15] PR-3 consisted of 224 amino

TABLE 1. Amino Acid Composition of Purified Proteinases

| | Residues/Molecule | | |
	PR-3	HLE[a]	Cat G[a]
Cys/2	8	6	6
Asx	20	24	17
Thr	11	7	12
Ser	7	13	15
Glx	20	18	23
Pro	16	10	13
Gly	32	28	21
Ala	17	24	13
Val	19	25	14
Met	2	2	4
Ile	11	11	11
Leu	19	20	16
Tyr	4	3	5
Phe	12	9	10
His	9	4	6
Lys	4	1	3
Arg	*11*	22	*31*
Trp	2	2[b]	nd[c]

[a] From ref. 15.
[b] Determined spectrophotometrically.
[c] nd = not determined.

acid residues per molecule. The composition was similar to that of HLE and cathepsin G, the greatest difference being the comparatively lower amount of arginine. The arginine content of PR-3 was one half that of HLE and approximately one third that of cathepsin G.

A comparison of the first 20 residues of PR-3[16] with other serine proteinases shows that the first four N-terminal residues (ile[1] to gly[4]) were identical to that of HLE,[17] whereas the eight-residue stretch from pro[9] to ala[16] was identical to that of Cat G,[18] murine granzymes A to F,[19-22] rat mast cell proteases,[23,24] and human lymphocyte proteases.[25] Over the first 20 residues, PR-3 has approximately 60% homology to HLE,[17] 70% to Cat G,[18] and 75% to human lymphocyte proteases.[25]

When compared with the granzyme family of serine proteases, PR-3 exhibited strong homology to granzyme B and less to granzymes A and C through F and to rat mast cell proteases. The first 20 residues of the PR-3 sequence are essentially identical to those recently published for c-ANCA antigen and AGP7.[26,27]

In light of the initial studies suggesting that PR-3 could induce emphysema, its elastinolytic activity was evaluated. Assays employing ^{3}H-elastin (TABLE 2) demonstrated that PR-3 hydrolyzed more elastin than did HLE (w/w) at pH 6.5, that its activity at pH 7.4 was ~50% that of HLE, and that it was much less active than was HLE at pH 8.9.

We assessed the esterolytic and amidolytic activity of PR-3 against selected chromogenic and fluorogenic substrates. The data are summarized in TABLE 3. Using monomeric p-nitrophenyl ester substrates, PR-3 hydrolyzed the monomeric p-nitrophenyl esters of alanine and valine. PR-3 showed no activity against ester substrates containing leucine at the p1 site. We next tested the amidolytic activity of PR-3 against four other HLE-specific oligopeptide substrates. Of the three tripeptide substrates (Suc-Ala-Ala-Ala-NA, Suc-Ala-Ala-Val-NA, and Suc-Ala-Pro-Ala-MCA) and one tetrapeptide substrate (MeO-Suc-Ala-Ala-Pro-Val-NA) tested, only Suc-Ala-Pro-Ala-MCA and MeO-Suc-Ala-Ala-Pro-Val-NA were hydrolyzed. PR-3 showed no activity against other oligomeric amide substrates having Leu at the p1 site.

We next focused on the interaction of PR-3 with the human plasma proteinase inhibitors α_1-PI, α_2-M, and α_1-Achy utilizing the most efficiently hydrolyzed substrate Boc-ala-ONp. PR-3 was inhibited by α_1-PI and α_2-M under the experimental conditions employed; PR-3 was not inhibited by α_1-Achy.

The uniqueness of PR-3 was further documented when purified PR-3, HLE, and cathepsin G were tested for their ability to react with monoclonal antibodies raised against PR-3. A representative result using anti-HPR-3 monoclonal antibody H-66 is shown in FIGURE 1. Only PR-3 reacted with H-66. Similar results were obtained using three other monoclonal antibodies.

DISCUSSION

PR-3 was originally described by Baggiolini et al.[28] as an α-naphthyl acetate esterase present in azurophilic granules of human PMNL. Its electrophoretic

TABLE 2. Degradation of Elastin by Purified PMNL Proteinase[a]

	^{3}H-Elastin (cpm released/5 mg ^{3}H-Elastin/6 h)		
	pH 6.5	pH 7.4	pH 8.9
PR-3	1710 ± 50[b]	2,250 ± 30	2,100 ± 120
HLE	1,190 ± 240	4,470 ± 300	18,770 ± 640
Cathepsin G	nd[c]	408[d]	890 ± 50

[a] From ref. 6.

[b] All values are corrected for background activity by subtracting radioactivity in buffer controls. Background activity was always <10% of most active fraction. Values are expressed as mean ± standard error.

[c] nd = not determined.

[d] Mean of duplicate determinations.

TABLE 3. PR-3 Specificity on Synthetic Substrates

Substrate						Substrate Conc. (mM)	Activity
p_5	p_4	p_3	p_2	p_1^-	p_1'		
					Boc-Ala-ONp	0.1	++
					Boc-Val-ONp	0.1	+
					Boc-Leu-ONp	0.1	−
					Suc-Ala-NA	1.0	−
				Suc-Ala-Ala-Ala-NA		1.0	−
			Suc-Ala-Ala-Val-NA			0.3	−
	MeO-Suc-Ala-Ala-Pro-Val-NA					0.1	+
		Z--Gly-Gly-Leu-NA				0.2	−
					Lys-Ala-MCA	0.2	−
			Suc-Ala-Pro-Ala-MCA			0.1	+
					Leu-MCA	0.2	−

ABBREVIATIONS: Boc = butoxycarbonyl; MeO-Suc = methoxy succinyl; Suc = Succinyl; MCA = 4-Methyl-Coumaryl-7-amide; NA = 4-Nitroanilide; ONp = *p*-nitrophenyl ester; Z = benzyloxy carbonyl.

mobility in nondenaturing gels was distinctly slower than that of HLE or Cat G. We subsequently purified PR-3 and demonstrated that it was an elastolytic neutral serine proteinase that caused extensive tissue damage and emphysema after intratracheal instillation into hamsters. The present studies further define the structural and functional features of PR-3.

We demonstrated that the pI of the major band of PR-3 was 9.1,[6] and a minor band with a pI of 8.9 was also observed in the isoelectric focusing gels. The pI of HLE is known to be near 11 and that of cathepsin G is reported to be greater than 11.[29] Analysis of the amino acid composition (TABLE 1) indicates that the lower pI of PR-3 compared to that of HLE and cathepsin G is due largely to the lower content of the basic amino acid arginine in PR-3.

The partial primary structure of PR-3 reported herein provides insight into the possible role of PR-3 *in vivo* and its mechanism of action. PR-3 shows NH_2-terminal sequence similarity to proteases of the trypsin superfamily with strongest homology to HLE,[17] cathepsin G,[18] and human lymphocyte protease.[25] The first 20 residues of the PR-3 sequence are identical to those recently published for AGP7, suggesting that one role of PR-3 may be that of an antimicrobial protein.[27] Of note, the antimicrobial activity of AGP7, similar to that of Cat G, was insensitive to di-isopropyl fluorophosphate, implying a nonproteolytic mechanism for the microbicidal action.[30,31]

The first 20 residues of PR-3 are also identical to the 29,000-dalton target antigen for the antineutrophil cytoplasmic autoantibodies (c-ANCA) associated with Wegener's granulomatosis, a disorder characterized by systemic necrotizing vasculitis.[32] We have further established the specificity of c-ANCA for PR-3 by showing that Wegener's granulomatosis sera reacted specifically with purified PR-3 in an ELISA and that the c-ANCA staining pattern of PMNLs observed by indirect immunofluorescence microscopy was blocked by PR-3-specific monoclonal antibodies.[33] The role of PR-3 in the pathobiochemistry of this devastating disorder is currently not known.

As previously reported,[6] PR-3 hydrolyzed elastin (TABLE 2), hemoglobin, and azocasein. Preliminary studies in our laboratory have demonstrated that it also degrades fibronectin, vitronectin, laminin, and type IV collagen, but it has no or minimal activity against collagen type I or III. Thus, similar to HLE and Cat G, PR-3 appears to have broad substrate specificity.

To gain insight into the preferred sites of peptide bond hydrolysis by PR-3, we examined a panel of synthetic substrates. Our analysis showed that PR-3 prefers substrates with either Ala or Val residues at the p1 site and pro at the p2 site, suggesting that its specificity with synthetic substrates is similar to but distinct from that of HLE. The preference for neutral aliphatic amino acid (ala and val) observed with oligopeptide substrates confirmed similar results obtained with insulin chains.

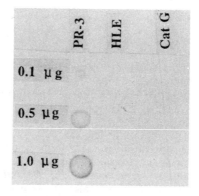

FIGURE 1. Dot blot comparing reactivity of mouse anti-PR-3 monoclonal antibody to PMNL neutral serine proteinases. The indicated amounts of purified neutral serine proteinases were spotted directly on a nitrocellulose membrane, followed by immunodetection using anti-PR-3 (1 : 1000 dilution) and alkaline phosphatase conjugated to rabbit anti-mouse IgG for detection.

PR-3 is inhibited by PMSF or DFP, but not by metallo, thiol, or carbonyl proteinase inhibitors,[6] establishing it as having a serine active residue. The present studies indicate that plasma proteinase inhibitors α_1-PI and α_2-M, but not α_1-Achy, are effective inhibitors of PR-3 *in vitro*. Thus, both the catalytic properties and the inhibitory profiles are distinct from those of other PMNL serine proteinases. Immunoblot studies using anti PR-3 monoclonal antibodies (FIG. 1) indicated that PR-3 is also immunologically distinct from HLE and cathepsin G.

In summary, the present studies extend current knowledge about PR-3, describing its partial primary structure and catalytic properties. PR-3 is capable of degrading a broad range of matrix proteins. Its catalytic properties and inhibitory profile are distinct from those of other PMNL proteinases. The finding that PR-3 in the antigen for c-ANCA suggests that it may play an important role in the pathophysiology of Wegener's granulomatosis.

REFERENCES

1. REILLY, C. F. & J. TRAVIS. 1980. Biochim. Biophys. Acta **621**: 147–157.
2. CHAPMAN, H. A. & O. L. STONE. 1984. Biochem. J. **222**: 721–728.
3. BOUDIER, C., C. HOLLE & J. G. BIETH. 1981. J. Biol. Chem. **256**: 10256–10258.
4. LUCEY, E. C., P. J. STONE, R. BREUER, T. G. CHRISTENSEN, J. D. CALORE, A. CANTANESE, C. FRANZBLAU & G. L. SNIDER. 1985. Am. Rev. Respir. Dis. **132**: 362–366.
5. REILLY, C. J., Y. FUKUNAGA, J. C. POWERS & J. TRAVIS. 1984. Hoppe-Seyler's Z. Physiol. Chem. **365**: 1131–1135.
6. KAO, R. C., N. G. WEHNER, K. M. SKUBITZ, B. H. GRAY & J. R. HOIDAL. 1988. J. Clin. Invest. **82**: 1963–1973.
7. HARTREE, E. F. 1972. Anal. Biochem. **48**: 422–427.
8. SIMPSON, R. J., M. R. NEUBERGER & T. Y. LIU. 1976. J. Biol. Chem. **251**: 1936–1940.
9. TARR, G. E., J. F. BEECHER, M. BELL & D. J. McKEAN. 1978. Anal. Biochem. **84**: 622–627.
10. GRAY, W. R., A. F. LUQUE, B. M. OLIVERA, J. BARRETT & L. J. CRUZ. 1981. J. Biol. Chem. **256**: 4734–4740.
11. STONE, P. J., C. FRANZBLAU & H. M. KAGAN. 1982. Methods Enzymol. **82**: 588–605.
12. BARRETT, A. J. 1981. Methods Enzymol. **80**: 561–565.
13. VIRCA, G. D. & J. TRAVIS. 1984. J. Biol. Chem. **259**: 8870–8874.
14. HAWKES, R. 1986. Methods Enzymol. **121**: 484–491.
15. TRAVIS, J., R. BAUGH, P. J. GILES, D. JOHNSON, J. BOWEN & C. F. REILLY. 1978. *In* Neutral Proteases of Human Polymorphonuclear Leukocytes. K. Havemann & A. Janoff, eds.: 118–128. Urban & Schwarzenberg. Baltimore-Munich.
16. RAO, N. V., N. G. WEHNER, B. C. MARSHALL, W. R. GRAY, B. H. GRAY & J. R. HOIDAL. J. Biol. Chem., in press.
17. SINHA, S., W. WATOREK, S. KARR, J. GILES, W. BODE & J. TRAVIS. 1987. Proc. Natl. Acad. Sci. USA **84**: 2228–2232.
18. SALVESEN, G., D. FARLEY, J. SHUMAN, A. PRZYBYLA, C. REILLY & J. TRAVIS. 1987. Biochemistry **26**: 2289–2293.
19. GERSHENFELD, H. K. & I. L. WEISSMAN. 1986. Science **232**: 854–858.
20. LOBE, C. G., B. B. FINLAY, W. PARANCHYCH, V. H. PAETKAU & R. C. BLEACKLEY. 1986. Science **232**: 858–861.
21. JENNE, D. E., C. REY, D. MASSON, K. K. STANLEY, J. HERZ, G. PLAETINCK & J. TSCHOPP. 1988. J. Immunol. **140**: 318–323.
22. JENNE, D. E., C. REY, J.-A. HAEFLIGER, B.-Y. QIAO, P. GROSCURTH & J. TSCHOPP. 1988. Proc. Natl. Acad. Sci. USA **85**: 4814–4818.
23. TRONG, H. L., D. C. PARMELEE, K. A. WALSH, H. NEURATH & R. G. WOODBURY. 1987. Biochemistry **26**: 6988–6994.
24. WOODBURY, R. G., N. KATUNUMA, K. KOBAYASHI, K. TITANI & H. NEURATH. 1978. Biochemistry **17**: 811–819.
25. SCHIMDT, J. & C. WEISSMANN. 1987. J. Immunol. **139**: 250–256.
26. NILES, J. L., R. T. McCLUSKEY, M. F. AHMAD & M. A. ARNAOUT. 1989. Blood **74**: 1888–1893.
27. WILDE, C. G., J. L. SNABLE, J. E. GRIFFITH & R. W. SCOTT. 1990. J. Biol. Chem. **265**: 2038–2041.
28. BAGGIOLINI, M., U. BRETZ & M. E. FEIGENSON. 1978. Agents Actions **8**: 3–11.

29. TRAVIS, J., P. J. GILES, L. PORCELLI, C. F. REILLY, R. BAUGH & J. POWERS. 1980. Ciba Found. Symp. **75:** 51–68.
30. GABAY, J. E., R. W. SCOTT, D. CAMPANELLI, J. GRIFFITH, C. WILDE, M. N. MARRA, M. SEEGER & C. F. NATHAN. 1989. Proc. Natl. Acad. Sci. USA **86:** 5610–5614.
31. ODEBERG, H. & I. OLSSON. 1975. J. Clin. Invest. **56:** 1118–1124.
32. VAN DER WOUDE, F. J., N. RASMUSSEN, S. LOBATTO, A. WIIK, H. PERMIN, L. A. VAN ES, M. VAN DER GIESSEN, G. K. VAN DER HEM & T. H. THE. 1985. Lancet **i:** 425.
33. JENNETTE, J. C., J. R. HOIDAL & R. J. FALK. 1990. Blood **75:** 2263–2265.

Elastin Degradation by Mononuclear Phagocytes[a]

STEVEN D. SHAPIRO,[b] EDWARD J. CAMPBELL,[c]
HOWARD G. WELGUS,[d] AND ROBERT M. SENIOR[b,e]

Divisions of Respiratory and Critical Care[b] and Dermatology[d]
Department of Medicine
Jewish Hospital at Washington University Medical Center
St. Louis, Missouri 63110

[c]Division of Respiratory, Critical Care and
Occupational Pulmonary Medicine
Department of Medicine
University of Utah Health Science Center
Salt Lake City, Utah 84132

Macrophages participate in a variety of phagocytic, immune, and inflammatory processes. By secreting proteinases, proteinase inhibitors, and cytokines that affect proteinase production by other cells, macrophages also play a role in the degradation and remodeling of the extracellular matrix.[1] Cigarette smoking is associated with a marked increase in the number of alveolar macrophages recovered from bronchoalveolar lavage,[2] and histologic examination of smokers' lungs demonstrates infiltration of macrophages within and surrounding the small airways where destructive changes in emphysema are most prominent.[3] These observations and the close association between smoking and emphysema suggest that human alveolar macrophages play a role in the pathogenesis of pulmonary emphysema. Because degradation of lung parenchymal elastic fibers appears to be a central feature in the development of emphysema, interest has become directed at the elastolytic activity of alveolar macrophages.[4,5] This article reviews information relating to the potential of macrophages and macrophage precursors to promote elastin degradation through the production and release of elastolytic enzymes. It should be recognized, however, that alveolar macrophages can affect the elastase-antielastase balance of the lungs in diverse ways (TABLE 1).

ELASTASE EXPRESSION DURING HUMAN MONONUCLEAR PHAGOCYTE DEVELOPMENT

Human mononuclear phagocytes synthesize and release proteinases capable of degrading extracellular matrix. The spectrum of such proteinases varies with the developmental stage of the cells.[6] We used human promonocyte-like cells, U937, to study the profile of elastases and other matrix-degrading proteinases

[a] This work was supported by USPHS grant HL29594.
[e] Address for correspondence: Robert M. Senior, M.D., Respiratory and Critical Care Division, Jewish Hospital at Washington University Medical Center, 216 S. Kingshighway, St. Louis, Missouri 63110.

produced during mononuclear phagocyte differentiation. Upon exposure to phorbol esters, such as 12-o-tetradecanoyl-phorbol-13-acetate (TPA), U937 cells acquire characteristics of more mature mononuclear phagocytes.[7] We found that U937 cells contain neutrophil elastase and cathepsin G; however, after exposure to TPA, there is a delayed (12–16 hour) but sharp decline in the cellular content of these serine proteinases, with minimal activity after 72 hours (FIG. 1). These findings reflect changes in gene expression. Nuclear runoff assays in basal U937 cells demonstrate active transcription of cathepsin G; exposure to TPA markedly down-regulates transcription, probably by stimulating the synthesis of a suppressor protein.[8]

In contrast to their production of serine proteinases, undifferentiated U937 cells do not produce interstitial collagenase and they secrete only small amounts of the tissue inhibitor of metalloproteinases (TIMP), whereas TPA-differentiated U937 cells show high levels of synthesis of collagenase and TIMP following a 12–16-hour lag time.[9] These changes in production of collagenase and TIMP are matched by similarly delayed increases in steady-state mRNA. Production of 92-kD type IV collagenase, a metalloenzyme with potent gelatinolytic activity in

TABLE 1. Effects of Alveolar Macrophages on Elastase-Antielastase Balance[a]

Increased elastase activity	
Release of elastase activity:	Metalloelastase, cysteinyl elastase (cathepsin L), serine (neutrophil) elastase
Inactivation of α_1-antiproteinase:	Oxidation by activated reactive oxygen species, proteolysis by metalloelastase and cathepsin L
Recruitment of neutrophils:	Secretion of chemoattractants
Stimulation of elastase release from neutrophils:	Secretion of neutrophil secretagogue activity
Decreased elastase activity	
Release of elastase inhibitors:	Alpha$_1$-antiproteinase, alpha$_2$-macroglobulin, tissue inhibitor of metalloproteinases (TIMP), cystatin C
Receptor-mediated endocytosis of elastases:	Neutrophil elastase, elastase-alpha$_2$-macroglobulin complexes

[a] Adapted from Senior and Kuhn.[5]

addition to its capacity to degrade basement membrane collagen, also increases markedly with TPA-induced differentiation of U937 cells.[10] Thus, the differentiated U937 cells have a profile of metalloproteinase production similar to that of normal alveolar macrophages.

Takahashi et al.[11] studied the expression of the neutrophil elastase gene in HL-60 cells, a human myelomonocytic cell line that can be induced to differentiate along monocytic or neutrophilic pathways. In the basal state, HL-60 cells were found to express neutrophil elastase mRNA, and cells induced along the myelocytic lineage with dimethyl sulfoxide expressed increased amounts of neutrophil elastase steady-state mRNA. Maturation along the monocytic lineage with phorbol esters, however, resulted in decreased neutrophil elastase mRNA levels. No elastase mRNA was detected in mature neutrophils, monocytes, or macrophages.

Campbell et al.[12] found that normal peripheral blood monocytes contain immunoreactive neutrophil elastase and cathepsin G, which are rapidly released in response to degranulating agents. The amount of each enzyme is approximately 6% of that in neutrophils (FIG. 2). Further analysis revealed that a small subpopu-

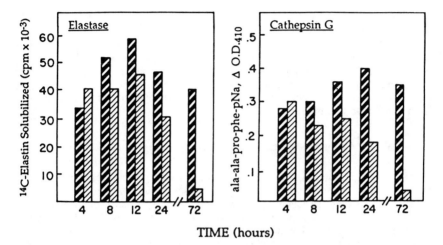

FIGURE 1. Time course of U937 cell elastase and cathepsin G activities after incubation with TPA (*thick lines,* ◪) or no treatment; control (*thin lines,* ▨). Elastase and cathepsin G activities were determined by measuring radioactivity following exposure of cell extracts at neutral pH with ^{14}C-labeled insoluble elastin[36] and succinyl-alanyl-alanyl-prolyl-phenyl-alanyl-*p*-nitroanilide,[37] respectively. (Reproduced from Senior *et al.*[6] with permission.)

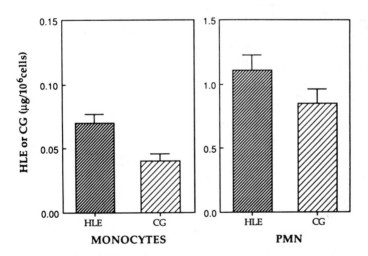

FIGURE 2. Immunoreactive human leukocyte elastase (HLE) and cathepsin G (CG) in extracts of monocytes and neutrophils. Whole cells were extracted with 10 mM phosphate, 0.02% Triton X-100, 1.0 M NaCl, pH 7.4, and subjected to competitive binding ELISA. Data are the means of triplicate assays of lysates from each donor: for monocytes and neutrophils, $n = 23$ and 19 donors, respectively. The heights of the bars represent mean values ± SEM. Note that the amounts of HLE and CG were similar in each cell type; the monocyte contents of HLE and CG were 6.2% and 5.2% of neutrophil content, respectively. (Reproduced from Campbell *et al.*[12] with permission from J. Immunol.)

lation of monocytes (15–20% of total) accounts for the enzymes. Most of the cells contain little or none of the enzymes.

Overall, these results with monocyte precursors and monocytes demonstrate that the genes for the serine proteinases neutrophil elastase and cathepsin G are expressed transiently during monomyelocytic development in cells destined to be monocytes, and that a subpopulation of monocytes resemble neutrophils in having granules containing neutrophil elastase and cathepsin G that can be released from the cells on stimulation.

Macrophage Elastase Activities

Mature human macrophages do not contain serine proteinases, but they have the capacity to secrete several metalloproteinases capable of degrading extracellular matrix (TABLE 2). Despite the broad substrate specificity of these metalloenzymes, however, none has been shown to possess substantial elastolytic activity. In fact, the ability of human macrophages to degrade elastin at all has been controversial for years. In the following section we summarize information about a well-characterized macrophage elastase, the metalloelastase produced by murine macrophages.

22-kD Murine Metalloelastase

Werb and Gordon[13] were the first to identify a macrophage-derived elastase. They used mouse peritoneal macrophages to generate conditioned media and demonstrated elastase activity using elastin-agar plates. The inhibition profile of the elastase activity did not fit with a serine proteinase, and it exhibited a narrower range of substrate specificity than did serine elastases. Banda and Werb[14] subsequently isolated the enzyme, confirmed it to be a metalloproteinase that requires zinc for activity, and found that it has a molecular weight of 22 kD. Kettner and Shaw[15] determined its peptide bond specificity using the beta-chain of insulin as a reference substrate. The enzyme was shown to be an endopeptidase that cleaves peptide bonds with leucine contributing to the amino group. Similar macrophage metalloelastase activity has been found in varying degrees in other species including the cow,[16] monkey,[17] and rabbit,[13] but not in humans.

In addition to its direct elastolytic activity, murine macrophage 22-kD metalloelastase might augment elastin degradation indirectly by catalytically degrading and inactivating alpha$_1$-antiproteinase, the major physiologic inhibitor of neutrophil elastase.[18] The cleavage affects a single peptide bond between Pro^{357} and Met^{358} of alpha$_1$-antiproteinase, resulting in 4.2-kD and 48-kD fragments that remain associated but alter the structure of the protein.[19] The 4.2-kD fragment is chemotactic for neutrophils, further suggesting that metalloelastase might promote elastin degradation in multiple ways.[20]

We have extended the characterization of murine metalloelastase using enzyme purified from media conditioned by P388D1 murine macrophage-like cells (Senior, Griffin, and Welgus, unpublished observations). The amino acid sequence of the amino-terminus shows similarity with three human metalloproteinases: interstitial collagenase, 72-kD type IV collagenase, and stromelysin.[21] Also, similar to other metalloenzymes, this proteinase is secreted as a zymogen. By casein-PAGE analysis of freshly harvested conditioned medium, the proenzyme has a molecular mass of about 36 kD, and the proenzyme can be converted to the 22-kD active form by treatment with trypsin. It is less effectively converted by organomercurials.

HUMAN MACROPHAGE-MEDIATED ELASTOLYSIS

Internalized Neutrophil Elastase

In contrast to murine macrophages, detecting and characterizing elastase activity with human macrophages have been problematic. Early studies consisting of assays of cell extracts gave negative results. Later approaches using macrophage-conditioned medium led to the finding of elastase activity, but characterization of the elastase has proven difficult. Rodriguez *et al.*[22] found elastolytic activity in conditioned media from alveolar macrophages obtained from bronchoalveolar lavage of smokers but not nonsmokers. The source of the elastase activity was unknown; however, its capacity to attack the synthetic peptide substrate succinyl-L-alanyl-L-alanyl-L-alanine-p-nitroanilide (SLAPN) indicated

TABLE 2. Macrophage Metalloproteinases

Proteinase	Proenzyme Molecular Weight (kD)	Substrates
Interstitial collagenase	52,57	Collagens types I–III, VII, X, alpha$_1$-antiproteinase
Stromelysin	57,60	Proteoglycans, laminin, fibronectin, alpha$_1$-antiproteinase
Type IV collagenase/ gelatinase	92,72	Collagen types IV, V, denatured collagens, fibronectin
Elastase (murine)	36	Elastin, proteoglycans, type IV collagen, fibronectin, alpha$_1$-antiproteinase

that it was likely to be a serine proteinase. Other investigators have shown that the elastase activity of conditioned media from human alveolar macrophages has a similar profile to that of neutrophil elastase, and that elastolysis is not inhibited by cycloheximide.[23] Moreover, additional studies indicated that macrophages can internalize neutrophil elastase via receptor-mediated binding and subsequently release the enzyme still in an active form.[24,25] Therefore, neutrophil elastase internalized by macrophages could be responsible for macrophage-mediated elastolysis without invoking the synthesis of an elastase by macrophages.

Potential Role of Cathepsin L

Problems in detecting elastase activity in the early studies of cultured human alveolar macrophages may have been a result of proteinase inhibitors simultaneously secreted into the conditioned media employed. To circumvent this potential complication, Chapman and associates[26,27] used an experimental system consisting of living alveolar macrophages cultured in contact with adherent radiolabeled elastin. They showed that human alveolar macrophages can degrade insoluble elastin and that the elastin-degrading activity persists in serum-containing media. These investigators also demonstrated that human macrophages synthesize cathepsin L, a cysteine proteinase that cleaves collagen, proteoglycans, and elastin. On the basis of inhibition by the synthetic inhibitor Z-phenylalanine-phenylalanine-diazomethylketone, cathepsin L appeared to be at least partly responsible for the observed elastolysis.[28] Cathepsin L is synthesized as a proenzyme of 29 kD and stored in lysosomes, where it is converted to a 25-kD species that is active at an acid pH. Although newly synthesized enzyme molecules are directed into lysosomes, the chaneling process is imperfect because small amounts of the precursor forms of the enzyme are released into the media by human alveolar macrophages in culture.[29] Like metalloelastase, cathepsin L cleaves alpha$_1$-antiproteinase.[30]

Indirect Role of Plasminogen and Plasminogen Activator

Human alveolar macrophages exhibited minimal elastolytic activity when cultured with a more complex, elastin-rich matrix deposited by rat smooth muscle cells *in vitro*. The addition of plasminogen markedly augmented elastolytic activity.[31] Presumably, tissue plasminogen activator acted on plasminogen to form plasmin, which in turn degraded fibrin and glycoproteins that covered the elastin substrate. However, activation of proelastases is another possible mechanism. These findings corroborate work by Jones and Werb[32] with mouse macrophages and elastin-containing matrices. The synergism between macrophage elastases and other proteinases to degrade complex matrices more efficiently suggests that, unlike neutrophil elastase, the spectrum of substrates attacked by macrophage elastase(s) may be limited and that elastin degradation in extracellular matrix by macrophages requires combined activity of a number of proteinases.

Metalloelastase Activity

To further investigate solubilization of elastin by human macrophages we adopted the procedures of Chapman and associates.[33] Accordingly, human alveolar macrophages and murine macrophages (P388D1 cell line) were incubated on radiolabeled elastin substrate for varying lengths of time, and elastin degradation was measured by release of [³H] elastin peptides.[33] During the first 24 hours in culture, elastin degradation by either human or murine macrophages was minimal; however, the increase in elastolysis over the next 48 hours was large (FIG. 3). Despite the elastolysis, elastase activity was not found in the conditioned medium. Elastin degradation required close association between the cell and elastin substrate. When cells were cultured on micropore filters placed between the cells and the labeled substrate, or when plates were only partially coated with elastin and cells were plated adjacent to the substrate, elastolysis was minimal compared to that of cells plated directly on elastin (FIG. 4). These findings suggest that direct contact of cells with the substrate effectively excluded inhibitors, and they raise the possibility that contact with elastin has an inductive effect on elastase production by macrophages.

Elastin degradation produced by human macrophages and P388D1 cells had similar inhibitor profiles (FIG. 5A), consistent with metalloelastase activity. They

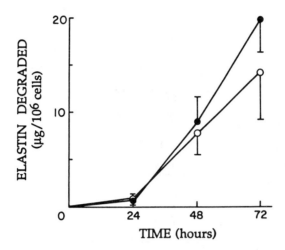

FIGURE 3. Time course of degradation of radiolabeled elastin adherent to tissue culture wells by human alveolar macrophages (*open circles*) and P388D1 cells (*closed circles*). Human alveolar macrophages were recovered by saline bronchoalveolar lavage from healthy smokers.[38] Cells were washed twice in Hanks' solution and plated in Dulbecco's modified Eagle medium (DMEM) with 10% fetal bovine serum. P388D1 murine macrophage-like cells were also cultured in DMEM with 10% fetal bovine serum. Plates contained [³H] elastin dried on the inside bottom of the wells. Elastin degradation was measured as release of [³H] into the conditioned media.[39] Human alveolar macrophages were collected from five separate bronchoalveolar lavages, and six separate experiments were performed on P388D1 cells. Each experiment was performed in triplicate. Values are mean ± SEM. (Reproduced from Senior *et al.*[33] with permission from Am. Rev. Respir. Dis.)

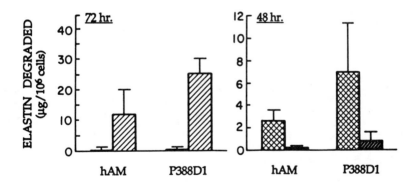

FIGURE 4. Effects of cell contact with elastin on the degradation of radiolabeled elastin by human alveolar macrophages and P388D1 cells. (*Left panel*) Cells were plated on Millicell inserts, which were then placed into tissue culture wells coated with elastin (*solid columns*), or plated directly on elastin-coated wells (*hatched columns*); four experiments. (*Right panel*) Cells were plated on tissue culture wells that were partially coated with elastin, with the cells plated either on the elastin (*cross-hatched columns*) or adjacent to it (*hatched columns*); three experiments. Each experiment was done in triplicate. Values are mean ± SEM. (Reproduced from Senior *et al.*[33] with permission from Am. Rev. Respir. Dis.)

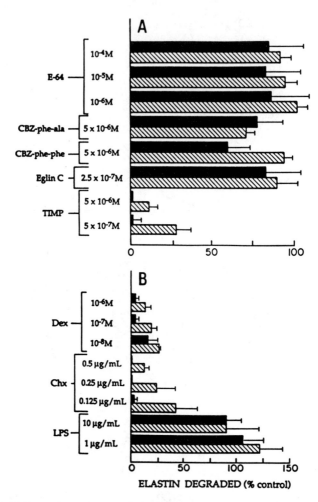

FIGURE 5. Effects of proteinase inhibitors (*panel* **A**), dexamethasone (Dex), cyclohexi-mide (Chx), and lipopolysaccharide (LPS) (*panel* **B**) on the degradation of radiolabeled elastin by human alveolar macrophages (*hatched columns*) and P388D1 cells (*solid columns*). Each inhibitor was tested in triplicate at the indicated concentration in four to eight separate experiments from separate lavages. Values are mean ± SEM, with the results expressed as the percentage of degradation observed in parallel control cultures. (Reproduced from Senior *et al.*[33] with permission from Am. Rev. Respir. Dis.)

were both inhibited in a dose-dependent manner by TIMP, but inhibitors of cysteine proteinases including cystatin C, epoxysuccinyl-leucyl-agmatine (E-64), CBZ-phe-phe-CHN$_2$, and CBZ-phe-ala-CHN$_2$ had little effect. The serine proteinase inhibitor eglin C was completely ineffective. Of note, not only was elastase activity not found in the medium, but also the medium was found capable of inhibiting isolated murine metalloelastase. Because nontoxic concentrations of cycloheximide drastically reduced elastolytic activity (FIG. 5B), elastin degradation appeared to be dependent on protein synthesis. Dexamethasone also diminished elastin degradation (FIG. 5B), a finding perhaps explained by steroid-induced suppression of elastase synthesis, because with other metalloproteinases, dexamethasone has been shown to block transcription.[34]

Thus, human alveolar macrophages plated on elastin appear to synthesize and secrete a metalloenzyme similar to the murine metalloelastase. These results agree with the findings of Chapman *et al.*[27] that human macrophages can degrade elastin, but differ in that our work demonstrated a prominent role for metalloproteinase activity rather than for cysteine proteinase activity. The reason for the discrepancy is not clear; however, it may be important that Chapman *et al.* studied elastolytic activity only during the first 24 hours of *in vitro* culture.

SUMMARY

During their development, mononuclear phagocytes express a changing profile of proteinases that may participate in the degradation of elastin and other extracellular matrix components. Neutrophil elastase is produced and stored in azurophil-like granules in immature mononuclear phagocytes. Monocytes contain small amounts of neutrophil elastase but do not synthesize the enzyme. Macrophages neither synthesize nor contain neutrophil elastase, but they can internalize and secrete scavenged neutrophil elastase. Human alveolar macrophages synthesize cysteine proteinases including cathepsin L, a lysosomal enzyme with elastolytic activity at an acidic pH. Macrophages from several animal species synthesize an approximately 22-kD metalloelastase that, in the mouse, is secreted as a zymogen of about 36 kD. In addition to its direct elastolytic properties, this metalloelastase may also promote elastolysis by cleaving alpha$_1$-antiproteinase and thus protecting neutrophil elastase from inhibition. A human counterpart of this enzyme has not yet been purified; however, the elastolytic activity of human macrophages appears to depend predominantly on the activity of one or more metalloproteinases.

Because elastin is intertwined with other matrix components in natural matrices, degradation of elastin *in vivo* probably involves cooperation of multiple proteinases to uncover macromolecules that mask the elastic fibers. Degradation of matrix may be localized to pericellular sites, where proteinases are protected from inhibitors and where potentially surface-bound enzymes may be concentrated.[35] Complete breakdown of matrix may be completed within the cells after partially cleaved molecules are internalized.

Growth and remodeling of the extracellular matrix must involve highly coordinated interactions between cells, cytokines, proteinases, proteinase activators and inhibitors, as well as the matrix itself. The intrapulmonary process resulting in emphysema probably involves equally complex interactions. Mononuclear phagocytes accumulate in large numbers in the lung in response to cigarette smoking, and they may play a role in the pathogenesis of the alveolar septal injury that characterizes pulmonary emphysema.

Note added in proof: In recent studies, we have found that the 92-kD type IV collagenase released by alveolar macrophages is elastolytic (Senior *et al.* 1990. J.

Cell Biol. **111:** 16a). These data suggest that this enzyme accounts for at least part of the metalloelastase activity of human macrophages.

ACKNOWLEDGMENT

The authors express appreciation to Gail L. Griffin for invaluable assistance in the unpublished studies cited.

REFERENCES

1. WERB, Z. 1989. Proteinases and matrix degradation. *In* Textbook of Rheumatology, 3rd ed. W. N. Kelley, E. D. Harris, S. Ruddy & C. B. Sledge, eds.: 300–321. W. B. Saunders. Philadelphia.
2. HUNNINGHAKE, G. W., J. E. GADEK, O. KAWANAMI, V. J. FERRINS & R. G. CRYSTAL. 1979. Inflammatory and immune processes in the human lung in health and disease: Evaluation by bronchoalveolar lavage. Am. J. Pathol. **97:** 149–198.
3. NIEWOEHNER, D. E., J. KLEINERMAN & D. B. RICE. 1974. Pathologic changes in the peripheral airways of young cigarette smokers. N. Engl. J. Med. **291:** 755–758.
4. NIEWOEHNER, D. E. 1988. Cigarette smoking, lung inflammation, and the development of emphysema. J. Lab. Clin. Med. **111:** 15–27.
5. SENIOR, R. M. & C. KUHN. 1988. The pathogenesis of emphysema. *In* Pulmonary Diseases and Disorders, 2nd ed. A. P. Fishman, ed.: 1209–1218. McGraw-Hill. New York.
6. SENIOR, R. M., N. L. CONNOLLY, E. J. CAMPBELL & H. G. WELGUS. 1987. Human mononuclear phagocytes produce different proteinases at different stages of differentiation: Observations with TPA-differentiated U937 cells. *In* Pulmonary Emphysema and Proteolysis 1986. J. C. Taylor & C. Mittman, eds.: 219–226. Academic Press. Orlando.
7. HARRIS, P. & P. RALPH. 1985. Human leukemic models of myelomonocytic development: A review of the HL-60 and U937 cell lines. J. Leukocyte Biol. **37:** 407–422.
8. HANSON, R. D., N. L. CONNOLLY, D. BURNETT, E. J. CAMPBELL, R. M. SENIOR & T. J. LEY. 1990. Developmental regulation of the human cathepsin G gene in myelomonocytic cells. J. Biol. Chem. **265:** 1524–1530.
9. WELGUS, H. G., N. L. CONNOLLY & R. M. SENIOR. 1986. 12-o-tetradecanoyl-phorbol-13-acetate-differentiated U937 cells express a macrophage-like profile of neutral proteinases: High levels of secreted collagenase and collagenase inhibitor accompany low levels of intracellular elastase and cathepsin. G. J. Clin. Invest. **77:** 1675–1681.
10. WELGUS, H. G., E. J. CAMPBELL, J. D. CURY, A. Z. EISEN, R. M. SENIOR, S. M. WILHELM & G. I. GOLDBERG. 1990. Neutral metalloproteinases produced by human mononuclear phagocytes: Enzyme profile, regulation and expression during cellular development. J. Clin. Invest. **86:** 1496–1502.
11. TAKAHASHI, H., T. NUKIWA, P. BASSET & R. G. CRYSTAL. 1988. Myelomonocytic cell lineage expression of the neutrophil elastase gene. J. Biol. Chem. **263:** 2543–2547.
12. CAMPBELL, E. J., E. K. SILVERMAN & M. A. CAMPBELL. 1989. Elastase and cathepsin G of human monocytes. Quantification of cellular content, release in response to stimuli, and heterogeneity in elastase-mediated proteolytic activity. J. Immunol. **143:** 2961–2968.
13. WERB, Z. & S. GORDON. 1975. Elastase secretion by stimulated macrophages. J. Exp. Med. **142:** 361–377.
14. BANDA, M. J. & Z. WERB. 1981. Mouse macrophage elastase. Purification and characterization as a metalloproteinase. Biochem. J. **193:** 589–605.
15. KETTNER, C. & E. SHAW. 1981. The specificity of macrophage elastase on the insulin B-chain. Biochem. J. **195:** 369–372.

16. VALENTINE, R. & G. L. FISHER. 1984. Characteristics of bovine alveolar macrophage elastase. J. Leuk. Biol. **35:** 449–457.
17. DeCREMOUX, H., W. HORNBECK, M. C. JAURAND, J. BIGNON & L. ROBERT. 1978. Partial characterisation of elastase-like enzyme secreted by human and monkey alveolar macrophages. Am. J. Pathol. **125:** 171–177.
18. BANDA, M. J., E. CLARK & Z. WERB. 1980. Limited proteolysis by macrophage elastase inactivates human alpha-1-proteinase inhibitor. J. Exp. Med. **152:** 1563–1570.
19. BANDA, M. J., E. J. CLARK, S. SINHA & J. TRAVIS. 1987. Interaction of mouse macrophage elastase with native and oxidized human alpha-1-antiproteinase. J. Clin. Invest. **79:** 1314–1317.
20. BANDA, M. J., A. G. RICE, G. L. GRIFFIN & R. M. SENIOR. 1988. α-1-proteinase inhibitor is a neutrophil chemoattractant after proteolytic inactivation by macrophage elastase. J. Biol. Chem. **263:** 4481–4484.
21. COLLIER, I. E., S. M. WILHELM, A. Z. EISEN, B. L. MARMER, G. A. GRANT, H. L. SELTZER, A. M. KRONBERGER, C. HE, E. A. BAUER & G. I. GOLDBERG. 1988. H-ras oncogene-transformed human bronchial epithelial cells (TBE-1) secrete a single metalloprotease capable of degrading basement membrane collagen. J. Biol. Chem. **263:** 6579–6587.
22. RODRIGUEZ, R. J., R. R. WHITE, R. M. SENIOR & E. A. LEVINE. 1977. Elastase release from human alveolar macrophages: Comparison between smokers and nonsmokers. Science **198:** 313–314.
23. HINMAN, L. M., C. A. STEVENS, R. A. MATTHAY & J. B. L. GEE. 1980. Elastase and lysozyme activities in human alveolar macrophages. Effect of cigarette smoking. Am. Rev. Respir. Dis. **121:** 263–271.
24. CAMPBELL, E. J., R. R. WHITE, R. M. SENIOR, R. J. RODRIGUEZ & C. KUHN. 1979. Receptor-mediated binding and internalization of leukocyte elastase by alveolar macrophages in vitro. J. Clin. Invest. **64:** 824–833.
25. CAMPBELL, E. J. 1982. Human leukocyte elastase, cathepsin G, and lactoferrin: A family of neutrophil granule glycoproteins which bind to an alveolar macrophage receptor. Proc. Natl. Acad. Sci. USA **79:** 6941–6945.
26. CHAPMAN, H. A., JR., O. L. STONE & Z. VAVRIN. 1984. Degradation of fibrin and elastin by intact human alveolar macrophages in vitro. Characterization of a plasminogen activator and its role in matrix degradation. J. Clin. Invest. **73:** 806–815.
27. CHAPMAN, H. A., JR. & O. L. STONE. 1984. Comparison of live human neutrophil and alveolar macrophage elastolytic activity in vitro. Relative resistance of macrophage elastolytic activity to serum and alveolar proteinase inhibitors. J. Clin. Invest. **74:** 1693–1700.
28. REILLY, J. J., R. MASON & R. YEE. 1989. Synthesis and processing of an elastolytic enzyme, cathepsin L, by human alveolar macrophages. Biochem. J. **257:** 493–498.
29. IMORT, M., M. ZUHLSDORT, U. FEIGE, A. HASILIK & K. VON FIGURA. 1983. Biosynthesis and transport of lysosomal enzymes in human monocytes and macrophages. Biochem. J. **214:** 671–678.
30. JOHNSON, D. A., A. J. BARRETT & R. W. MASON. 1986. Cathepsin L inactivates α-1-proteinase inhibitor by cleavage in the reactive site region. J. Biol. Chem. **261:** 14748–14751.
31. CHAPMAN, H. A., JR., J. J. REILLY & L. KOBZIK. 1988. Role of plasminogen activator in degradation of extracellular matrix protein by live human alveolar macrophages. Am. Rev. Respir. Dis. **137:** 412–419.
32. JONES, P. A. & Z. WERB. 1980. Degradation of connective tissue matrices by macrophages. II. Influence of matrix composition on proteolysis of glycoproteins, elastin, and collagen by macrophages in culture. J. Exp. Med. **152:** 1527–1536.
33. SENIOR, R. M., N. L. CONNELLY, J. D. CURY, H. G. WELGUS & E. J. CAMPBELL. 1989. Elastin degradation by human alveolar macrophages: A prominent role of metalloproteinase activity. Am. Rev. Respir. Dis. **139:** 1251–1256.
34. FRISCH, S. M. & H. E. RULEY. 1987. Transcription from the stromelysin promoter is induced by interleukin-1 and repressed by dexamethasone. J. Biol. Chem. **262:** 16300–16304.

35. WERB, Z., D. F. BAINTON & P. A. JONES. 1980. Degradation of connective tissue matrices by macrophages. III. Morphological and biochemical studies on extracellular, pericellular, and intracellular events in matrix proteolysis by macrophages in culture. J. Exp. Med. **152:** 1537–1553.
36. SENIOR, R. M., E. J. CAMPBELL, J. A. LANDIS, F. R. COX, C. KUHN & H. S. KOREN. 1982. Elastases derived from human monocyte-like cells: Comparisons with elastases derived from human monocytes and neutrophils and murine macrophage-like cells. J. Clin. Invest. **69:** 384–393.
37. SENIOR, R. M. & E. J. CAMPBELL. 1984. Cathepsin G in human mononuclear phagocytes: Comparison between monocytes and U937 monocyte-like cells. J. Immunol. **132:** 2547–2551.
38. SENIOR, R. M., E. J. CAMPBELL & B. VILLAGER. 1981. Obtaining and culturing alveolar macrophages. *In* Methods for Studying Mononuclear Phagocytes. D. O. Adams, P. J. Edelson & H. S. Koren, eds.: 69–83. Academic Press. New York.
39. BANDA, M. J., H. F. DOVEY & Z. WERB. 1981. Elastinolytic enzymes. *In* Methods for Studying Mononuclear Phagocytes. D. O. Adams, P. J. Edelson & H. S. Koren, eds.: 603–618. Academic Press. New York.

Neutrophil Proteinases

Caution Signs in Designing Inhibitors Against Enzymes with Possible Multiple Functions

JAMES TRAVIS, ADAM DUBIN, JAN POTEMPA,
WIESLAW WATOREK, AND ANNA KURDOWSKA

Department of Biochemistry
University of Georgia
Athens, Georgia 30602

Abnormal connective tissue turnover in man has been suggested to be associated with an imbalance between endogenous proteinases and their controlling inhibitors.[1] These enzymes are apparently primarily derived from neutrophils that are attracted to inflammatory sites, degranulate, and degrade damaged tissue. In the presence of sufficient proteinase inhibitors there appears to be tight regulation of such activities. When functional inhibitor levels are low, however, tissue damage may be exacerbated, leading to further connective tissue damage. The primary enzyme that apparently is responsible for tissue damage is neutrophil elastase (HNE) (and also, perhaps, cathepsin G). The rationale for this is strong because (1) the proteinase can degrade connective tissue proteins including elastin, collagen, and proteoglycan,[2] and (2) deficiencies in alpha$_1$-proteinase inhibitor (α_1-PI), the regulating inhibitor of this enzyme, have been correlated with the development of pulmonary emphysema.[3] For this reason significant efforts have been made to develop both protein and nonprotein inhibitors of HNE for therapy, to reduce uncontrolled tissue damage. The purpose of this report is to examine the possible multiple functions of HNE and other neutrophil proteinases, to describe briefly the properties of their controlling inhibitors, and to indicate what properties new inhibitors should have in order for them to be considered safe therapeutic agents.

ELASTOLYTIC ENZYMES

During the past several years emphasis has been placed on the role of HNE in connective tissue damage. When we consider that the concentration of the enzyme inside a neutrophil is between 0.036 and 0.058 mM (FIG. 1) and that the normal elimination rate is approximately 250 mg per day, it is easy to understand why the controlling inhibitor of HNE, α_1-PI, is present in such high concentrations in plasma[4] and also why individuals with a genetic deficiency are more susceptible to increased tissue damage, especially emphysema. However, what has not really been noted and is, indeed, startling is that HNE is a rather poor elastolytic enzyme. This is best seen from the data in TABLE 1 which clearly shows that when porcine pancreatic elastase is used as a standard for comparison of a series of animal and bacterial elastolytic proteinases, HNE ranks at the very bottom. Therefore, we can only conclude that its ability to degrade elastin is a minor, perhaps overstated function and that HNE must normally be involved in

Diameter of Single Cell	12–14 microns
Volume of Single Cell	$9.0–14.4 \times 10^{-13}$ liters
Enzyme in Single Cell	5.2×10^{-17} moles

Molar Concentration of Elastase in a Single Neutrophil

0.036–0.058 mM

FIGURE 1. Elastase concentration in human neutrophils.

the degradation of proteins in a nondiscriminant manner. However, as will be described, we suggest that HNE may have other functions that have been overlooked during the development of inhibitors to this enzyme.

FUNCTIONS OF HUMAN NEUTROPHIL ELASTASE

HNE is known to play a major role in the degradation of proteins ingested by neutrophils during phagocytosis. Indeed, this is believed to be its major role. However, it is likely that other functions may be attributed to this enzyme.

One possibility is that the ability of HNE to degrade connective tissue proteins may not be associated solely with tissue damage. Data are slowly accumulating that suggest that HNE, and probably cathepsin G as well, may be involved in the movement of the neutrophil through tissues during chemotactic responses.[5,6] This is supported by evidence of receptors (binding sites?) for HNE on the surface of the neutrophil.[7] Thus, it is possible that HNE which escapes from this cell may become bound to its external surface in a functional form and be utilized to degrade matrix proteins encountered during neutrophil migration. Although this is only a theoretical argument at the moment, some further biochemical data that are based on the isozyme patterns known to exist in both HNE and cathepsin G offer support. Examination of the carbohydrate moiety of the individual protein forms of these enzymes indicates the presence of significantly different types of structures with the minor forms of each proteinase containing bi-antennary complex carbohydrate side chains. The major forms of HNE and cathepsin G, on the other hand, have truncated high-mannose polysaccharide side chains (W. Watorek, unpublished observations) usually associated with lysosomal enzymes. Our suggestion is that these latter forms are primarily involved in phagocytosis, whereas the minor forms, which contain secretory carbohydrate side chains, may be transported to the cell surface where they can be utilized in cell movement. In effect, the carbohydrate side chain differences may allow for dual roles for HNE and cathepsin G, one involving phagocytic degradation and the other cell migration.

NEUTROPHIL ELASTASE AND THE ACUTE PHASE RESPONSE

It was recently established that HNE and HNE : α_1-PI complexes can stimulate further inhibitor production in monocytes and macrophages.[8] This suggests that HNE may play some role in the acute phase response. However, individuals who are genetically deficient in α_1-PI (phenotype Z) and who have liver disease also have a significant acute phase reaction,[9] suggesting that HNE must be functioning by a different mechanism to produce this response. Our data suggest that the probable mode of action of HNE is through interactions with a different inhibitor, namely, alpha$_1$-antichymotrypsin (α_1-Achy).

It is clear that HNE cannot be inhibited by α_1-Achy.[10] Rather, we recently found that this inhibitor is exceedingly sensitive to HNE, being inactivated rapidly by cleavage after the P_1-reactive site leucine residue. However, this is not the only proteinase capable of carrying out α_1-Achy inhibitor inactivation. As shown in FIGURE 2, the reactive site loop of this inhibitor is readily attacked by proteinases from a number of different sources, each of which causes complete inhibitor inactivation. These data suggest that in individuals deficient in α_1-PI some free HNE may be utilized to turn over functional α_1-Achy and perhaps other plasma proteinase inhibitors. Recently,[11] we found that both proteolytically inactivated α_1-Achy and α_1-Achy : cathepsin G complexes could stimulate the production of IL-6 in fibroblasts and that IL-6 would then, in turn, cause the increased production of acute phase proteins by HEP G2 cells. These data thus suggest that a primary function of both HNE and cathepsin G is to stimulate second-message cytokine production during inflammatory reactions, primarily through either α_1-Achy complex formation or inactivation. A schematic of this approach is shown in FIGURE 3.

PROTEINASE III AND TISSUE DEGRADATION

All of the foregoing data suggest multiple functions for neutrophil proteinases. These are summarized in TABLE 2 which also suggests that one other function of

TABLE 1. Comparative Elastinolytic Activities of Mammalian and Bacterial Proteinases

Enzyme	Elastinolytic Activity (%)[a]	Pathogen
Porcine pancreatic elastase	100	—
Dog pancreatic elastase	199	—
Human pancreatic elastase	17	—
Human neutrophil elastase	7	—
Horse leukocyte elastase	80	—
Human cathepsin G	2	—
Human cathepsin L	50	—
Flavobacterium immotum elastase	1,190	No
Streptomyces fradiae elastase	818	No
Bacillus thermoproteolyticus	419	No
Bacillus subtilis alkaline proteinase	80	No
Streptomyces griseus proteinase	53	No
Pseudomonas aeruginosa elastase	41	No
Staphylococcus aureus elastase	16	Yes
Flavobacterium meningosepticum proteinase	+	Yes
Clostridium histolyticum proteinase	+	Yes
Vibrio vulnificus proteinase	+	Yes
Aspergillus oryzae proteinase	+	No

[a] Elastinolytic activity is expressed as the percentage of porcine pancreatic elastase; + = no quantitative data available for comparison.

Sequence:

- Ala - Val - Lys - Ile - Thr - Leu - Leu - Ser - Ala - Leu - Val
      6   7   2   1,2 3   4,5

[1] Reactive Center for Cathepsin G Complex Formation

[2] Inactivation Sites for Human Neutrophil Elastase

[3] Inactivation Site for *S. marcescens* Metalloproteinase

[4] Inactivation Site for *S. aureus* Metalloproteinase

[5] Inactivation Site for *P. aeruginosa* Elastase

[6] Inactivation Site for Porcine Pancreatic Elastase

[7] Inactivation Site for Human Pancreatic Trypsin

FIGURE 2. Reactive site loop of alpha$_1$-antichymotrypsin.

some of these enzymes involves bacteria killing.[12] Included in this list is proteinase III, a newly described enzyme that is apparently present in a concentration near that of HNE and cathepsin G within neutrophil granules.[13] The function of this enzyme is in some dispute because it has been suggested to be important in both abnormal connective tissue degradation and bacteria killing.[14] Evidence for

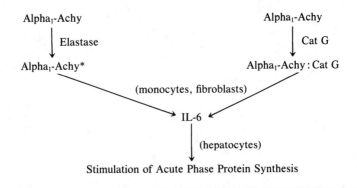

FIGURE 3. Role of alpha$_1$-antichymotrypsin in the acute phase reaction.

TABLE 2. Possible Functions of Human Neutrophil Proteinases

Function	Enzyme		
	HNE	Cat G	Pr III
Protein degradation	Yes	Yes	Yes
Bactericidal activity	No	Yes	Yes
Inhibitor turnover	Yes	?	?
Acute phase involvement	Yes	Yes	?
Chemotaxis	?	?	?

the latter is very strong; however, it is doubtful if this enzyme plays any major role in the development of emphysema, as has been suggested.[13] In fact, our evidence (TABLE 3) indicates that the enzyme is not important as an elastolytic enzyme because it is not inactivated by either elastase or cathepsin G chloromethyl ketone inhibitors.[13] Yet, either an elastase-specific synthetic inhibitor, MeO-suc-ala-ala-pro-val-chloromethyl ketone, or α_1-PI, can reduce the elastin-degrading activity of neutrophil granule extracts by 85%–88%. Therefore, the role of proteinase III certainly must be minor in elastin degradation. Indeed, its properties make it more like cathepsin G in terms of weak proteinase activity and strong bactericidal activity.

DESIGNING ELASTASE INHIBITORS

As suggested in TABLE 2, HNE may have multiple functions. Therefore, the production of new inhibitors to regulate its activity should be considered with some caution. If all of the naturally occurring elastase inhibitors in humans, including α_1-PI, the bronchial mucus inhibitor,[15] and the recently described cytosolic inhibitor,[16] are examined, a specific set of properties is found to exist in each, which new inhibitors should probably include. These are listed as follows:

1. All inhibitors should be oxidation sensitive. This is perhaps the most important property of all and allows for regulation of activity at or near the cell surface. Indeed, it is likely that during neutrophil migration, a halo of oxidizing agents provided by either the respiratory burst or the myeloperoxidase reaction exists around the cell as it moves towards its target. Consider what might happen if a nonoxidizable inhibitor was utilized as a therapeutic agent.[6]

2. All inhibitors should have a high degree of specificity for HNE. That is,

TABLE 3. Effect of Inhibitors on Neutrophil Neutral Proteinase Activity

Inhibitor	Substrate	Hydrolysis (% Control)
Cbz-gly-leu-ala-CMK	*E. coli*	20
Ac-ala-ala-ala-CMK	Elastin	14
Meo-suc-ala-ala-pro-val-CMK	Elastin	12
Z-gly-leu-phe-CMK[a]	Elastin	94
Alpha$_1$-PI	Elastin	15
Alpha$_1$-Achy	Elastin	98
DFP	Elastin	2

[a] CMK = chloromethylketone.

there should be a major drop in second order association rates for other proteinases which can be complexed. Considering the turnover rate of HNE in humans, the load of a new inhibitor will have to be relatively high. Therefore, interaction with other proteolytic systems will have to be extremely low.

3. Inhibitors should be rather large in size. Here, there may be some debate. However, the bronchial mucus inhibitor with a molecular weight around 10,000 kD would appear to be the smallest desired. If smaller molecules are devised, they may be able to saturate all tissues and perhaps enter neutrophils and other cells. Chemotaxis would be a difficult process if the area contains a battlefield of proteinase inhibitors. Larger molecules would diffuse more slowly, thus reducing their effective concentrations in and around cells. Of course, this may be a mute point if the inhibitor being considered is oxidation sensitive.

4. The use of human inhibitors should be encouarged. Not only can they be inactivated to signal for acute phase protein synthesis, but also complexes with HNE can easily be removed through receptor-mediated binding. What happens with nonhuman or synthetic inhibitor-HNE complexes has to be seriously questioned.

The most important question is what will happen if HNE function is significantly reduced by the use of synthetic, nonoxidizable, elastase inhibitors? Will any or all of the presumed functions suggested herein and summarized in TABLE 2 be compromised? Clearly, the pharmaceutical industry has a major challenge ahead. Yet, these and other potential functions for neutrophil proteinases must seriously be evaluated during inhibitor development.

REFERENCES

1. JANOFF, A. 1972. Annu. Rev. Med. **23:** 177–202.
2. STARKY, P. M. & A. J. BARRETT. 1976. Biochem. J. **155:** 255–263.
3. ERIKSSON, S. 1965. Acta Med. Scand. 177 (Suppl. 432): 20–24.
4. TRAVIS, J. & G. S. SALVESEN. 1983. Annu. Rev. Biochem. **52:** 655–709.
5. HORNEBECK, W., J. SOLEILHAC, J. TIXIER, E. MOCZAR & L. ROBERT. 1987. Cell Biochem. & Function **5:** 113–122.
6. BANDA, M. J., A. G. RICE, G. L. GRIFFIN & R. M. SENIOR. 1989. J. Biol. Chem. **263:** 4481–4484.
7. TOST, A. & W. HOLLE. 1986. J. Clin. Chem. Clin. Biochem. **24:** 299–308.
8. PERLMUTTER, D., J. TRAVIS & P. PUNSAL. 1988. J. Clin. Invest. **81:** 1774–1780.
9. PERLMUTTER, D., M. SCHIESINGER, J. PIERCE, P. PUNSAL & A. SCHWARTZ. 1989. J. Clin. Invest. **84:** 1555–1561.
10. TRAVIS, J., J. BOWEN & R. BAUGH. 1978. Biochemistry **17:** 5651–5657.
11. KURDOWSKA, A. & J. TRAVIS. 1990. J. Biol. Chem. **265:** 21023–21026.
12. BANGALORE, N., J. TRAVIS, V. ONUNKA, J. POHL & W. SHAFER. 1990. J. Biol. Chem. **265:** 13584–13588.
13. KAO, R., N. WEHNER, K. SKUBITZ, B. GRAY & J. HOIDAL. 1988. J. Clin. Invest. **82:** 1963–1973.
14. GABAY, J., R. SCOTT, D. CAMPANELLI, J. GRIFFITH, C. WILDE, M. MARRA, M. SEEGER & C. NATHAN. 1989. Proc. Natl. Acad. Sci. USA **86:** 5610–5614.
15. THOMPSON, R. & K. OHLSSON. 1986. Proc. Natl. Acad. Sci. USA **83:** 6692–6699.
16. REMOLD-O'DONNELL, E., J. NIXON & R. ROSE. 1989. J. Exp. Med. **169:** 1071–1086.

Role of Enzyme Receptors and Inhibitors in Regulating Proteolytic Activities of Macrophages[a]

HAROLD A. CHAPMAN, JR.[b]

Department of Medicine
Brigham and Women's Hospital and
Harvard Medical School
Boston, Massachusetts 02115

Proteolytic enzymes released by inflammatory cells are part of the orderly response of the lung to potentially injurious agents. In addition, circumstantial evidence links abnormalities in protease regulation to the pathobiology of several lung diseases, including emphysema.[1-5] Thus, the expression and regulation of proteases by inflammatory cells have received much attention. The expression and regulation of the proteolytic activities of macrophages differ in several important respects from those of other myeloid cells such as neutrophils and mast cells. These differences and their implications for the regulation of connective tissue turnover mediated by macrophages are the subject of this report.

Secretion of newly synthesized proteins by cells occurs primarily along two secretory pathways.[6,7] The first is a regulated secretory pathway in which newly synthesized proteins are initially channeled and then concentrated and stored for subsequent release in response to cellular stimulation. The morphologic hallmark of this pathway is the presence within cells of dense secretory granules, as is well described in neutrophils, mast cells, and eosinophils. Channeling of proteins into such granules depends on recognition markers on the particular proteins, and secretion occurs in response to specific cellular signals. The second secretory pathway is a constitutive one along which newly synthesized proteins are released without significant concentration soon after they are synthesized. The synthesis of each secretory protein entering this pathway, and perhaps their rate of secretion, may be regulated, but the secretory process itself is thought to occur without additional signaling.

A third major pathway along which some proteins are channeled after their synthesis is the pathway leading to accumulation of acidic hydrolases in lysosomes. Diversion of newly synthesized proteins into this pathway depends on the acquisition by the targeted proteins, posttranslationally, of mannose-6-phosphate residues and the subsequent recognition of these residues by mannose-6-phosphate receptors.[8] These receptors traffic for the most part between lysosomal vacuoles and the Golgi apparatus and thereby concentrate acidic hydrolases made by the cells into lysosomes.[9] Although evidence exists that lysosomal hydrolases

[a] This work was supported by HL35505 and a grant from the Council for Tobacco Research and performed during the tenure of Dr. Chapman as a Career Investigator of the American Lung Association.

[b] Address for correspondence: Harold A. Chapman, M.D., Respiratory Division, Brigham and Women's Hospital, 75 Francis St., Boston, MA 02115.

can be released directly from lysosomes,[10] these granules are not simply secretory storage granules, as their constituents have major functions inside the cell; once released into the neutral pH environment of the extracellular space, these enzymes are largely inactive. In this regard the release of lysosomal enzymes into zones of adhesion between cell surfaces and protein substrates, such as bone matrix or elastin, may be more similar to the process of secondary lysosome formation than to secretory granule secretion. The functional and morphologic properties of these pathways for protein movement after their synthesis have been the subject of recent reviews.[11,12]

Available evidence indicates that macrophages, unlike monocytes, neutrophils, and mast cells, do not concentrate proteases into secretory storage granules.[13] Macrophages do have a greatly expanded lysosomal compartment compared to that of other myeloid cells. Dense granules within macrophages are largely lysosomal in nature.[14] Most, if not all, nonlysosomal proteins destined for secretion are secreted soon after synthesis.[13,15] Therefore, whereas neutrophils and mast cells regulate their expression of secreted proteases not only by synthesis of the particular enzymes but also by their subsequent concentration and storage in specific secretory granules, macrophages must use other mechanisms to regulate their expression of secretory proteases. These distinctions from other myeloid cells have important implications for both the regulation of the proteolytic activities of macrophages and for the process by which macrophages use proteases to mediate the important biologic activities of these cells.

SUMMARY OF PROTEASES, PROTEASE INHIBITORS, AND PROTEASE RECEPTORS EXPRESSED BY HUMAN MONONUCLEAR PHAGOCYTES

Studies over several years from a number of laboratories have resulted in an accumulating list of proteases made by human mononuclear phagocytes (summarized in TABLE 1). These enzymes include all major mechanistic classes of proteases and range in activity from those with quite restricted proteolytic activity, such as Factor VII, to enzymes with potent and broad substrate specificity, such as cathepsin L. Of particular interest is the observation that for most of the proteases synthesized by monocytes/macrophages, the cells also synthesize a specific inhibitor. Furthermore, in virtually every situation where the cells synthesize a protease, the cells also express a specific receptor for that protease. As is also indicated in TABLE 1, macrophages are known to express several proteolytic enzymes that they do not actually synthesize. In these cases as well, the expression and regulation of the enzymes may depend on the concurrent expression by the cells of specific receptors and the availability of plasma-derived protease inhibitors. For example, human monocytes/macrophages are reported to express surface receptors for plasminogen, Factor Xa, and thrombin.[16–18]

The pattern of concomitant expression of enzyme and enzyme inhibitor distinguishes macrophages from other myeloid cells in which proteases are largely made, stored, and released after stimulation in the absence of protease inhibitors. It seems plausible that the absence of a regulated secretory pathway in macrophages underlies, at least in part, this difference in the pattern of protein expression. The concomitant synthesis of a receptor and an inhibitor for each protease by macrophages may represent an alternative, dual mechanism for compartmentalizing proteolytic activity. Macrophages could be viewed as using receptors and inhibitors to subserve functions similar to those for which neutrophils and other granulocytic cells use azurophilic and specific granules. This organization of the cell's secretory machinery could be expected to allow macrophages a greater adaptability to environmental signaling than that afforded by the prepackaging of enzymes. This is consonant with the known long life span of macrophages and the versatility of their secretory capabilities as compared to those of other myeloid cells.[14] Examples of the combined function of inhibitors and receptors in compartmentalizing macrophage proteolytic activities related to connective tissue turnover are discussed subsequently herein; similar physiology seems applicable to the regulation of neutral proteases initiating coagulation and complement pathways[19] (TABLE 1).

TABLE 1. Summary of Known Proteases, Protease Inhibitors, and Receptors for Proteases Expressed by Human Mononuclear Phagocytes

Protease	Protease[a] Inhibitor	Protease[b] Receptor	References
A. Enzymes Synthesized by Mononuclear Phagocytes			
Urokinase	PAI-1, PAI-2	Yes	26,44,45,35
Factor VII	?	Yes	20,46
Factor B	Factor H	Yes	47,48
C1	C1 Inhibitor	Yes (C1q)	47–50
Collagenase(s)	TIMP	?	51,52
Cathepsin B	Cystatin C	Yes	53–55
Cathepsin L	Cystatin C	Yes	54
Cathepsin D	?	Yes	53
Neutrophil[c]			
Elastase	Alpha$_1$-antitrypsin	Yes	56–58
Cathepsin G			
B. Enzymes Expressed but Not Synthesized by Mononuclear Phagocytes			
Plasminogen	Alpha$_2$-antiplasmin	Yes	18
Factor Xa	Antithrombin 111	Yes	16
Thrombin	Antithrombin 111	Yes	17

[a] PAI-1, PAI-2, and TIMP refer to plasminogen activator inhibitor types 1 and 2 and tissue metalloprotease inhibitor, respectively. In addition to the inhibitors listed in the table, monocytes in culture reportedly synthesize, release, and bind alpha$_2$-macroglobulin, a type 111 cystatin.[24]

[b] For this discussion a receptor is defined as a specific and saturable binding site on a cell membrane with no implied signal transducing function.

[c] Made by monocyte progenitors only.

It should be noted that the set of protease receptors expressed by monocytes/macrophages cannot be depicted accurately as a single class of receptors. Several of the receptors, such as those for tissue factor, plasminogen, and urokinase, are endocytized relatively slowly, whereas others, such as neutrophil elastase/cathepsin G receptors, have a fast turnover.[20–22] A major function of receptors of the tissue factor class is to provide cell surface or intracellular anchoring sites, whereas the function of receptors such as the elastase receptor may be found in enzyme clearance. Recent evidence suggests that, in addition to providing anchoring sites, receptors for serine proteases, such as the uPA receptor, may also transduce intracellular signals important to macrophage function.[23]

Although not shown in TABLE 1, macrophages also synthesize receptors that recognize protease/protease inhibitor complexes.[24,25] These receptors recognize determinants on the protease inhibitor component of the complex, and the specific protease inhibitors recognized are again known to be synthesized and secreted by macrophages, such as alpha$_2$-macroglobulin and alpha$_1$-antitrypsin. These receptors also appear to function as a mechanism for clearing inactivated protease.

EXTRACELLULAR MATRIX DEGRADATION BY HUMAN MACROPHAGES: A COOPERATIVE PROCESS INVOLVING NEUTRAL AND ACIDIC PROTEASES

Previous studies have established that human alveolar macrophages can degrade purified elastin by a contact-dependent process relatively resistant to plasma protease inhibitors.[26,27] Scanning electron photomicrographs illustrating purified bovine ligamentum elastin adherent to a plastic tissue culture surface (panel A), and the close contact between macrophages and this elastin in coculture (panels B and C) are shown in FIGURE 1. The extent of degradation and the linear kinetics, at least in our experience, of the degradation of this substrate (^{3}H-bovine ligamentum elastin) by live macrophages in coculture are shown in FIGURE 2. The enzymatic mechanisms leading to elastin degradation are unsettled but depend, in part, on whether the elastin is in a purified form (FIG. 1A) or is part of a complicated extracellular matrix. Panel D of FIGURE 1 illustrates elastin within a matrix created by mixing purified elastin and fibrin. In the case of either the fibrin/elastin matrix or a natural extracellular matrix containing elastin, degradation by macrophages depends on both an active plasminogen activator (uPA) and cysteine proteases that are directly elastolytic.[26-30] Recent studies by Senior and colleagues[27] also implicate a metalloprotease in elastin degradation by human alveolar macrophages, although the enzyme remains to be defined. The function of the uPA/plasmin enzyme system appears to be to degrade elastin-associated proteins and thereby to allow close contact between macrophages and elastin. Presumably, the cells are then capable of acidifying a substrate-cell surface interface, allowing lysosomal granules that contain elastolytic proteases to migrate to the interface and mediate elastolysis. Direct evidence for such substrate-cell surface acidification and for migration of lysosomal granules to zones of contact has been obtained in both rat osteoclasts and murine peritoneal macrophages.[31-34] No data are available in humans.

ROLE OF RECEPTORS AND INHIBITORS IN COMPARTMENTALIZING ENZYMES MEDIATING MATRIX DEGRADATION BY MACROPHAGES

TABLE 1 shows that macrophages synthesize uPA as well as specific receptors and inhibitors for uPA. Normal alveolar macrophages express ~190,000 surface receptors for uPA per cell with an average Kd of 5 nM.[35] When stimulated with endotoxin, these cells also synthesize specific inhibitors of uPA.[36] Stimulation of monocytes with cytokines is reported to up-regulate receptor number.[37] Recent studies of the interaction of uPA, the uPA receptor, and uPA inhibitors on both blood monocytes and alveolar macrophages shed new light on the physiologic role of these proteins. Complexes of uPA and PA inhibitors bind to uPA receptors more slowly and with much lower affinity than does uncomplexed uPA.[35,38] Conversely, once uPA is bound to cell-surface uPA receptors, the active enzyme is protected from inhibition by the PA inhibitors.[34,38,39] Because the turnover of

FIGURE 1. Scanning electron photomicrographs of bovine ligamentum elastin in coculture with macrophages. **(A)** Purified elastin adherent to tissue culture plastic (magnification × 1,000). **(B)** Macrophages adherent to the elastin (magnification × 1,000). **(C)** High power view illustrating the close and extensive contact between macrophages and elastin (magnification × 2,000). **(D)** A matrix composed of purified fibrin and elastin (magnification × 2000). Enzymes mediating degradation of this elastin depend on whether the elastin is in pure form **(A)** or part of a matrix **(D)**.

these receptors on macrophages is a matter of hours rather than minutes, these receptors provide a relatively protected environment for uPA activity in the presence of molar excesses of uPA inhibitors. Viewed in another way, the receptor and inhibitor cooperate in focusing uPA activity to the site of their primary function on macrophages, the cell surface. Prior studies show that similar principles apply to the regulation of plasmin activity. Plasminogen and plasmin bind to both cell surfaces and extracellular matrix proteins by a binding site separate from the active site of the enzyme.[18,40] Once bound, plasmin is protected from inhibition by its major inhibitor, alpha$_2$-antiplasmin. These results may explain why macrophages can degrade plasmin-sensitive substrates such as fibronectin and laminin even in the presence of serum that contains inhibitors for both uPA and plasmin. These data also underscore why plasminogen-dependent proteolysis requires close contact between cells and the extracellular matrix under physiologic conditions.

Macrophages compartmentalize their lysosomal cysteine protease activity by similar mechanisms. Cysteine proteases such as cathepsins B, H, and L, like

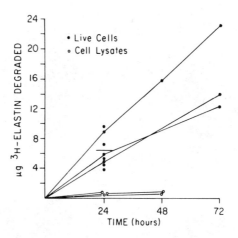

FIGURE 2. Degradation of ^{3}H-ligamentum elastin by human alveolar macrophages *in vitro*. Macrophages are cultured (10^6 cells/16 mm culture well) in the presence of 1% human serum and the elastin. At various times, as indicated in the figure, the media were sampled for soluble radioactivity. The specific activity of the ^{3}H-elastin was 1,200 cpm/μg protein. Each data point represents a mean of duplicate determinations of cell cultures from separate volunteers. The data show that only live cells (●) and not cells lysed at neutral pH (○) solubilized elastin.

other proteins targeted for lysosomes, acquire mannose-6-phosphate residues shortly after synthesis. The mannose-6-phosphate serves as a recognition system for receptors that channel these proteases into lysosomes. In addition, our recent studies show that macrophages synthesize a high affinity inhibitor of cysteine proteases, cystatin C.[41,42] Cystatin C antigen is not found in the lysosomal compartment but instead is continuously secreted along the constitutive secretory pathway of macrophages. Over 90% of the total cystatin C made by normal human alveolar macrophages during the first 24 hours *in vitro* is secreted. The inhibitor appears to be a major secretory product of these macrophages, accounting for 8–12% of the total protein secreted during this time. Available evidence suggests that one function of cystatin C is to localize the activity of cysteine proteases to their primary site of function, lysosomes. Although the bulk of cysteine proteases is targeted to lysosomes, a small fraction presumably escapes this channeling mechanism and travels along the constitutive secretory pathway until secreted into the medium. Cystatin C functions to inhibit such enzymes as judged by experiments showing that removal of cystatin C from the culture supernatant of alveolar macrophages uncovers latent cysteine protease activity.[41] Consequently, the combination of mannose-6-phosphate receptors and cystatin C appears to cooperate in compartmentalizing acid protease activity to the lysosomal compartment. Because secretory vesicles leaving the Golgi and migrating to the cell surface are reportedly acidic,[43] cystatin C may function to protect other secretory proteins from hydrolysis before their release from the cell. However, no direct evidence for such a function exists.

Interestingly, the amounts of cystatin C released by alveolar macrophages derived from cigarette smokers are ~50% less than those from nonsmoker macrophages, and there is a corresponding increase in the activity of secreted cysteine proteases by these cells.[41] Whether the secreted cysteine proteases contribute to extracellular matrix degradation by macrophages in an acidic pericellular microenvironment and whether cystatin C functions to modulate this process remain to be defined. Although polarized secretion is an established phenomenon in cells of the monocyte/macrophage lineage,[32] no data are available regarding the potential vectorial release of cystatin and/or cysteine proteases by macrophages.

SUMMARY AND CONCLUSIONS

Expression of proteases by neutrophils and other cells with a prominent regulated secretory pathway is determined largely by stimulus-response secretion of proteins prepackaged in high concentration. The regulated secretory pathway is apparently minor in macrophages, and instead proteases are either channeled into lysosomes or secreted constitutively. Posttranslational regulation of macrophage proteases then depends on compartmentalizing enzymes to their sites of primary function. Available data suggest that cells use both specific receptors and inhibitors to accomplish this. Viewed in this context protease inhibitors primarily function to inhibit enzyme not bound to their receptor. Consonant with this model of regulation, connective tissue turnover by macrophages is a contact-dependent process relatively resistant to exogenous macromolecular inhibitors. Although limited information is available regarding determinants that modulate matrix metabolism by human macrophages, this model suggests that determinants of adhesion and colocalization of enzyme and substrate would be as or more important than alterations of inhibitors in the microenvironment of the cell.

ACKNOWLEDGMENT

The author thanks Dr. Farhad Moatamed for assistance with the scanning electron microscopy.

REFERENCES

1. JANOFF, A., B. SLOAN, G. WEINBAUM, V. DAMIANO, R. A. SANDHAUS, J. ELIAS & P. KIMBEL. 1977. Experimental emphysema induced with purified human neutrophil

elastase, tissue localization of the instilled protease. Am. Rev. Respir. Dis. **115:** 461–478.

2. SENIOR, R. M., H. TEGNER, C. KUHN, K. OHLSSON, B. S. STARCHER & J. A. PIERCE. 1977. The induction of pulmonary emphysema with human neutrophil elastase. Am. Rev. Respir. Dis. **116:** 469–475.

3. KAO, R. C., N. G. WEHNER, K. M. SKUBITZ, B. H. GRAY & J. R. HOIDAL. 1988. Proteinase 3. A distinct human polymorphonuclear leukocyte proteinase that produces emphysema in hamsters. J. Clin. Invest. **82:** 1963–1973.

4. SITRIN, R. G., P. G. BRUBAKER & J. FANTON. 1987. Tissue fibrin deposition during acute lung injury in rabbits and its relationship to local expression of procoagulants and fibrinolytic activities. Am. Rev. Respir. Dis. **135:** 930.

5. REILLY, J. J., JR. & H. A. CHAPMAN, JR. 1988. Association between macrophage plasminogen activator activity and lung function in young cigarette smokers. Am. Rev. Respir. Dis. **138:** 1422–1428.

6. FARQUHAR, M. 1983. Multiple pathways of exocytosis, endocytosis, and membrane recycling: Validation of a Golgi route. Fed. Proc. **42:** 2407–2413.

7. BURGESS, T. L. & R. B. KELLY. 1987. Constitutive and regulated secretion of proteins. Ann. Rev. Cell. Biol. **3:** 243–293.

8. SLY, W. S., H. D. FISCHER, A. GONZALEZ-NORIEGA, J. H. GRUBB & M. NATOWICZ. 1981. Role of the 6-phosphomannosyl-enzyme receptor in intracellular transport and adsorptive pinocytosis of lysosomal enzymes. Methods Cell Biol. **23:** 191–212.

9. IMORT, M., M. ZUHLSDORT, V. FEIGE, A. HASILIK & K. VON FIGURA. 1983. Biosynthesis and transport of lysosomal enzymes in human monocytes and macrophages. Biochem. J. **214:** 671–678.

10. VAES, G. 1968. On the mechanism of bone resorption: The action of parathyroid hormone on the excretion and synthesis of lysosomal enzymes and on the extracellular release of acid by bone cells. J. Cell. Biol. **39:** 676–697.

11. KORNFELD, S. & I. MELLMAN. 1989. The biogenesis of lysosomes. Ann. Rev. Cell. Biol. **5:** 483–526.

12. FARQUHAR, M. G. 1985. Progress in unraveling pathways of Golgi traffic. Ann. Rev. Cell. Biol. **1:** 447–488.

13. TARTAKOFF, A. & P. VASSALLI. 1979. Comparative studies of intracellular transport of secretory proteins. J. Cell. Biol. **79:** 694–707.

14. FELS, A. O. S. & Z. A. COHN. 1986. The alveolar macrophage. J. Appl. Phys. **60:** 353–369.

15. WERB, Z. & J. R. CHIN. 1983. Apoprotein E is synthesized and secreted by resident and thioglycollate-elicited macrophages but not by pyran copolymer or bacillus calmette-guerin-activated macrophages. J. Exp. Med. **158:** 1272–1293.

16. TRACEY, P. B., M. S. ROHRBACH & K. G. MANN. 1983. Functional prothrombinase complex assembly on isolated monocytes and lymphocytes. J. Biol. Chem. **258:** 7264–7267.

17. GOODNOUGH, L. T. & H. SAITO. 1982. Specific binding of thrombin by human peripheral blood monocytes. J. Lab. Clin. Med. **99:** 873–884.

18. PLOW, E. F., D. E. FREANEY, J. PLESCIA & L. A. MILES. 1986. The plasminogen system and cell surfaces: Evidence for plasminogen and urokinase receptors on the same cell type. J. Cell. Biol. **103:** 2411–2420.

19. WALSH, P. N. 1981. Platelets and coagulation proteins. Fed. Proc. **40:** 20860–20869.

20. BROZE, G. J. 1982. Binding of Factor VII and VIIa to monocytes. J. Clin. Invest. **70:** 526–535.

21. CAMPBELL, E. J. 1982. Human leukocyte elastase, cathepsin G and lactoferrin: Family of neutrophil granule glycoproteins that bind to an alveolar macrophage receptor. Proc. Natl. Acad. Sci. **79:** 6941–6945.

22. STOPPELLI, M. P., A. CORTI, A. SOFFIENTINI, G. CASSANI, F. BLASI & R. K. ASSOIAN. 1985. Differentiation-enhanced binding of the amino-terminal fragment of human urokinase plasminogen activator to a specific receptor on U937 monocytes. Proc. Natl. Acad. Sci. **82:** 4939–4943.

23. NUSRAT, A. R. & H. A. CHAPMAN, JR. 1991. Autocrine role for urokinase in myeloid cell differentiation. J. Clin. Invest., in press.

24. HOVI, T., D. MOSHER & A. VAHERI. 1977. Cultured human monocytes synthesize and secrete alpha-2-macroglobulin. J. Exp. Med. **145:** 1580–1589.
25. PERLMUTTER, D. H., G. I. GLOVER, C. S. SCHASTEEN & R. J. FALLON. 1990. Identification of a serpin-enzyme complex receptor on human hepatoma cells and human monocytes. Clin. Res. **38:** 423A.
26. CHAPMAN, H. A., JR., O. L. STONE & Z. VAVRIN. 1984. Degradation of fibrin and elastin by human alveolar macrophages *in vitro*. Characterization of a plasminogen activator and its role in matrix degradation. J. Clin. Invest. **73:** 806–815.
27. SENIOR, R. M., N. L. CONNOLLY, J. D. CURY, H. G. WELGUS & E. J. CAMPBELL. 1989. Elastin degradation by human alveolar macrophages. A prominent role of metalloproteinase activity. Am. Rev. Respir. Dis. **139:** 1251–1256.
28. CHAPMAN, H. A., JR., J. J. REILLY & L. KOBZIK. 1988. Role of plasminogen activator in degradation of extracellular matrix protein by live human alveolar macrophages. Am. Rev. Respir. Dis. **317:** 412–419.
29. CHAPMAN, H. A., JR. & O. L. STONE. 1984. Comparison of live human neutrophil and alveolar macrophage elastolytic activities. Resistance of macrophage elastolysis to serum and alveolar proteinase inhibitors. J. Clin. Invest. **74:** 1693–1700.
30. MASON, R. W., D. JOHNSON, A. J. BARRET & H. A. CHAPMAN, JR. 1986. Elastolytic activity of human cathepsin L. Biochem. J. **122:** 925–927.
31. BARON, R., L. NEFF, D. LOUVARD & P. J. COURTOY. 1985. Cell mediated extracellular acidification and bone resorption: Evidence for a low pH in resorbing lacunae and localization of a 100 KD lysosomal membrane protein at the osteoclast ruffled border. J. Cell. Biol. **101:** 2210–2222.
32. WERB, Z., R. TAKEMURA, P. E. STENBERG & D. F. BAINTON. 1989. Directed exocytosis of secretory granules containing apolipoprotein E to the adherent surface and basal vacuoles of macrophages spreading on immobile immune complexes. Am. J. Pathol. **134:** 661–670.
33. AGGELER, G. & Z. WERB. 1982. Initial events during phagocytosis by macrophages viewed from outside and inside the cell. Membrane-particle interaction and clathrin. J. Cell. Biol. **94:** 613–623.
34. ETHERINGTON, D. J., D. PUGH & I. A. SILVER. 1981. Collagen degradation in an experimental inflammatory lesion: Studies on the role of macrophages. Acta Biol. Med. Germ. **40:** 1625–1636.
35. CHAPMAN, H. A., JR., P. BERTOZZI, L. SAILOR & A. R. NUSRAT. 1990. Alveolar macrophage urokinase receptors localize enzyme activity to the cell surface. Am. J. Phys.: Lung Cell Molec. Physiol. In press.
36. CHAPMAN, H. A., JR. & O. L. STONE. 1985. Human alveolar macrophages secrete a fibrinolytic inhibitor: Induction with endotoxin. Am. Rev. Respir. Dis. **132:** 569–575.
37. KIRCHHEIMER, J. C., Y.-H. NONG & H. G. REMOLD. 1988. IFN-gamma, tumor necrosis factor-alpha, and urokinase regulate the expression of urokinase receptors on human monocytes. J. Immunol. **141:** 4229–4234.
38. KIRCHHEIMER, J. C. & H. G. REMOLD. 1989. Functional characteristics of receptor-bound urokinase on human monocytes: Catalytic efficiency and susceptibility to inactivation by plasminogen activator inhibitors. Blood **74:** 1396–1402.
39. CHAPMAN, H. A., JR., Z. VAVRIN & J. B. HIBBS, JR. 1982. Macrophage fibrinolytic activity: Identification of two pathways of plasmin formation by intact cells and of a plasminogen activator inhibitor. Cell **28:** 653.
40. KNUDSEN, B. S., R. L. SILVERSTEIN, L. L. LEUNG, P. C. HASPEL & R. L. NACHMAN. 1986. Binding of plasminogen to extracellular matrix. J. Biol. Chem. **261:** 10765–10771.
41. CHAPMAN, H. A., J. J. REILLY, R. YEE & A. GRUBB. 1990. Identification of cystatin C, a cysteine proteinase inhibitor, as a major secretory product of human alveolar macrophages in vitro. Am. Rev. Respir. Dis. **141:** 698–705.
42. BARRETT, A. J., M. E. DAVIES & A. GRUBB. The place of human gamma-trace (cystatin C) amongst the cysteine proteinase inhibitors. Biochem. & Biophysic. Res. Comm. **120:** 631–636.
43. MELLMAN, I., R. FUCHS & A. HELENIUS. 1986. Acidification of the endocytic and exocytic pathways. Ann. Rev. Biochem. **55:** 603–700.

44. VASSALLI, J. D., J. M. DAYER, A. WOHLWEND & D. BELIN. 1984. Concomitant secretion of prourokinase and of a plasminogen activator-specific inhibitor by cultured human monocytes-macrophages. J. Exp. Med. **159:** 1653–1668.
45. VASSALLI, J. D., D. BACCINO & D. BELIN. 1985. A cellular binding site for the Mr 55,000 form of the human plasminogen activator, uorkinase. J. Cell. Biol. **100:** 86–92.
46. CHAPMAN, H. A., JR., D. FAIR, C. ALLEN & O. L. STONE. 1985. Human alveolar macrophages synthesize Factor VII *in vitro*. Possible role in interstitial lung disease. J. Clin. Invest. **75:** 480.
47. COLE, F. S., W. J. MATTHEWS, T. H. ROSSING *et al.* 1983. Complement biosynthesis by human alveolar macrophages. Clin. Immunol. Immunopathol. **27:** 153.
48. WHALEY, K. 1980. Biosynthesis of the complement components and the regulatory proteins of the alternative pathway by human peripheral blood monocytes. J. Exp. Med. **151:** 501.
49. BENSA, J. C., A. REBOUL & M. G. COLOMB. 1983. Biosynthesis in vitro of complement components C1q, C1s, and C1 inhibitor by resting and stimulated human monocytes. Biochem. J. **216:** 385–392.
50. ARVIEUX, J., A. REBOUL, J. C. BENSA & M. G. COLOMB. 1984. Characterization of the C1q receptor on a human macrophage cell line, U937. Biochem. J. **218:** 547–555.
51. WELGUS, H. G., E. J. CAMPBELL, Z. BAR-SHAVIT, R. M. SENIOR & S. L. TEITELBAUM. 1985. Human alveolar macrophages produce a fibroblast-like collagenase and collagenase inhibitor. J. Clin. Invest. **76:** 219–224.
52. HIBBS, M. S., J. R. HOIDAL & A. H. KANG. 1985. Secretion of a metalloproteinase by human alveolar macrophages which degrades gelatin and native Type V collagen. J. Cell. Biol. **101:** 216A.
53. MORTLAND, B. 1985. Cathepsin B activity in human blood monocytes during differentiation *in vitro*. Scand. J. Immunol. **22:** 9–16.
54. REILLY, J. J., JR., R. W. MASON, P. CHEN, V. SUKATME, L. JOSEPH, R. YEE & H. A. CHAPMAN, JR. 1989. Synthesis and processing of cathepsin L, an elastase, by human alveolar macrophages. Biochem. J. **257:** 493–498.
55. WARFEL, A. H., D. ZUCKER-FRANKLIN, B. FRANGIONE & J. CJISO. 1987. Constituitive secretion of cystatin C (gamma-trace) by monocytes and macrophages and its downregulation after stimulation. J. Exp. Med. **166:** 1912–1917.
56. PERLMUTTER, D. H., F. S. COLE, P. KILBRIDGE, T. H. ROSSING & H. R. COLTEN. 1985. Expression of the antiproteinase inhibitor gene in human monocytes and macrophages. Proc. Natl. Acad. Sci. **82:** 795–799.
57. SENIOR, R. M., E. J. CAMPBELL, J. A. LANDIS, F. R. COX, C. KUHN, III & H. S. KOREN. 1982. Elastase of U-397 monocytelike cells. Comparison with elastases derived from human monocytes and neutrophils and murine macrophagelike cells. J. Immunol. **132:** 2547.
58. CAMPBELL, E. J., E. K. SILVERMAN & M. A. CAMPBELL. 1989. Elastase and cathepsin G of human monocytes. J. Immunol. **143:** 2961–2968.

Role of Antileukoprotease in the Human Lung

J. A. KRAMPS,[a] A. RUDOLPHUS, J. STOLK,
L. N. A. WILLEMS, AND J. H. DIJKMAN

Department of Pulmonology
University Hospital
Leiden, The Netherlands

Considerable evidence exists that pulmonary emphysema, characterized by destruction of alveolar structures, is the result of an unrestrained action in the lung of proteinases, especially elastase originating from neutrophils.[1-3] In a small number of patients with emphysema the disease is linked to a genetic severe deficiency of alpha$_1$-proteinase inhibitor (α_1-PI), the major physiologic role of which is the inactivation of neutrophil elastase.[4]

The most common type of emphysema, however, is the centrilobular lesion characterized by involvement of respiratory bronchioles early in the disease.[5] Many studies have established that the primary risk factor for centrilobular emphysema is cigarette smoking.[6] Cigarette smoke induces an increase in the number of neutrophils in the lung and an even more dramatic increase in the number of alveolar macrophages.[7,8] The presence of these cells may result in a focal increase in the concentrations of proteinases, including neutrophil elastase. In addition, the inflammatory cells are stimulated by cigarette smoke to produce oxidants that are capable of destroying antielastase activity.[9] Although several lines of evidence support the hypothesis that emphysema in smokers is the result of an acquired local imbalance between elastase and antielastase,[9] it has become clear that smokers form a very heterogeneous group regarding the presence and severity of the disease, suggesting that the pathogenesis is multifactorial.[10]

In addition to α_1-PI, which is plasma derived, at least one other inhibitor of neutrophil elastase may be of significance in the pathogenesis of emphysema. This is the locally produced antileukoprotease (ALP), which is also called secretory leukocyte proteinase inhibitor.[11] This report focuses on the biochemical characteristics of ALP and its potential role in the lung under normal and inflammatory conditions.

BIOCHEMICAL CHARACTERISTICS OF ANTILEUKOPROTEASE

ALP is a potent, reversible inhibitor of several serine proteinases,[12] and it can be found in mucus secretions including those from the nose, lung, and reproductive tract.[13,14] Its most plausible physiologic role is to inhibit neutrophil elastase, because a highly stable complex ($K_i = 3 \times 10^{-10}$ M) is rapidly ($k_{ass} = 6.4 \times 10^6$ $M^{-1}s^{-1}$) formed between ALP and this enzyme.[15] The inhibitor is a nonglycosyl-

[a] Address for correspondence: J. A. Kramps, Ph.D., Department of Pulmonology (C3-P), University Hospital, NL-2300 RC Leiden, The Netherlands.

ated, highly basic (pI $\approx$ 10) acid-stable monomeric protein consisting of 107 amino-acid residues (mol. mass 11,726 daltons). The N-terminal (residues 1–56) and the C-terminal (residues 57–107) halves represent two distinct domains, showing an internal sequence homology of 35%.[16,17] The two domains each contain four disulfide bridges, the positions of which are completely conserved (FIG. 1).

On the basis of the arrangement and number of disulfide bridges in the domains, ALP can be assigned to a new family of serine proteinase inhibitors.[18] It was postulated, based on a slight homology with the active site of Kazal-type inhibitors and on the localization in the ALP molecule at which limited proteolytic cleavage occurs, that residues Leu19-Arg20-Tyr21 in the first NH₂-terminal domain represent trypsin inhibitory activity, whereas residues Leu72-Met73-Leu74 in the second COOH-terminal represent the chymotrypsin- and elastase-inhibitory activities.[16,17] X-ray structure analyses showed that the binding site of chymotrypsin and most likely also that of elastase are indeed defined by Leu72-Met73 in the second domain of ALP.[19] Concerning the localization of trypsin inhibitory activity, recent results strongly suggest that this activity is also represented by the second domain (to be discussed).[20]

The functional inhibitory activity of ALP can be affected by reactive oxidants.[21-23] By investigating the effect of stimulated human polymorphonuclear leukocytes (PMNs) on the activity of ALP, we recently observed a rapid loss of elastase-inhibitory activity for which myeloperoxidase-derived oxidants were found to be responsible.[20,23] This loss of activity probably occurs as a result of oxidation of the methionine residue that is involved in the inhibitory activity of the second domain. Therefore, ALP may be inactivated by oxidants in a way similar to that observed to occur with α_1-PI, the reactive center of which also contains a methionine residue.[9]

In a recently performed study,[20] additional evidence was obtained on the involvement of methionine in the inhibitory activities of ALP. It was found that modification of methionine in the ALP molecule by selective reagents like cis-platinumdiamminedichloride (FIG. 2), N-chlorosuccinimide, or the myeloperoxidase-H₂O₂-Cl⁻ system resulted in a dose-dependent inactivation of the elastase- and chymotrypsin-inhibitory activities. Besides affecting the activities located in the second domain, the methionine-selective agents also affected, in an almost identical dose-dependent manner, the trypsin-inhibitory activity (FIG. 2). However, the first NH₂-terminal domain, which was postulated to represent the activity against trypsin,[16,17] does not contain any methionine residue. Consequently, involvement of methionine in the trypsin-inhibitory activity indicates localization of this activity in the second domain.

Additional observations support inhibition of trypsin by the second domain; that is, it was observed that elastase, which binds to the second domain, is able to displace trypsin from the inhibitor molecule.[20] This indicates that the binding sites

FIGURE 1. Schematic representation of antileukoprotease, showing the arrangement of disulfide bridges in the two domains. *Black dots* indicate half-cystine residues. The proposed reactive sites in the inhibitor molecule are indicated by *arrows*. It can be seen that the positions of the four disulfide bridges are highly conserved in the two domains.

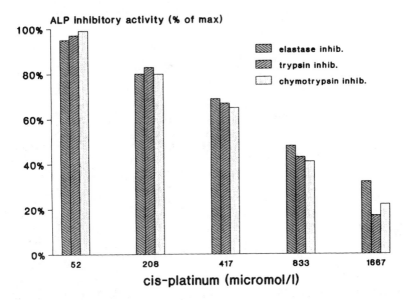

FIGURE 2. Dose-related effect of *cis*-platinumdiamminedichloride (*cis*-platinum) on the inhibitory activity of antileukoprotease (ALP). ALP (0.7 μM) was incubated with increasing amounts of the methionine-selective reagent *cis*-platinum. Thereafter, inhibitory activities of ALP against neutrophil elastase, trypsin, and chymotrypsin were measured using synthetic para-nitroanilide substrates. The inhibitory activities were expressed relative to the activity of ALP which was not incubated with *cis*-platinum.

of the two enzymes are identical or located close to each other in the ALP molecule. Moreover, the recombinant second domain of ALP, expressed in *Escherichia coli* using the gene fragment encoding amino acids 49–107, was able to inhibit elastase, chymotrypsin, as well as trypsin activity.[20,24]

The observations just mentioned strongly suggest that all known inhibitory activities of ALP are located in the second COOH-terminal domain and that methionyl residues are involved, making ALP susceptible to oxidative inactivation. Inactivation of ALP by oxidants generated by stimulated PMNs[23] may contribute to a local disturbance of the protecting shield against elastase-mediated tissue destruction.

So far there is no evidence of any functional inhibitory activity that can be ascribed to the NH₂-terminal domain of ALP. Instead of inhibiting a specific proteinase we may speculate that this first domain is involved in specific interactions with other types of proteins.

PHYSIOLOGIC ROLE OF ANTILEUKOPROTEASE IN THE LOWER RESPIRATORY TRACT

In quantitative terms, α_1-PI and ALP are the main components of the antielastase screen in the lung,[25] although it was recently shown that other partially characterized low molecular weight inhibitors[26,27] and hitherto undefined inhibitors are also present in variable amounts.[28-30] The origin and role of the latter inhibitors are so far unknown.

Local Production of ALP in Central and Peripheral Airways of the Human Lung

Immunohistochemical and immunocytochemical investigations on tissue sections clearly showed that ALP, in contrast to the plasma-derived α_1-PI, is produced locally.[31] At the bronchial level, ALP has been localized in nonciliated cells of the epithelial lining and in serous cells of submucosal glands. In bronchioles, ALP-producing cells are present and identified as Clara cells and goblet cells.[11]

Immunochemical quantifications of ALP relative to α_1-PI in lavage fluid obtained from peripheral airspaces revealed ALP/α_1-PI molar ratios of approximately 0.1.[30,32] In contrast, the molar concentration of ALP in tracheobronchial secretions exceeds that of α_1-PI by a factor of 3 or more,[25,29] suggesting that ALP is the major inhibitor of neutrophil elastase in the large conducting airways.

Serial measurements of ALP concentrations in sputum of patients with recurrent or chronic bronchial infections revealed intraindividual ALP levels that were negatively correlated with daily sputum production, suggesting that the absolute amount of ALP produced in these airways is rather constant. However, between patients with comparable severity of disease, ALP production was found to vary over a 10-fold range (0.5–9 mg/24 h).[33] The presence of relatively large quantities of ALP in bronchi, together with the observation that secretory cell metaplasia can be induced by neutrophil-derived lysosomal proteinases, led to the suggestion that a proteinase/proteinase-inhibitor imbalance may play a role in secretory cell hyperplasia and metaplasia of human chronic bronchitis.[34]

The molar quantity of ALP in the peripheral airspaces of approximately 10% of that of α_1-PI[30,32] means an ALP level of 0.3 μM ($\approx$4 μg/ml), assuming an α_1-PI level in the alveolar epithelial lining fluid of 3.4 μM.[35] Although its efficiency *in vivo* can be predicted from this concentration and from the kinetic constants of ALP,[36] it may be postulated, on the basis of the molar excess of α_1-PI over ALP, that ALP is of minor importance in protecting the peripheral part of the lung. Several recent observations, however, do suggest that ALP plays a role in the defense against destructive processes in distal airways.

Investigating the extracellular localization of ALP in human lungs by immunofluorescence and immunoelectron microscopic techniques, we were able to demonstrate ALP in the parenchymal matrix of the alveolar wall. Most interestingly, the extracellular ALP was localized mainly in association with elastin fibers.[37,38] If pulmonary emphysema indeed is the result of an unrestrained activity of neutrophil elastase, leading to degradation of elastin fibers in the alveolar septa, the observed immunohistochemical association of ALP with these fibers strongly indicates that this inhibitor is a modulating factor in the pathogenesis of the disease.

Is ALP Able to Protect the Elastin Fibers against Elastase-Mediated Destruction?

In a study on human lungs Damiano and coworkers[39] found, by immunostaining, neutrophil elastase localized extracellularly in association with interstitial elastic fibers. The quantity of elastin-bound elastase correlated with the local severity of emphysema, supporting the role of this enzyme in the pathogenesis of the disease. Concerning the antielastase screen in the lung, it is of importance that elastase inhibitors are active against elastase that is bound to elastin. Investigating the potency of α_1-PI and ALP to inhibit elastin-bound elastase, Bruch and Bieth[40] observed that 25% of the bound elastase was resistant to the inhibitory effect of α_1-PI. In contrast, ALP was able to inhibit efficiently all elastin-associated elastase activity including the fraction not inhibited by α_1-PI. The difference might be attributed to distinct mechanisms of inhibition by the two inhibitors. In a very recent study Morrison *et al.*[41] indeed found evidence for this. It was demonstrated that α_1-PI enhances the dissociation of elastase from elastin and inhibits the enzyme in solution. In the presence of α_1-PI, elastin-bound elastase remains fully active. In contrast, ALP is able to inhibit elastase while remaining bound to elastin. As suggested by Gauthier *et al.*[42] and recently also by Morrison *et al.*,[41] it is possible that the major physiologic role of ALP is to inhibit elastin-bound elastase, whereas that of α_1-PI is to inhibit elastase activity in solution. Our light and electron microscopic observations of the association of interstitial ALP with elastin fibers may therefore be explained by binding of ALP to elastin-bound elastase.[37,38] However, direct interaction of ALP with elastin fibers by, for example, electrostatic interactions is also possible.[43]

What is the Potency of ALP, Relative to α_1-PI, in Inhibiting PMN-Mediated Extracellular Matrix Degradation?

Breakdown of lung-connective tissue may occur when PMNs migrate into the interstitium and adhere firmly to insoluble matrix molecules.[44,45] By this, a cell-substrate interface may be formed which limits access of proteinase inhibitors and permits extracellular proteolysis. In an attempt to mimic these events and to establish the efficacy of ALP and α_1-PI, the effect of these inhibitors was studied on the degradation of insoluble fibrinogen matrix mediated by firmly attached PMNs.

Attachment of PMNs onto the fibrinogen matrix was achieved by incubation of the cells with tumor necrosis factor α (TNFα) which is a macrophage-derived product.[46] Elastase-specific degradation of fibrinogen was measured after stimulation of attached PMNs in the absence or presence of various amounts of either of the two inhibitors. Fibrinogen degradation was monitored by a sensitive radioimmunoassay highly specific for Aα(1–21), which is a cleavage product generated by neutrophil elastase.[47] In this study, we observed a 50% inhibition of the Aα(1–21) peptide formation at ALP and α_1-PI concentrations of 85 $\pm$ 30 nM ($n = 7$, $\bar{x} \pm$ SD) and 220 $\pm$ 98 nM ($n = 7$, $\bar{x} \pm$ SD), respectively (FIG. 3). Based on these so-called IC50 values, it can be concluded that, on a molar basis, ALP is significantly more potent than is α_1-PI in inhibiting PMN-mediated matrix degradation.[48]

To evaluate the effect of reactive oxygen species (ROS) on the potency of ALP and α_1-PI as measured in the aforementioned PMN-fibrinogen test system, we performed additional experiments using PMNs from patients with chronic granulomatous disease (CGD). PMNs from these patients lack a functional membrane-associated NADPH oxidase and therefore are unable to produce ROS. When compared to normal PMNs, PMNs from patients with CGD were found to degrade approximately twice as much fibrinogen as measured by Aα(1–21) formation. Most likely, lysosomal enzymes released from normal PMNs are partly inactivated by the cell's own ROS.[49] Notwithstanding this increased formation of Aα(1–21) by PMNs from patients with CGD, lower concentrations of ALP (50 $\pm$

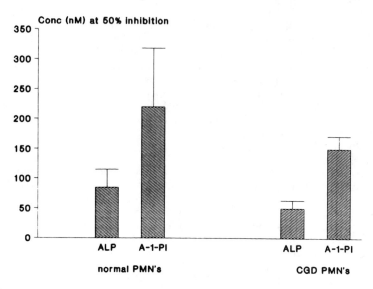

FIGURE 3. Concentrations of antileukoprotease (ALP) and alpha$_1$-proteinase inhibitor (A-1-PI) at which 50% inhibition occurred of PMN-mediated fibrinogen degradation (IC50: mean ± SD). PMNs (1.10^6) isolated from peripheral blood were firmly attached onto a fibrinogen matrix.[46] Thereafter, the cells were stimulated by cytochalasin-B and formyl-Met-Leu-Phe in the presence of increasing amounts of either ALP or A-1-PI (inhibitor concentrations ranging from 0.05–0.9 μM). After an incubation of 1 hour, fibrinogen degradation was determined by performing a radioimmunoassay specific for Aα(1-21), which is an elastase-specific peptide released from fibrinogen.[47] From the results of these experiments IC50 values were calculated. CGD-PMNs were isolated from blood of patients with chronic granulomatous disease. These PMNs are characterized by an inability to produce reactive oxygen species. Using normal PMNs, an IC50(A-1-PI)/IC50(ALP) ratio of 2.6 was calculated. Using CGD-PMNs this ratio amounted to 3.0.

13 nM) and α_1-PI (150 ± 21 nM) were needed to obtain 50% inhibition of Aα(1–21) formation when results are compared with those obtained with normal PMNs (FIG. 3). However, using PMNs from patients with CGD, again ALP was found to be more potent in inhibiting Aα(1–21) formation than was α_1-PI (FIG. 3). The results of these series of experiments confirm our previous observation that ALP, similar to α_1-PI, is susceptible to oxidative inactivation.[9,23] This explains the lower IC50 values obtained when experiments are performed with PMNs from patients with CGD (FIG. 3). Interestingly, irrespective of whether the experiments were performed with PMNs obtained from either normal individuals or those with CGD, IC50(ALP) values were approximately three times as low as IC50(α_1-PI) values (FIG. 3). So, the observed difference in potency between ALP and α_1-PI cannot be explained by differences in susceptibility to oxidative inactivation. Obviously, the difference in IC50 values of ALP and α_1-PI is explained by the difference in molecular mass of the two inhibitors (12,000 vs 52,000 daltons), making ALP more accessible to the interface between firmly adhered PMNs and the insoluble matrix. The relative high potency of ALP to inhibit PMN-mediated matrix degradation may be of importance for the *in vivo* role of ALP to protect the connective tissue matrix of the lung.

Evidence for the involvement *in vivo* of ALP in limiting the processes playing a role in destruction of pulmonary tissue has been obtained by Willems *et al.*[50] investigating human surgical lung specimens using morphometric methods. It was found that bronchiolar inflammation and destruction of alveoli attached to these bronchioles parallel the local number of ALP-containing epithelial cells. From this we postulate that an increase of ALP-containing cells is part of the general defense against inflammatory and destructive processes in distal human airways, leading to higher levels of proteinase inhibitor to minimize tissue damage.

PROTECTION BY ANTILEUKOPROTEASE OF EXPERIMENTALLY INDUCED EMPHYSEMA IN HAMSTERS

To investigate the *in vivo* potency of ALP in inhibiting elastase-mediated parenchymal destruction, we instilled ALP intratracheally in hamsters, 1 hour prior to the instillation of human neutrophil elastase.[51] Emphysema in lung-tissue sections was measured by the mean linear intercept (L_m), reflecting the average distance between alveolar walls. The results indicate that the increase in L_m, as induced by 420 μg human neutrophil elastase, can substantially be inhibited by 35% and 70% when ALP has been instilled at a dose of 365 and 730 μg, respectively (TABLE 1). Contrary to this, the elastase-induced bronchial secretory cell metaplasia, which develops in hamster lungs, was only minimally affected by ALP. The results were comparable to those obtained by Lucey *et al.*[52] who used recombinant ALP in the same hamster model.

Similar to what was observed in the human lung,[37,38] we found, using specific immunofluorescence staining, that at least part of intratracheal-instilled ALP in the hamster lung selectively associates with elastic fibers in the alveolar interstitium (FIG. 4). We speculate that it is this elastin-associated ALP that makes a major contribution to the protection against elastase-induced parenchymal destruction in these hamsters.

Quite recently it was observed by Starcher and Williams[53] that repeated intratracheal instillations of lipopolysaccharide (LPS) in mice cause emphysema, most likely mediated by elastase from the animal's own neutrophils. These neutrophils are recruited into the airways as a result of LPS instillation. In hamsters we were able to induce emphysema by instilling *E. coli* LPS (500 μg) twice a week for 4 weeks.[54] This treatment resulted in an increase of the mean linear intercept (L_m)

TABLE 1. Effect of Intratracheal Instilled ALP on Human Neutrophil Elastase (HNE)-Induced Emphysema in Hamsters as Measured by Mean Linear Intercept (L_m)

1st/2nd Instillation[a]	L_m (μm) (mean ± SEM)	Number of Animals Tested
PBS/HNE (420 μg)	95.1 ± 2.0	10
ALP (365 μg)/HNE (420 μg)	85.6 ± 2.7	13
ALP (730 μg)/HNE (420 μg)	76.9 ± 1.3	12
PBS/PBS[b]	68.6 ± 0.5	13
ALP (730 μg)/PBS	67.2 ± 0.7	14

[a] Time between first and second instillation, 1 hour. Instilled volume, 200 μg.
[b] PBS = phosphate-buffered saline solution.

of 25–30%. In a preliminary investigation we studied the efficacy of recombinant ALP (rSLPI, provided by Synergen, Inc., Boulder, Colorado) to inhibit LPS-induced emphysema. The inhibitor, instilled twice at a dose of 0.5 mg during and 7 hours after each LPS treatment, was able to inhibit the L_m increase by 33%. From these preliminary results we conclude that the LPS-mediated damage is caused, at least partly, by endogenous elastase from the hamster neutrophils. Moreover, ALP is effective in inhibiting LPS-induced parenchymal lesions.

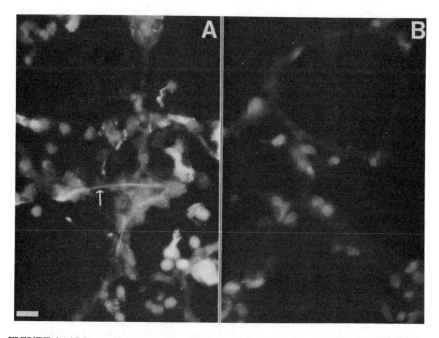

FIGURE 4. ALP-specific immunofluorescence staining of peripheral lung tissue from hamsters. **(A)** Tissue of a lung lavaged and processed 7 hours after intratracheal instillation of 500 μg antileukoprotease. Lavage (3 × 5 ml) was performed to wash out ALP from the airway lumen. Elastic fibers clearly show positive staining (*arrow*). Diffuse interstitial staining and staining in alveolar macrophages are also observed. **(B)** Tissue of a hamster not having received ALP (negative control). The immunostaining was applied to frozen sections fixed with paraformaldehyde-lysine-periodate, using ALP-specific rabbit IgG and FITC-conjugated swine anti-rabbit IgG.

CONCLUSION

Starting from the elastase/elastase-inhibitor imbalance hypothesis of the pathogenesis of emphysema, several lines of evidence have been obtained suggesting that ALP in the human lung is a compound that may limit the development of this disease. ALP, which is highly effective in inhibiting neutrophil elastase-mediated matrix degradation, is produced by bronchiolar cells localized closely to the region where centrilobular emphysema starts to develop. In human lungs the number of ALP-producing bronchiolar cells has been found to correlate with local inflammation and alveolar destruction, indicating involvement of this inhibitor in these processes. In the parenchymal interstitium, the inhibitor is found closely associated with elastin fibers, thereby strongly suggesting that ALP may play a role in protecting the fibers from elastase-mediated degradation. Finally, ALP, administered to hamsters by intratracheal instillation, has been effective in inhibiting emphysema induced by either human neutrophil elastase or LPS.

Although ALP was found to be a better inhibitor than α_1-PI of matrix degradation mediated by firmly attached PMNs, both inhibitors are clearly sensitive to oxidation. Consequently, oxidative inactivation may be a mechanism contributing to a local imbalance between elastase and elastase inhibitors and therefore may also play a role in the pathogenesis of emphysema. Most likely these and additional processes which are still unknown will interact and determine the final outcome of whether emphysema will develop and how rapidly the disease will progress.

Irrespective of the evidence that ALP plays a protective role in the peripheral airways, its role in the pathogenesis of emphysema is not clear as long as it is not demonstrated that emphysema in non-α_1-PI-deficient individuals is the result of unrestrained elastase activity.

Being a potent inhibitor of neutrophil elastase, ALP may be a very useful therapeutic agent in the management of emphysema in α_1-PI-deficient individuals. It is a stable, small, nonglycosylated protein of human origin that can be produced by recombinant DNA techniques.[55,56] It might even be possible to synthesize the COOH-terminal half of the molecule, showing full inhibitory activity[24,56] or to synthesize mutants that are resistant to oxidation.

REFERENCES

1. JANOFF, A. 1985. Elastases and emphysema. Current assessment of the protease-antiprotease hypothesis. Am. Rev. Respir. Dis. **132:** 417–433.
2. WEISSLER, J. C. 1987. Southwestern internal medicine conference: Pulmonary emphysema: Current concepts of pathogenesis. Am. J. Med. Sci. **293:** 125–138.
3. NIEWOEHNER, D. E. 1988. Cigarette smoking, lung inflammation, and the development of emphysema. J. Lab. Clin. Med. **111:** 15–27.
4. PERLMUTTER, D. H. & J. A. PIERCE. 1989. The α_1-antitrypsin gene and emphysema. Lung Cell. Mol. Physiol. **1:** L147–L162.
5. NIEWOEHNER, D. E. 1986. New messages from morphometric studies of chronic pulmonary disease. Semin. Respir. Med. **8:** 140–146.
6. SOBONYA, R. E. & B. BURROWS. 1983. The epidemiology of emphysema. Clin. Chest Med. **4:** 351–358.
7. HUNNINGHAKE, G. W. & R. G. CRYSTAL. 1983. Cigarette smoking and lung destruction. Accumulation of neutrophils in the lungs of cigarette smokers. Am. Rev. Respir. Dis. **128:** 833–838.
8. NIEWOEHNER, D. E., J. KLEINERMAN & D. P. RICE. 1974. Pathologic changes in the peripheral airways of young cigarette smokers. N. Engl. J. Med. **291:** 755–758.
9. JANOFF, A. 1983. Biochemical links between cigarette smoking and pulmonary emphysema. J. Appl. Physiol.: Respir. Environ. Exercise Physiol. **55:** 285–293.
10. MASON, R. J., A. S. BUIST, E. B. FISHER, J. A. MERCHANT, J. M. SAMET & C. H. WELSH. 1985. Cigarette smoking and health. Am. Rev. Respir. Dis. **132:** 1133–1136.
11. DE WATER, R., L. N. A. WILLEMS, G. N. P. VAN MUIJEN, C. FRANKEN, J. A. M. FRANSEN, J. H. DIJKMAN & J. A. KRAMPS. 1986. Ultrastructural localization of bronchial antileukoprotease in central and peripheral human airways by a gold-

labeling technique using monoclonal antibodies. Am. Rev. Respir. Dis. **133:** 882–890.

12. SMITH, C. E. & D. A. JOHNSON. 1985. Human bronchial leucocyte proteinase inhibitor. Rapid isolation and kinetic analysis with human leucocyte proteinases. Biochem. J. **225:** 463–472.

13. KRAMPS, J. A., C. FRANKEN & J. H. DIJKMAN. 1984. ELISA for quantitative measurement of low-molecular-weight bronchial protease inhibitor in human sputum. Am. Rev. Respir. Dis. **129:** 959–963.

14. FRANKEN, C., C. J. L. M. MEIJER & J. H. DIJKMAN. 1989. Tissue distribution of antileukoprotease and lysozyme in humans. J. Histochem. Cytochem. **37:** 493–498.

15. BOUDIER, C. & J. G. BIETH. 1989. Mucus proteinase inhibitor: A fast acting inhibitor of leucocyte elastase. Biochim. Biophys. Acta **995:** 36–41.

16. THOMPSON, R. C. & K. OHLSSON. 1986. Isolation, properties, and complete amino acid sequence of human secretory leukocyte protease inhibitor, a potent inhibitor of leukocyte elastase. Proc. Natl Acad. Sci. USA **83:** 6692–6696.

17. SEEMÜLLER, U., M. ARNHOLD, H. FRITZ, K. WIEDENMANN, W. MACHLEIDT, R. HEINZEL, H. APPELHANS, H.-G. GASSEN & F. LOTTSPEICH. 1986. The acid-stable proteinase inhibitor of human mucous secretions (HUSI-I, antileukoprotease) FEBS-Letters **199:** 43–48.

18. LASKOWSKI, M. & I. KATO. 1980. Protein inhibitors of proteinases. Ann. Rev. Biochem. **49:** 593–626.

19. GRÜTTER, M. G., G. FENDRICH, R. HUBER & W. BODE. 1988. The 2.5 Å X-ray crystal structure of the acid-stable proteinase inhibitor from human mucous secretions analyzed in its complex with bovine α-chymotrypsin. EMBO J. **7:** 345–351.

20. KRAMPS, J. A., C. VAN TWISK, H. APPELHANS, B. MECKELEIN, T. NIKIFOROV & J. H. DIJKMAN. 1990. Proteinase inhibitory activities of antileukoprotease are presented by its second COOH-terminal domain. Biochim. Biophys. Acta **1038:** 178–185.

21. OHLSSON, K., U. FRYKSMARK & H. TEGNER. 1980. The effect of cigarette smoke condensate on α₁-antitrypsin, antileukoprotease and granulocyte elastase. Eur. J. Clin. Invest. **10:** 373–374.

22. CARP, H. & A. JANOFF. 1980. Inactivation of bronchial mucous proteinase inhibitor by cigarette smoke and phagocyte-derived oxidants. Exp. Lung Res. **1:** 225–237.

23. KRAMPS, J. A., C. VAN TWISK, E. C. KLASEN & J. H. DIJKMAN. 1988. Interactions among stimulated human polymorphonuclear leucocytes, released elastase and bronchial antileukoprotease. Clin. Sci. **75:** 53–62.

24. MECKELEIN, B., T. NIKIFOROV, A. CLEMEN & H. APPELHANS. 1990. The location of inhibitory specificities in human mucus proteinase inhibitor (MPI): Separate expression of the COOH-terminal domain yield an active inhibitor of three different proteinases. Protein. Eng. **3:** 215–220.

25. TEGNER, H. 1978. Quantitation of human granulocyte protease inhibitors in nonpurulent bronchial lavage fluids. Acta Otolaryngol. **85:** 282–289.

26. KRAMPS, J. A. & E. C. KLASEN. 1985. Characterization of a low molecular weight antielastase isolated from human bronchial secretion. Exp. Lung Res. **9:** 151–165.

27. KUEPPERS, F. & B. J. BROMKE. 1983. Protease inhibitors in tracheobronchial secretions. J. Lab. Clin. Med. **101:** 747–757.

28. TETLEY, T. D., S. F. SMITH, A. J. WINNING, J. M. FOXALL, N. T. COOKE, G. H. BURTON, E. HARRIS & A. GUZ. 1989. The acute effect of cigarette smoking on the neutrophil elastase inhibitory capacity of peripheral lung lavage from asymptomatic volunteers. Eur. Respir. J. **2:** 802–810.

29. MORRISON, H. M., J. A. KRAMPS, S. C. AFFORD, D. BURNETT, J. H. DIJKMAN & R. A. STOCKLEY. 1987. Elastase inhibitors in sputum from bronchitic patients with and without α₁-proteinase inhibitor deficiency: Partial characterization of a hitherto unquantified inhibitor of neutrophil elastase. Clin. Sci. **73:** 19–28.

30. BOUDIER, C., A. PELLETIER, A. GAST, J.-M. TOURNIER, G. PAULI & J. G. BIETH. 1987. The elastase inhibitory capacity and the α₁-proteinase inhibitor and bronchial inhibitor content of bronchoalveolar lavage fluids from healthy subjects. Biol. Chem. Hoppe-Seyler **368:** 981–990.

31. WILLEMS, L. N. A., J. A. KRAMPS, R. DE WATER, T. STIJNEN, G. J. FLEUREN, C. FRANKEN & J. H. DIJKMAN. 1986. Evaluation of antileukoprotease in surgical lung specimens. Eur. J. Respir. Dis. **69:** 242–247.

32. KRAMPS, J. A., C. FRANKEN & J. H. DIJKMAN. 1988. Quantity of antileukoprotease relative to α_1-proteinase inhibitor in peripheral airspaces of the human lung. Clin. Sci. **75:** 351–353.

33. DIJKMAN, J. H., J. A. KRAMPS & C. FRANKEN. 1986. Antileukoprotease in sputum during bronchial infections. Chest **89:** 731–736.

34. SNIDER, G. L. 1986. Experimental studies on emphysema and chronic bronchial injury. Eur. J. Respir. Dis. **69** (suppl 146): 17–35.

35. WEWERS, M. D., A. CASOLARO, S. E. SELLERS, S. C. SWAYZE, K. M. MCPHAUL, J. T. WITTES & R. G. CRYSTAL. 1987. Replacement therapy for α-antitrypsin deficiency associated with emphysema. N. Engl. J. Med. **316:** 1055–1062.

36. BIETH, J. G. 1980. Pathophysiological interpretation of kinetic constants of protease inhibitors. Bull. Eur. Physiopathol. Respir. **16** (suppl): 183–195.

37. WILLEMS, L. N. A., C. J. M. OTTO-VERBERNE, J. A. KRAMPS, A. A. W. TEN HAVE-OPBROEK & J. H. DIJKMAN. 1986. Detection of antileukoprotease in connective tissue of the lung. Histochemistry **86:** 165–168.

38. KRAMPS, J. A., A. H. T. TE BOEKHORST, J. A. M. FRANSEN, L. A. GINSEL & J. H. DIJKMAN. 1989. Antileukoprotease is associated with elastin fibres in the extracellular matrix of the human lung. An immunoelectron microscopic study. Am. Rev. Respir. Dis. **140:** 471–476.

39. DAMIANO, V. V., A. TSANG, U. KUCICH, W. R. ABRAMS, J. ROSENBLOOM, P. KIMBEL, M. FALLAHNEJAD & G. WEINBAUM. 1986. Immunolocalization of elastase in human emphysematous lungs. Clin. Invest. **78:** 482–493.

40. BRUCH, M. & J. G. BIETH. 1986. Influence of elastin on the inhibition of leucocyte elastase by α_1-proteinase inhibitor and bronchial inhibitor. Biochem. J. **238:** 269–273.

41. MORRISON, H. M., H. G. WELGUS, R. A. STOCKLEY, D. BURNETT & E. J. CAMPBELL. 1990. Inhibition of human leukocyte elastase bound to elastin. Relative ineffectiveness and two mechanisms of inhibitory activity. Am. J. Respir. Cell Mol. Biol. **2:** 263–269.

42. GAUTHIER, F., U. FRYKSMARK, K. OHLSSON & J. G. BIETH. 1982. Kinetics of the inhibition of leukocyte elastase by the bronchial inhibitor. Biochim. Biophys. Acta **700:** 178–183.

43. LONKEY, S. A. & H. WOHL. 1983. Regulation of elastolysis soluble elastin by human leukocyte elastase: Stimulation by lysine-rich ligands, anionic detergents, and ionic strength. Biochemistry **22:** 3714–3720.

44. CAMPBELL, E. J. & M. A. CAMPBELL. 1988. Pericellular proteolysis by neutrophils in the presence of proteinase inhibitors: Effects of substrate opsonization. J. Cell Biol. **106:** 667–676.

45. WEISS, S. J. 1989. Tissue destruction by neutrophils. N. Engl. J. Med. **320:** 365–376.

46. HANLON, W. A., J. STOLK, P. DAVIES & R. J. BONNEY. 1989. Tumor necrosis factor α (rTNFα) facilitates the attachment of polymorphonuclear leukocytes to fibrinogen (abstr.). J. Leuk. Biol. **46:** 297.

47. MUMFORD, R. A., O. F. ANDERSEN, J. BOGER, S. BONDYS, R. J. BONNEY, D. BOULTON, P. DAVIES, J. DOHERTY, D. FLETCHER, K. HAND, J. MAO, H. WILLIAMS & M. E. DAHLGREN. 1989. Development of a direct, sensitive and specific RIA for a primary human leukocyte elastase (HLE) generated fibrinogen cleavage product, Aα(1–21). Am. Rev. Respir. Dis. **139:** A574.

48. STOLK, J., W. A. HANLON, P. DAVIES, R. MUMFORD, M. E. DAHLGREN, W. B. KNIGHT, J. A. KRAMPS & R. J. BONNEY. 1989. The effect of antileukoprotease (ALP) and α_1-proteinase inhibitor (α_1PI) on the PMN-mediated degradation of fibrinogen. Am. Rev. Respir. Dis. **139:** A201.

49. KOBAYASHI, M., T. TANAKA & T. USUI. 1982. Inactivation of lysomal enzymes by the respiratory burst of polymorphonuclear leukocytes. Possible involvement of myeloperoxidase-H_2O_2-halide system. J. Lab. Clin. Med. **100:** 896–907.

50. WILLEMS, L. N. A., J. A. KRAMPS, T. STIJNEN, P. J. STERK, J. J. WEENING & J. H.

DIJKMAN. 1989. Antileukoprotease-containing bronchiolar cells. Relationship with morphologic disease of small airways and parenchyma. Am. Rev. Respir. Dis. **139:** 1244–1250.

51. RUDOLPHUS, A., J. A. KRAMPS & J. H. DIJKMAN. 1990. Intratracheal instillation of antileukoprotease (ALP) in hamsters: Clearance, parenchymal localization and efficacy on human neutrophil elastase (NE)-induced lung injury (abstr.). Am. Rev. Respir. Dis. **141:** A109.

52. LUCEY, E. C., P. J. STONE, R. BREUER, T. G. CHRISTENSEN, R. C. THOMPSON & G. L. SNIDER. 1989. Human secretory leukocyte-protease inhibitor (SLPI) prevents human neutrophil elastase-induced emphysema and secretory cell metaplasia (SCM) in the hamster. Am. Rev. Respir. Dis. **139:** A226.

53. STARCHER, B. & I. WILLIAMS. 1989. The beige mouse: Role of neutrophil elastase in the development of pulmonary emphysema. Exp. Lung Res. **15:** 785–800.

54. RUDOLPHUS, A., D. OSINGA, J. STOLK, D. FLETCHER, K. KEENAN & J. A. KRAMPS. 1990. Induction of emphysema and secretory cell metaplasia (SCM) by intratracheal instillations of E. coli lipopolysaccharide (LPS) in hamsters. A new animal model (abstr.). Am. Rev. Respir. Dis. **141:** A736.

55. HEINZEL, R., H. APPELHANS, G. GASSEN, U. SEEMÜLLER, W. MACHLEIDT, H. FRITZ & G. STEFFENS. 1986. Molecular cloning and expression of cDNA for human antileukoprotease from cervix uterus. Eur. J. Biochem. **160:** 61–67.

56. STETLER, G. L., C. FORSYTH, T. GLEASON, J. WILSON & R. C. THOMPSON. 1989. Secretion of active, full- and half-length human secretory leukocyte protease inhibitor by saccharomyces cerevisae. Biotechnology **7:** 55–60.

Secreted Inhibitors of Mammalian Metalloproteinases[a]

MICHAEL J. BANDA[b] AND ERIC W. HOWARD

Laboratory of Radiobiology and Environmental Health
University of California
San Francisco, California 94143-0750

Whether disruption of the proteinase-inhibitor balance is the cause or the result of emphysema has been the subject of continued debate and experimentation. However, because of the destruction and remodeling of the alveolar connective tissue matrix, the involvement of proteinases is certain. The role of the neutrophil serine elastase and its principal inhibitor alpha$_1$-proteinase inhibitor (α_1-PI) in this process is addressed in several articles in this volume.

Mammalian metalloproteinases have an established role in connective tissue turnover. Interstitial collagenase, stromelysin, M_r 72,000 gelatinase (type IV collagenase), and M_r 92,000 gelatinase (type V collagenase) are secreted by several types of cells and can, especially in combination, severely damage matrix or mediate matrix turnover. In emphysema an elastase is implicated. Of the mammalian metalloelastases, mouse macrophage elastase has been partially purified,[1] and macrophage metalloelastase activity has been described in bovine alveolar macrophages,[2] monkey alveolar macrophages,[3] and human macrophages.[4] The human macrophage elastinolytic activity may contribute to the destruction associated with emphysema.

The regulation of mammalian matrix-degrading metalloproteinases by endogenous protein inhibitors of metalloproteinase has been considered to be the result of the action of the tissue inhibitor of metalloproteinases[5] (TIMP, M_r 30,000), which was initially purified as the dermal collagenase inhibitor.[6] Although TIMP can inhibit most secreted mammalian metalloproteinases as well as the metalloproteinase activity from human alveolar macrophages, detailed studies of the kinetics of TIMP-metalloproteinase interaction have been limited to a few metalloproteinases, most notably interstitial collagenase.[7,8] Whether TIMP is a biologically important inhibitor of all mammalian metalloproteinases remains to be determined. Studies *in vitro* have shown that TIMP can regulate the invasiveness of tumor cells,[9] as well as chondrocytes and microvascular endothelial cells.[10] Antisense TIMP RNA has a similar effect on Swiss 3T3 cells.[11] Although the extent to which TIMP-mediated inhibition is biologically significant is not known, it is generally accepted that TIMP is an important secreted protein inhibitor of metalloproteinases, particularly of interstitial collagenase. It is certainly the best-characterized inhibitor of its type, and its entire amino acid sequence has been determined.[12]

[a] This work was supported by the Office of Health and Environmental Research, U.S. Department of Energy (contract no. DE-AC03-76-SF01012), a National Research Service Award from the NIEHS (5 T32 ES07106), and a grant from the National Institutes of Health (AR32746).

[b] Address for correspondence: Michael J. Banda, Laboratory of Radiobiology and Environmental Health, LR-102, University of California, San Francisco, CA 94143-0750.

In studies in which TIMP was found to regulate rabbit microvascular endothelial cell collagenase and stromelysin activities,[13,14] the existence of two metalloproteinase inhibitors that were lower in molecular weight than TIMP and did not cross-react with anti-TIMP antibodies was discovered. Those non-TIMP inhibitors were referred to as inhibitors of metalloproteinases 1 and 2 (IMP-1 and IMP-2). Inhibitors corresponding to TIMP, IMP-1, and IMP-2, and a third inhibitor, IMP-3, were detected in the medium conditioned by human glioma and astrocytoma cells.[15] Other inhibitors of metalloproteinases that appear to be distinct from TIMP have been detected in chick embryonic cartilage,[16] human polymorphonuclear leukocytes,[17] and bovine cartilage.[18,19] All of these inhibitors have a lower molecular weight than does human TIMP. It is not clear in all cases how the inhibitors are related to TIMP. In this study we examined the functional selectivity and physical characteristics of TIMP and one of the IMPs, IMP-2.

RESULTS

Characterization of IMP-2

IMP-2 (also known as TIMP-2) has been purified from medium conditioned by human melanoma cells with the use of gelatin-affinity chromatography.[20] We used the same technique to purify IMP-2 from human fibroblasts (HS-395). When the conditioned medium was passed over a gelatin-affinity column, IMP-2 and the M_r 72,000 and M_r 92,000 gelatinases appeared to bind specifically to the column and elute together. IMP-2 was separated from the two proteinases by reverse-phase chromatography. When the resulting IMP-2, now free of proteinase, was reapplied to the gelatin-affinity column, it did not bind. This suggests that the original binding of IMP-2 to the column was the result of its being in complex with the proform of gelatinase that binds to gelatin. The pure IMP-2 is an effective inhibitor of the active form of the M_r 72,000 gelatinase.[20] Thus, IMP-2 can bind to both the pro- and active forms of M_r 72,000 gelatinase.

Because the gelatinase recovered from the dissociation of IMP-2: gelatinase complexes remains active, it is possible that IMP-2-mediated inhibition *in vivo* is reversible. Whether reversible inhibition is biologically relevant remains to be established. This possible reversibility distinguishes the metalloproteinase inhibitors from the serpins, such as α_1-PI. If the serpin-serine proteinase complex is separated *in vitro,* neither the serpin nor the serine proteinase remains active; the serpin undergoes a proteolytic clip as part of the mechanism of inhibition. Neither TIMP nor IMP-2 is proteolyzed during inhibition of its target proteinase.

The amino-terminal sequence of IMP-2 is similar to that of TIMP (FIG. 1), but it does not match any sequence in TIMP exactly.[20] Therefore, IMP-2 is not a cleavage fragment of TIMP. The sequence similarity, however, does suggest that TIMP and IMP-2 may be part of a related family of proteinase inhibitors.

```
IMP-2    C-S-C-S-P-V-H-P-Q-Q-A-F-C-N-A-D-V-V-I-R-A-K-
TIMP     C-T-C-V-P-P-H-P-Q-T-A-F-C-N-S-D-L-V-I-R-A-K-
```

FIGURE 1. Amino-terminal sequences of human IMP-2 and human TIMP.

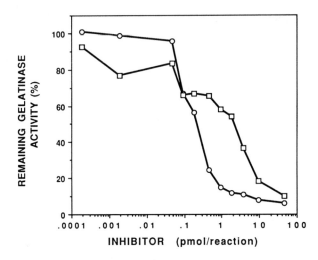

FIGURE 2. Inhibition of human M_r 72,000 gelatinase by human TIMP (□) and human IMP-2 (○). The M_r 72,000 gelatinase was activated with APMA and used at 1.4 pmol per reaction.

Selectivity of TIMP and IMP-2

On the basis of the kinetics of inhibition, the serpin family of proteinase inhibitors exhibits remarkable selectivity for a target serine proteinase. The most pertinent example is that the target for human α_1-PI is human neutrophil elastase. The object of the experiments to be described was to determine if the two purified metalloproteinase inhibitors, TIMP and IMP-2, can be distinguished by their ability to inhibit individual metalloproteinases selectively.

To determine the selectivity of TIMP and IMP-2, purified M_r 72,000 gelatinase was activated by incubation for up to 4 hours with 1 mM 4-aminophenylmercuric acetate (APMA) and then incubated separately with increasing concentrations of TIMP or IMP-2 for 30 minutes at ambient temperature. The remaining gelatinolytic activity was measured by means of a radioactive gelatin degradation assay. The results show that IMP-2 is better at inhibiting the M_r 72,000 gelatinase than is TIMP (FIG. 2). In a similar experiment, purified interstitial collagenase was activated with 1 mM APMA and then incubated separately with increasing concentrations of TIMP or IMP-2. The remaining collagenolytic activity was determined with a radiolabeled collagen fibril assay (FIG. 3). In contrast to the results with the M_r 72,000 gelatinase, TIMP was a better inhibitor of interstitial collagenase than was IMP-2.

When the same experiment was performed with a metalloelastase, the results were unclear. Partially purified mouse macrophage elastase was incubated separately with increasing concentrations of TIMP or IMP-2, and the remaining elastinolytic activity was measured with a radioactive insoluble elastin assay. Both inhibitors appeared to inhibit this proteinase with similar efficiency (FIG. 4). The lack of selectivity in this case may be due to the species difference in the origin of the elastase (mouse) and the inhibitors (human).

Highly invasive malignant cells are thought to use metalloproteinases to facili-

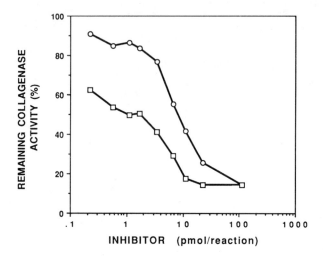

FIGURE 3. Inhibition of human interstitial collagenase by human TIMP (□) and human IMP-2 (O). The collagenase was activated with APMA and used at 19 pmol per reaction.

tate their transit across the basement membrane and the extracellular matrix. A mouse melanoma-derived cell line that forms tumors in the lungs, B16, is thought to have elastin-degrading capabilities. These cells can degrade the elastin and collagen in the matrix established by R-22 rat smooth muscle cells. When B16 cells were incubated on a radiolabeled R-22 matrix with increasing concentrations of IMP-2, their ability to degrade the matrix was partially inhibited (FIG. 5).

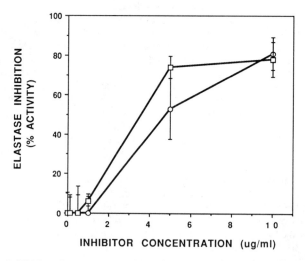

FIGURE 4. Inhibition of mouse macrophage elastase by human TIMP (□) and human IMP-2 (O).

Because B16 cells are derived from mice, the lack of full inhibition may be due to the species difference between the cells and the inhibitor. Another possibility is that IMP-2 alone cannot fully inhibit the battery of metallo- and nonmetalloproteinases that may be required for B16 cell-mediated matrix degradation. Nevertheless, IMP-2 may play a role in the regulation of tissue invasion.

DISCUSSION

It is well established that normal matrix cells, inflammatory cells, and tumor cells secrete a number of distinct but related metalloproteinases that are capable

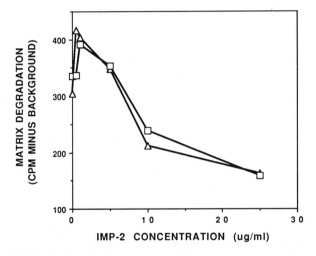

FIGURE 5. Inhibition of matrix degradation by B16 mouse melanoma cells (a gift of I. J. Fidler, M. D. Anderson Cancer Institute, Houston, Texas). Two strains of B16 cells, BL6 (△) and F1 (□), were cultured separately on a radiolabeled matrix secreted by R-22 rat smooth muscle cells (a gift of P. A. Jones, Cancer Center, USC School of Medicine, Los Angeles, California) with increasing concentrations of human IMP-2 for 24 hours. Matrix degradation was determined by measuring the release of radioactivity into the culture medium.

of degrading various components of the extracellular matrix. Recent studies suggest that the same cells also secrete one or more distinct but related inhibitors of metalloproteinases, the best characterized of which is TIMP. Three other secreted inhibitors of metalloproteinases, called IMPs, have also been discovered. Based on the entire amino acid sequence, TIMP and IMP-2 (TIMP-2) have 41% homology.[20] This supports the contention that TIMP and IMP-2 are members of a related family of metalloproteinase inhibitors. The sequence of the bovine equivalent of IMP-2 has also been reported.[21]

The data gathered to date suggest that IMP-2 may preferentially inhibit at least the M_r 72,000 gelatinase. Until the kinetics of inhibition by IMP-2 and TIMP are determined for several metalloproteinases, the selectivity of these inhibitors cannot be determined. However, it should be noted that little if any TIMP bound to and eluted from a gelatin-affinity column. Therefore, TIMP may not be involved in the *in vivo* regulation of at least the M_r 72,000 gelatinase.

TIMP is an excellent inhibitor of interstitial collagenase[8] ($K_i < 10^{-9}$), with tight binding kinetics[7] ($K_d = 1.4 \times 10^{-10}$). It is intriguing to speculate on the selectivity of TIMP versus IMP-2. Collagenase and stromelysin are structurally similar and may be preferentially inhibited by TIMP, whereas the two gelatinases are structurally similar and may be preferentially inhibited by IMP-2. It remains to be determined whether this speculation is supported by new data. However, it is proper to conclude that the degradation of extracellular matrix by metalloproteinases is regulated by more than simply the presence or absence of TIMP. As a family of metalloproteinases is involved in the turnover of matrix proteins, so a family of inhibitors of metalloproteinases, of which TIMP and IMP-2 are members, may be involved in the regulation of proteolysis.

ACKNOWLEDGMENTS

We are grateful to Susan Brekhus for help in preparing the manuscript and to Mary McKenney for editing it. We thank Karyn Bouhana and Elizabeth Bullen for technical assistance.

REFERENCES

1. BANDA, M. J. & Z. WERB. 1981. Mouse macrophage elastase. Purification and characterization as a metalloproteinase. Biochem. J. **193:** 589–605.
2. VALENTINE, R. & G. L. FISHER. 1984. Characteristics of bovine alveolar macrophage elastase. J. Leukocyte Biol. **35:** 449–457.
3. DECREMOUX, H., W. HORNBECK, M. C. JAURAND, J. BIGNON & L. ROBERT. 1978. Partial characterisation of elastase-like enzyme secreted by human and monkey alveolar macrophages. Am. J. Pathol. **125:** 171–177.
4. SENIOR, R. M., N. L. CONNOLLY, J. D. CURY, H. G. WELGUS & E. J. CAMPBELL. 1989. Elastin degradation by human alveolar macrophages: A prominent role of metalloproteinase activity. Am. Rev. Respir. Dis. **139:** 1251–1256.
5. MURPHY, G., M. B. McGUIRE, R. G. RUSSELL & J. J. REYNOLDS. 1981. Characterization of collagenase, other metalloproteinases and an inhibitor (TIMP) produced by human synovium and cartilage in culture. Clin. Sci. **61:** 711–716.
6. WELGUS, H. G., G. P. STRICKLIN, A. Z. EISEN, E. A. BAUER, R. B. COONEY & J. J. JEFFREY. 1979. A specific inhibitor of vertebrate collagenase produced by human skin fibroblasts. J. Biol. Chem. **254:** 1938–1943.
7. CAWSTON, T. E., G. MURPHY, E. MERCER, W. A. GALLOWAY, B. L. HAZLEMAN & J. J. REYNOLDS. 1983. The interaction of purified rabbit bone collagenase with purified rabbit bone metalloproteinase inhibitor. Biochem. J. **211:** 313–318.
8. WELGUS, H. G., J. J. JEFFREY, A. Z. EISEN, W. T. ROSWIT & G. P. STRICKLIN. 1985. Human skin fibroblast collagenase: Interaction with substrate and inhibitor. Collagen Relat. Res. **5:** 167–179.
9. MIGNATTI, P., E. ROBBINS & D. B. RIFKIN. 1986. Tumor invasion through the human amniotic membrane: Requirement for a proteinase cascade. Cell **47:** 487–498.
10. GAVRILOVIC, J., R. M. HEMBRY, J. J. REYNOLDS & G. MURPHY. 1987. Tissue inhibitor of metalloproteinases (TIMP) regulates extracellular type I collagen degradation by chondrocytes and endothelial cells. J. Cell Sci. **87:** 357–362.
11. KHOKHA, R., P. WATERHOUSE, S. YAGEL, P. K. LALA, C. M. OVERALL, G. NORTON & D. T. DENHARDT. 1989. Antisense RNA-induced reduction in murine TIMP levels confers oncogenicity on Swiss 3T3 cells. Science **243:** 947–950.

12. DOCHERTY, A. J. P., A. LYONS, B. J. SMITH, E. M. WRIGHT, P. E. STEPHENS, T. J. R. HARRIS, G. MURPHY & J. J. REYNOLDS. 1985. Sequence of human tissue inhibitor of metalloproteinases and its identity to erythroid-potentiating activity. Nature **318:** 66–69.

13. HERRON, G. S., M. J. BANDA, E. J. CLARK, J. GAVRILOVIC & Z. WERB. 1986a. Secretion of metalloproteinases by stimulated capillary endothelial cells. II. Expression of collagenase and stromelysin activities is regulated by endogenous inhibitors. J. Biol. Chem. **261:** 2814–2818.

14. HERRON, G. S., Z. WERB, K. DWYER & M. J. BANDA. 1986b. Secretion of metalloproteinases by stimulated capillary endothelial cells. I. Production of procollagenase and prostromelysin exceeds expression of proteolytic activity. J. Biol. Chem. **261:** 2810–2813.

15. APODACA, G., J. T. RUTKA, K. BOUHANA, M. E. BERENS, J. R. GIBLIN, M. L. ROSENBLUM, J. H. MCKERROW & M. J. BANDA. 1990. Expression of metalloproteinases and metalloproteinase inhibitors by fetal astrocytes and glioma cells. Cancer Res. **50:** 2322–2329.

16. YASUI, N., H. HORI & Y. NAGAI. 1981. Production of collagenase inhibitor by the growth cartilage of embryonic chick bone: Isolation and partial characterization. Coll. Relat. Res. **1:** 59–72.

17. MACARTNEY, H. W. & H. TSCHESCHE. 1983. The collagenase inhibitor from human polymorphonuclear leukocytes. Isolation, purification and characterisation. Eur. J. Biochem. **130:** 79–83.

18. BUNNING, R. A. D., G. MURPHY, S. KUMAR, P. PHILLIPS & J. J. REYNOLDS. 1984. Metalloproteinase inhibitors from bovine cartilage and body fluids. Eur. J. Biochem. **139:** 75–80.

19. MURRAY, J. B., K. ALLISON, J. SUDHALTER & R. LANGER. 1986. Purification and partial amino acid sequence of a bovine cartilage-derived collagenase inhibitor. J. Biol. Chem. **261:** 4154–4159.

20. STETLER-STEVENSON, W. G., H. C. KRUTZSCH & L. A. LIOTTA. 1989. Tissue inhibitor of metalloproteinase (TIMP-2): A new member of the metalloproteinase inhibitor family. J. Biol. Chem. **264:** 17374–17378.

21. DECLERCK, Y. A., T.-D. YEAN, B. J. RATZKIN, H. S. LU & K. E. LANGLEY. 1989. Purification and characterization of two related but distinct metalloproteinase inhibitors secreted by bovine aortic endothelial cells. J. Biol. Chem. **264:** 17445–17453.

Regulation of Elastin Gene Expression[a]

JOEL ROSENBLOOM,[b,c] MUHAMMAD BASHIR,[b]
HELENA YEH,[b] JOAN ROSENBLOOM,[b]
NORMA ORNSTEIN-GOLDSTEIN,[b] MICHAEL FAZIO,[d]
VELI-MATTI KAHARI,[d] AND JOUNI UITTO[d]

[b]Department of Anatomy and Histology and
Center for Oral Health Research
School of Dental Medicine
University of Pennsylvania

and

[d]Department of Dermatology
Jefferson Medical College
Philadelphia, Pennsylvania 19104

The elastic properties of many tissues of the vertebrate body such as the lung, larger blood vessels, and dermis are due mainly to the presence of extracellular matrix elastic fibers that are composed primarily of the protein elastin. Within the fibers, the individual elastin polypeptide chains are covalently connected by cross-linkages derived from the oxidation of lysine residues by a copper-requiring enzyme, peptidyl lysyl oxidase.[1-3] The extensive cross-linking results in great insolubility, and significant progress in the primary structure determination of elastin came about only after the isolation of a soluble polypeptide designated tropoelastin ($M_r \sim 70,000$) from copper-deficient or lathyritic animals.[4,5] In the analysis of tryptic peptides derived primarily from porcine and to a lesser extent from chick tropolelastin, two types of peptides were recovered: (1) small peptides of three or four residues rich in alanine and terminated by lysine, and (2) larger peptides of 15 to 81 residues rich in hydrophobic amino acids. On the basis of these data and consideration of the structure and number of cross-linkages found in elastin, a model was developed in which alanine, lysine-rich regions formed cross-linkages between tropoelastin molecules and these cross-link points alternated between hydrophobic regions.[6] Validation of this model and definition of its details were impeded by the lack of ordering of the tryptic peptides and the intractable nature of insoluble elastin to primary structure analysis. Furthermore, the difficulty of obtaining tropoelastin from copper-deficient or lathyritic animals has limited analysis of the protein to a relatively few species.

The application of recombinant DNA techniques has resulted in the determination of the complete amino acid sequences of human,[7] bovine,[8,9] and chicken[10] tropoelastins, as well as determination of the structure of the entire bovine[9,11] and human[7,12] elastin genes. This paper summarizes our current knowledge of the structure of the elastin gene and cDNA including consideration of the heterogene-

[a] This work was supported by National Institutes of Health Research Grants AR-20553, AR-28450, and AR-35229 and Grant 1-989 from the National Foundation-March of Dimes.
[c] Address for reprint requests: Joel Rosenbloom, Research Center in Oral Biology, School of Dental Medicine, University of Pennsylvania, Philadelphia, PA 19104.

ity observed in the mature mRNA due to alternative splicing of the primary transcript. These findings will be related to the structure of the protein. Finally, recent studies on the control of transcription of the elastin gene will be considered in detail.

cDNA CLONING AND SEQUENCE ANALYSIS

Although early attempts to clone elastin cDNA yielded only incomplete chick[13] and sheep cDNAs,[14] recent improved cloning techniques[15] have permitted the construction of cDNA clones encompassing essentially the entire length of elastin mRNA isolated from embryonic chick aorta,[10] bovine nuchal ligament,[8,9] and human fetal aorta.[7] In general, very good agreement is found at the nucleotide and encoded amino acid sequence levels between human and bovine; however, large differences, which will be discussed in more detail, exist between them and the chicken. There is also good agreement with the sequenced tryptic peptides of pig tropoelastin, all of which have been accounted for.[16] Although porcine peptides homologous to most of the human and bovine tropoelastins were found, the bovine and human tropoelastins were shown to contain two segments not found in the recovered porcine peptides: (1) sequence LPGGYGLPYTTGK (residues 213–225 in the human sequence), which is part of a tyrosine-rich region (other tyrosine residues in the human sequence are located at residues 206, 228, 230, and 244), and (2) the carboxy-terminal region extending from residue 739 to residue 786. The carboxy-terminal region of the protein is interesting because it contains two cysteines located at positions 6 and 11 from the carboxy terminus, and the segment is exceedingly basic because the sequence ends with four basic residues and another lysine is located at position 8 from the terminus. These features suggest that this portion of tropoelastin may interact strongly, possibly through disulfide bonds, with other matrix proteins such as those composing the microfibrillar component. Although the precise nature of the proteins composing the microfibrillar component is not known, it is very likely that these proteins are acidic and rich in cysteine. (For a review, see ref. 17.)

As predicted by the protein sequencing of tropoelastin,[6] the cDNA sequence analyses have demonstrated that the tropoelastin molecule consists, for the most part, of alternating hydrophobic domains responsible for the elastic properties and of lysine-rich domains that form cross-links between molecules. This segregation of domains is conveniently visualized when the amino acid sequences are analyzed for the distribution of hydrophobic and hydrophilic segments by methods such as that of Kyte and Doolittle.[18] A graphic display of such analyses (FIG. 1) permits ready identification of the lysine-rich potential cross-linking sequences, which project as relatively hydrophilic regions. In these cross-link domains the lysines usually occur in pairs, but in two instances three lysines are found near one another. These findings suggest that a given cross-link segment in the formation of a desmosine/isodesmosine serves to join only two tropoelastin molecules rather than four molecules which is a theoretic possibility. It is apparent that these potential cross-linking sequences are not uniformly distributed and occur at shorter intervals in the first 200 residues. In addition, the potential cross-linking sequences in the first 200 residues frequently contain a prolyl or other residues between the lysines instead of the usual alanines. The conformation of tropoelastin in the cross-linking segments containing alanyl residues is likely to be largely helical, and this conformation may be important in the alignment and condensa-

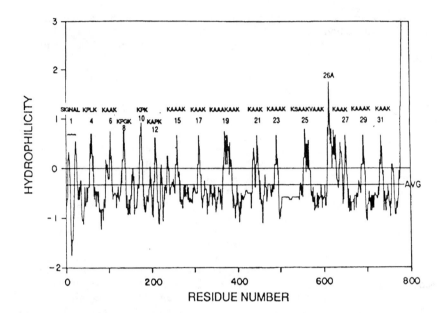

FIGURE 1. Hydrophilicity/hydrophobicity analysis of tropoelastin done by the method of Kyte and Doolittle.[20] The hydrophilic peaks correspond for the most part to potential cross-linking domains.

tion of the lysyl residues in desmosine formation. However, occasional differences in the number of alanines present in a given segment are found among the species, and these differences may lead to variation in the particular type of cross-link formed. In contrast, the presence of prolyl residues clearly disrupts helix formation, and the segments containing proline may participate in unusual cross-links. For reasons that are not clear, cross-links containing such sequences have not as yet been identified in protein analysis.[16] The two cross-linking sequences, KAAKAAK and KSAAKLAAK, that contain three lysine residues are found near the center of the molecule (human residues 375–382 and 558–567), and they may have a critical role in cross-linking the tropoelastin. The 200-residue amino-terminal segment of tropoelastin ends in a tyrosine-rich region whose function is unknown. It is possible that this region is involved in some fashion in the interaction of tropoelastin with other matrix macromolecules or alignment of the molecules within the fiber. Although the homology among the human, bovine, and porcine amino acid sequences is extremely strong, these three differ considerably from the chicken. Among the mammalian sequences, most substitutions are of a minor, conservative nature such as frequent interchange of hydrophobic amino acids or substitution of tyrosine for phenylalanine. However, some significant differences do exist. For example, the number of amino acids that may be found in a hydrophobic region may differ among species. Near the center of bovine, porcine, and chick tropoelastins a pentapeptide, GVGVP, is repeated 11 times, but this repeat segment is considerably different and more irregular in human tropoelastin. Similarly, in human tropoelastin, a hexapeptide, GVGVAP, is repeated seven times (residues 505–546) but only five times with conservative substitutions in bovine tropoelastin. Such repeating units have been postulated to confer an unusual conformation called a beta-spiral in portions of the tropoelastin molecule.[19] The observed variations suggest that a particular number of amino acids and a precise sequence in a given hydrophobic region are not critical to the adequate functioning of the molecule. The chicken sequence is quite homologous to the mammalian sequences for the first 302 residues and the last 57 residues. In the central portion, although some segments are homologous, major differences exist which appear to be due to some type of duplication and deletion events. The most striking of these differences is the occurrence in chicken tropoelastin of the repeating tripeptide $(GVP)_{12}$ which is not found in mammalian elastins. The presence of this repeating tripeptide suggests that elastin may have a distant evolutionary relationship to collagen. Somewhat surprisingly, rather extensive homology (~80%) exists among the bovine, ovine, and human sequences in the 3′ untranslated region, suggesting that this region may have a function either in stabilizing the mature mRNA or even in modulating translation (FIG. 2). Two polyadenylation signals, both of which are apparently used, are found 230 bp

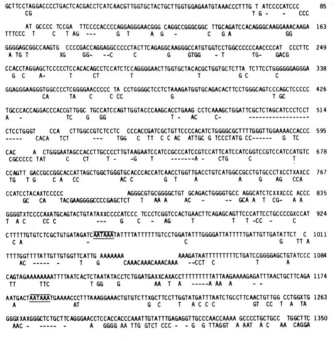

FIGURE 2. Comparison of the human and bovine 3′ untranslated region. The *top line* is the human nucleotide sequence. Differences in the bovine sequence are shown in the *second line*. Spaces have been inserted in the human sequence corresponding to insertions in the bovine gene. Deletions in the bovine gene are denoted by dashes. (X) indicates that the nucleotide could not be unequivocally determined in the human gene. Numbering counts the human sequence. Two polyadenylation consensus sequences AATAAA are delineated. The extensive homology between the two species breaks down about 100 nucleotides past the second polyadenylation signal.

apart. Both the human and bovine genes (see below) were sequenced for 750 bases 3' of the second signal sequence, but no other polyadenylation consensus signals were found.

STRUCTURE OF ELASTIN GENES

The entire bovine (9,11) and human (7,12) elastin genes, which are diagrammed in FIGURE 3, have been extensively characterized. Comparison of the bovine and human cDNA sequences with the genomic sequences has permitted definition of the exon:intron structure of these genes. Because all known cDNA sequences have been identified, the exons have now been renumbered starting at the 5' end, and homologous exons in the two genes have been numbered identically to maintain uniformity. The bovine elastin gene contains 36 exons with a total of 2,280 translated nucleotides, and the human gene 34 exons with 2,358 translated nucleotides. Each entire gene encompasses about 45 kb. Thus, a striking feature of both genes is the small size of the translated exons (27–186 bp) which are interspersed in large expanses of introns. The intron-to-coding ratio is about 20:1, which is very large even in comparison with that of other extracellular matrix proteins such as the fibrillar collagens which have ratios of 8:1.[20] Another important characteristic of the elastin gene is that coding sequences corresponding to hydrophobic and cross-link domains of the protein are found in separate exons, as indicated by the cross-hatched and filled boxes in FIGURE 3. With rare exceptions, these two types of exons alternate. Thus, the domain structure of the protein is a reflection of the gene in which functional domains are usually segregated into separate exons. Typical exons of each type are illustrated in FIGURE 4. Note that in the cross-link exons, two lysines are found near one another, and the exons frequently terminate with an aromatic amino acid. It has been suggested that an aromatic residue may modulate the activity of lysyl oxidase and that the adjacent lysine may remain unoxidized. Curiously, the exon encoding the signal sequence also encodes the amino acid found at the amino terminus of the secreted tropoelastin, so that there is not a distinct separation of these two regions of the protein in the gene.

Although the exons are all multiples of three nucleotides and glycine is found usually at the exon-intron junctions, the exons do not exhibit any regularity in size as is found in the fibrillar collagen genes. It should also be noted that exon-intron borders always split codons in the same way in both genes. Thus, at the 5' border of an exon, the second and third nucleotides of a codon are included, whereas the first nucleotide of a codon is found at the 3' border. This consistent structure is an important feature because it permits extensive alternative splicing of the primary transcript in a cassette-like fashion while maintaining the reading frame (see below). Sequences homologous to bovine exons 34 and 35 have not been found in the human gene despite very extensive sequencing of human genomic clones independently isolated from two human genomic libraries. At present, it is impossible to discern what functional difference may exist between the bovine and human tropoelastins because of this variation.

INTRON STRUCTURE

Remarkably, the introns of both the human and bovine elastin genes are GC rich to nearly the same extent (60% GC) as are the exons, and they contain several

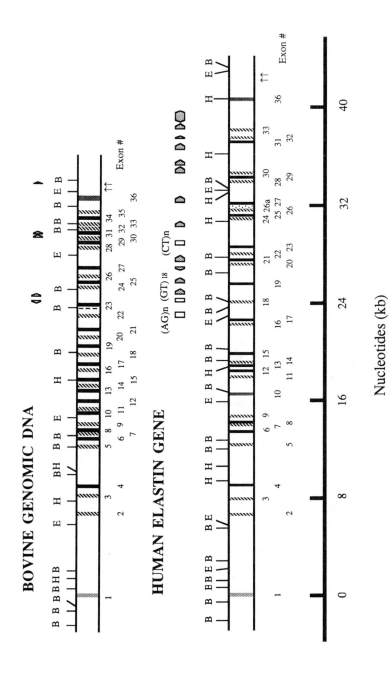

FIGURE 3. Diagram of the bovine and human elastin genes. Exons are not drawn to scale and have been numbered starting at the 5' end of the bovine gene, because all known coding sequences have been identified. Homologous exons in the two genes have been numbered identically to maintain uniformity. Exons encoding potential cross-link domains (■); exons encoding hydrophobic domains (▨); segments containing repetitive DNA (⬢). B = *BamHI*, E = *EcoRI*; H = *HindIII*. (↑↑) = polyadenylation sites.

Exon 31   Crosslink Exon (39 bp)

Human	ag	GT	ATA	CCT	CCA	GCT	GCA	GCC	GCT	AAA	GCA	GCT	AAA	TAC	G gt
Bovine			G G	T			T					C		T	

Human			Gly	Ile	Pro	Pro	Ala	Ala	Ala	Ala	Lys	Ala	Ala	Lys	Tyr
Bovine				Val	Ser										Phe

Exon 32   Hydrophobic Exon (54bp)

Human	ag	GT	GCT	GCT	GGC	CTT	GGA	GGT	GTC	CTA	GGG	GGT	GCC	GGC
Bovine										A				CA

Human		Gly	Ala	Ala	Gly	Leu	Gly	Gly	Val	Leu	Gly	Gly	Ala	Gly
Bovine													Gly	Gln

Exon 36   Translated Region (41 bp)

Human	ag	GT	GGG	GCC	TGC	CTG	GGG	AAA	GCT	TGT	GGC	CGG	AAG	AGA	AAA	TGA
Bovine									T C						G	

Human		Gly	Gly	Ala	Cys	Leu	Gly	Lys	Ala	Cys	Gly	Arg	Lys	Arg	Lys	Trm
Bovine									Ser							

FIGURE 4. Exons found at the 3' end of the bovine and human elastin genes. Note the presence of two cysteine residues near the carboxy terminus. The cross-linking exon contains two lysines separated by alanines.

elastin-like sequences such as GPGGVGALGG or GGAGG. However, these sequences do not contain the consensus exon-intron splice junctions, and they may represent the remnants of unspliced exons during evolution.

Abundance of Repetitive Sequences of the Alu Family

In the human genome, Alu sequences constitute 3–6% of the total mass of DNA and about 3×10^5 copies are present per haploid genome.[21] A consensus sequence for this family of repetitive DNA has been derived by Deininger et al.[22] from the sequence analysis of 10 different Alu sequences. Alu repeats are approximately 300 bp in length, consisting of two head-to-tail left and right repeat monomeric units. The longer right unit contains an insertion of about 30 bp. Alu sequences have some features of transposable elements and pseudogenes in that they are often flanked by direct repeats and have A-rich sequences at their 3' ends.

Each of the Alu repeats in the elastin gene contains the structural features of the classical Alu sequence to one degree or another. For example, FIGURE 5 displays the sequence of one of the repeats which contains a single left and right monomer, a 3' A-rich segment, and flanking direct repeats. All repetitive sequences do not contain direct repeats, and all monomer units are not of the same size and are not found strictly in pairs of left and right units. It should be noted that although most of the repeats are oriented in a 5' to 3' direction (FIG. 3), one paired unit is in the opposite orientation. In comparison to their overall abundance in the human genome, Alu sequences are found at a frequency of about four times the expected value in the elastin gene. In addition to Alu repeats, rather long stretches composed of either alternating purines or alternating pyrimidines occur. Several repetitive sequence elements, whose positions within the introns are diagrammed in FIGURE 3, are also found in the bovine gene.

Although the function, if any, of these repetitive elements remains to be determined, their presence particularly in the human gene raises questions concerning the genetic stability of the elastin gene in the population. In other human genes, such as that for the low density lipoprotein receptor,[23] human hemoglobin,[24] or the nonglobin gene cluster,[25] deletions apparently mediated by recombination between repetitive sequences have occurred, resulting in hereditary diseases. In addition, evidence for genomic instability in regions of human DNA enriched in Alu repeat sequences has been presented by Calabretta *et al.*[26] Limited Southern analyses have revealed several restriction length polymorphisms in the human gene, and it will be of considerable interest to determine whether significant polymorphism, possibly mediated by similar mechanisms, is generally found in the population.

ALTERNATIVE SPLICING OF ELASTIN mRNA

When the nucleotide sequences of several bovine[8,9] and human[7] cDNAs were determined, it became apparent that the sequences for all the cDNAs of each species were not identical. The differences were due to the presence of particular segments in some clones which were absent in other clones; but, for each species, sequences that were common to all the clones were identical. The composite structures for the human and bovine cDNA clones are diagrammed in FIGURE 6 in which nucleotide segments found to be missing in particular clones are designated with arrows. In the human cDNA, such segments correspond to exons 22, 23, 24, 26A, 32, and 33. In the bovine cDNA, such segments correspond to exons 13, 14, 27, 30, 32, and 33. In most cases, the splicing event either includes or deletes an exon in a cassette-like fashion. However, in two of the human cases alternative splicing occurs within an exon in which the exon sequence participates as either a 5' donor (exon 26A) or a 3' acceptor (exon 24).

Inasmuch as the omitted sequences are located between canonical splice junction boundaries, it is unlikely that cloning artifacts caused the observed variations. Consequently, such differences could be due to the existence of more than one gene in the genome or to allelic variation in a single gene. All available genetic evidence suggests that there is only one elastin gene per genome.[7,10] Additional evidence for one elastin gene is the perfect sequence identity, except for the

```
TAGTGAGGGGGATTGGCTGGGCXTGGTGGCCTCACGCCTGTAATCCCAGCACTTTGGGAG

GCCTAGGTGGGTGGATCAACTTGAGGTCCAAGGAGTTCGAGACCCAGTCTGGTCAAACAT

GGTGAACCCTGTCTCTACTAAAAAAAAATGGCAAAAATTAGCCAAACGTGGTGGACGCCTG

TAATCCCAGCTACTCGGGAGGCTGAGGCGGGAGAATCACTGGAGCCTGGGAAGCGGAGGT

TGCAGTGAGCCAAGATCGCACCACTGCACTCCAGCCTGGGTGACAGAGCAAGACCCCATC

TCAAAAAAATAATAATAAAATAAAATATAAAAAATTATATAGTGGGGGGGAT
```

FIGURE 5. Representative example of the Alu family sequence found in the human elastin gene. This particular sequence is found at about coordinate 41 in FIGURE 3. Direct repeats at the 5' and 3' end are underlined.

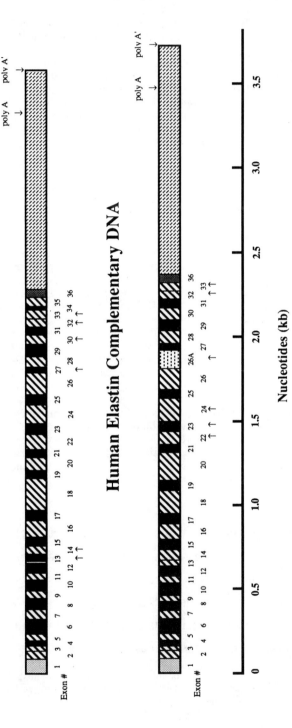

FIGURE 6. Composite diagrams of bovine and human cDNA clones. The cDNA is divided into exons which are numbered. Exons encoding hydrophobic sequences (□); exons encoding potential cross-linking sequences (■); exon encoding most of the signal sequence (▨). *Arrows* mark the exons subject to alternative splicing.

omitted sequences, among the cDNAs. Some variation in sequence would be expected, at least in wobble positions that involve no amino acid substitutions, if more than one gene existed. Because in both the human and bovine cases the cDNAs were constructed from mRNA isolated from a single fetus, and because at least three distinct cDNAs were observed in each species, the most likely explanation for the variability in the cDNA is alternative splicing of a single primary transcript.[27] Both hydrophobic and cross-link domains are affected, so that two cross-link domains may be brought into apposition (deletion of exon 30) or the interval between cross-link domains may be increased (deletion of exon 27). It is not possible to predict the functional consequences of these variations at present, although clearly a tighter or looser network could be produced.

S1 Nuclease Analysis

Sequencing of cDNA neither yields quantitative information on the frequency of particular splicing events nor does it guarantee that all events have been detected. Such information is essential to provide a description of the potential coding sequences found in the ensemble of mature mRNA molecules and hence in the encoded tropoelastin. To shed light on these important points, we carried out S1 nuclease protection experiments using mRNA isolated from the nuchal ligament of different aged bovine fetuses, neonate calves, and adult cows. Appropriate cloned cDNA restriction fragments, which collectively encompassed the entire length of the elastin mRNA, were end-labeled with ^{32}P at their 5' or 3' ends and used as probes. After hybridization to the isolated RNA, the RNA:DNA duplexes were digested with S1 nuclease and the digestion products analyzed by quantitative autoradiography. In addition to the control and experimental digest mixtures, a range of known concentrations of the particular labeled probes used in the individual experiments was also electrophoresed to be used as an internal standard in quantitative densitometry. Representative autoradiographs are illustrated in FIGURES 7 and 8. These experiments clearly demonstrated that only exon 33 is frequently subject to alternative splicing (deleted 45–55% of the time), whereas exons 13, 14, 27, 30, and 32 are infrequently spliced out in all the RNA samples from the different aged cows. An interesting observation was that the RNA from adult cows displayed a different splicing pattern from that of fetal and neonatal RNA. In one region encompassing exons 23–27, a low level of splicing out of every exon was observed in the RNA from the adult, whereas no such splicing events were seen in RNA at any of the other ages.

Isoforms of Tropoelastin

Significant variation will exist in the size and amino acid sequence of the tropoelastin if all the different mRNA molecules are translated, and such variation could explain the finding of at least two forms of tropoelastin in several species.[28-31] It will be important to validate this explanation and also to determine whether the splicing pattern is developmentally or tissue regulated and whether functional differences exist among the tropoelastin molecules. The finding of a varied pattern in the adult suggests that the precise type of monomer units synthesized in the adult could vary from that synthesized during growth. If a different type of splicing pattern were to occur in human tissues, particularly in pathologic situations such as atherosclerosis, it may contribute to the disease process.

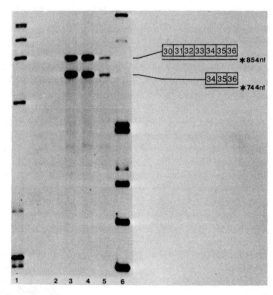

FIGURE 7. Alternative splicing of exon 33. The probe was 5' end-labeled and hybridized to 25 μg of bovine nuchal ligament RNA and the hybridization mixture digested with S1 nuclease. The digestion products were then electrophoresed on 6% polyacrylamide gels. Control *lane 1* contained no RNA; *lane 2*, 270-day fetal RNA; *lane 3*, 14-day neonate; *lane 4*, adult; *lane 5*, 1-kb ladder markers.

CHARACTERIZATION OF THE ELASTIN PROMOTER

To define mechanisms controlling transcription of the elastin gene, analyses of the region flanking the start of translation were carried out and experiments performed to identify elements important for promoter activity.

Sequence Analysis of the 5'-Flanking Segment

Both strands of 2,200 base pairs located 5' of the ATG codon have been sequenced, revealing several interesting features[12,32] (FIG. 9). No canonical TATA box was found, and although two CAAT sequences were found, their locations raise doubts concerning their functional significance. The most proximal one at −57 is extremely close to apparent initiation sites, and the distal one at −599 is quite far from the major initiation sites (see below). The 5'-flanking region is generally G+C rich (66%) and contains a remarkable number of potential binding sites for transcription regulatory factors. These include SP1 and AP2 binding sites, glucocorticoid-responsive elements, TPA-responsive elements, and cyclic AMP-responsive elements (CRE). There is also an extended sequence of alternating guanine and pyrimidine residues (−225 to −275), a type of sequence that may be associated with Z DNA and that may be involved in the regulation of transcription.[33] Collectively, these features have been associated with the promoters of so-called "housekeeping genes" as well as a few other genes.[34]

S1 Protection and Primer Extension Analyses

The absence of a TATA box in the putative promoter region suggested the possibility of multiple sites of transcription initiation. To test this possibility and to provide support for the belief that the isolated 5'-flanking segment contained the promoter, S1 protection and primer extension analyses were carried out (FIG. 10).[12] These two different types of analyses were remarkably consistent with one another and clearly indicated that elastin mRNA transcription in the human fetal aorta is initiated at multiple sites. The agreement between the two methods also suggests that the apparent initiation sites are not artifactually produced. Three major clusters of initiation were found centered around nucleotides −(7,8), −(15,16), and −(32,33). Four other minor initiation sites were observed between −45 and −80. The S1 probes extended 648 nucleotides 5' of the ATG codon, and a weak signal not visible in FIGURE 10, but seen in the original autoradiograph was found at −195. Furthermore, no splice junction consensus sequences are present in the region of the putative initiation sites, suggestive of an intervening sequence. Cumulatively, these data suggest that no intron is located 5' of the ATG initiation codon and that the promoter region is in fact located within the isolated 2,200-base pair segment.

A recent survey has identified 14 promoters that lack the usual TATA canoni-

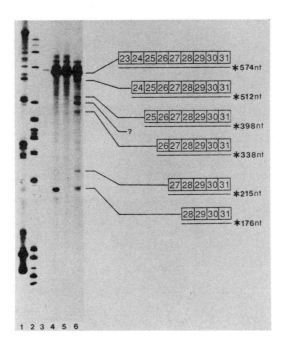

FIGURE 8. Alternative splicing of exons 23–31. The probe was 5' end-labeled and hybridized to 25 µg bovine nuchal ligament total RNA. After S1 nuclease digestion, the products were electrophoresed on 6% polyacrylamide gels. *Lane 1,* 123 nt ladder markers; *lane 2,* phi-X markers; *lane 3,* control, no RNA; *lane 4,* 270-day fetal RNA; *lane 5,* 14-day neonate; and *lane 6,* adult. The origin of the band labeled with a question mark is unknown.

cal sequence and other features frequently associated with most previously characterized promoters.[34] Many of these 14 promoters control transcription of genes encoding enzymes for metabolic reactions carried out in many cell types, and therefore they have been described as "housekeeping genes." In contrast, many of the earlier characterized genes with conventional promoter features encoded proteins whose expression was tissue specific. Some widely expressed genes lack a TATA box and also contain CpG islands (regions of high G+C content and a high frequency of CpG dinucleotides) in their 5'-flanking regions. However, no tissue-specific genes with 5'-CpG islands lack a TATA box, and at present the elastin promoter appears to represent an unusual case. It is clear from studies in different species that elastin expression is regulated with respect to both tissue type and developmental period. It is equally clear that expression is largely con-

```
-2260 CTCGAGAAGA GAGGGCTCCA GCTCCCCACA GTAGCCCCTG CCTTCCTCCT TCCCAGGCAG -2201
-2200 GCGGAGGCAC GCAGATATAC CATTGACTTC CCCTCCCCTG CAGCAGGCAC ATCCTGGGCA -2141
-2140 TCGAGCTTCA GACCCTGCCC CTGAGCAGCC CCTAACCCCA CCAACAAAGG GTGGCTTGGG -2081
-2080 GGGGCTTTCA CCCCAGCATA ATCTCCATCA GCTACCCTCA AAGCACCCCC AAATAAACAC -2021
-2020 ACACCGTAAG TAAGAGCTGT ACACTGGCTG TGTGCGTACA TCTTCAAGAC AATTCTCCCA -1961
-1960 GCATGCCCCT ACCTTCCAAA ATTCCAGAGC TGCTCCCTCC AAAGACCCAG GGAAAAGGAA -1901
-1900 GGGTTTGTCC AGGGTCCTGG GGTGGCCCCG TATAGACCAA AGCCTGATAA CTGTCCTAGA -1841
-1840 AGCAGAGTAC TTGCAGAGCG AGTGACGGCA ACTGTGGTAT TGACACCAGT CCTAGCACCA -1781
-1780 GCTGAACACA GAGCATTTTT GATCTAGCAG AAATACAAGA CCACGTTGTA TTTGTCTTTG -1721
-1720 CAATAATCTC TTAGCTAGGA ATACTGATCA CCTGTAGACA GATAAGGAAA CTGATGCTCT -1661
-1660 GTGGAGAGGT TTTCCTACCA GAAAGGCTAG AGCCAGAAAT TTACTTCTAG GTCCACCAAT -1601
-1600 ACCTGCCTTT GACCAATGCC TGCATTTGAC CTTTCCACGC TGAGCCACCC CTGCTGGCAC -1541
-1540 TCCAGACTGC CACAGTGCTC CTGCCTCCAC AAGGGGTCTT TAACTCATCC CTCGGAGCCA -1481
-1480 TCGTGGTGCA GGGAAAAGCC CACAGGGCGT GTGGCTTCCA TGCTGTTCCC TGACTGGCTG -1421
-1420 TGACCTAGGA CAAGGAACAA GTTTCCCTCT CCTATTCTCT AGGTCTCACA TTTCTTCTCC -1361
-1360 TCTAGCAGTA GTGGGAAGTG AGGGGTGGGG GACACGACCC TCCCCTGTTC CATCCCACAC -1301
-1300 TCCAACCCCC AAAATCCCCC AGGGTCCCCG TCCAGCTCAG TCCTGGGGGC AGAAATGCAG -1241
-1240 AGTTCTCCAG GAACGTGGTC CCAGCTGTTT CAGTGCAGGC CGCCCCCTCC TGGCCACCAG -1181
-1180 CGGAATGTCA GCCTTCCCAG AGGGGCCGGG AGAACAGCAG TCGAGAAGCT CCCAGACTGG -1121
-1120 TGTGGGCGCT AGCTGTGCTC AGCGTGGGGA TGGGAGGTGA CCCAGTGATA ATGGGAAGCT -1061
-1060 GGGCTGCCTG TCAGTCTGTG GGGGGCTCCC ACCTCCCTGT TCCCCCACAG GGCACCTGGG -1001
-1000 GATCCAGCCT GATTTTTACC AGACCTGCGG CCTGCATGGG GCTGGGTATA GGGCTGTGAC -941
-940 CTTGACCCAT GCAGAATAGA ACCCTGTGTG TCGGGATCCT CCATGTGCTC CAGATGCCCC -881
-880 TGGGGACAGC ACCAACATGG CCTTAACTCC CAAGCCATTC CCCTGCCTCT AACCCCCTGG -821
-820 CATCTGCAGG CATCCACCCC AGACCCACCC AACACCTCCT CCCCAGCTTC AGGCGCTAGG -761
-760 CAGAGACCTT GGCCCCTGCA GAATGCAGCC CTGTCCAGGG TCCCCTACCT TCCCCCCAGA -701
-700 TCCCTCCCAG AGCAATACCA ACCCGGGCCT ACCTTCCAGG CCATTCAACC TGCAGCCCCC -641
-640 CGGCCTCTGT AGACATCGCA CCCCCCAAAC CCCCAGACCT GCCCAATGCC TCCCTCCCC -581
-580 AGCTTTGGGC AGAACCTGTC TCTAGCCAGA CCTGGGGGTG TTGGGGAGTC TGGAGGGCCG -521
-520 GGGTGGGGGC TGAGGCGCGG GACAGCTGGC CCGTATCCTC ACACTGGGCC CAGCCGGACG -461
-460 GGCGGGGGCC TGGCCACTCG GGCCTTGGCT GGGGCTGGGA TTTTTGGCCT GGCCGCCAGG -401
-400 CCCTCCCTTC TGCTTCCTCT CCCGAGGGCT GTCCTGGCAG AGGCCCCCCT CGCTCTTTCT -341
-340 GGCGGGGAACA GGGCCAGCAG CGAAAGAACA GTCGCAGAGG GAAAGCGGGA AAGAGATGGG -281
-280 GGAAAGTGTG TGTGTGTGAG TGTGTGCTTG TGTGCATGTG TGTGCGTGTG TGTGTCTCAAG -221
-220 AAAAAAGCTC GCAGTCCAGC AGCCCGGGCC TGGGAGGCTT GTGAGCCGGG CCTTTCGTAA -161
-160 TTGTCCCCTC CGCGCGGGCC CCTCCCCAG GCCTCCCCCC TCTCCCGCCC TCCCGCCCGC -101
-100 CGTCTCTCCC TCCCTCTTTC CCTCACAGCC GACGAGGCAA CAATTAGGCT TTGGGGATAA -41
-40 AACGAGGTGC GGAGAGCGGG CTGGGGCATT TCTCCCCGAG
```

FIGURE 9. Nucleotide sequence of the 5'-flanking region of the human elastin gene. The nucleotides have been numbered upstream from the ATG translation initiation site. The putative SP1 binding sites are boxed (□), and the putative AP2 binding sites are outlined (⊏⊐). The glucocorticoid-responsive elements (⊔), TPA-responsive elements (⌊⌋), and CRE elements (⌐) are indicated in the figure.

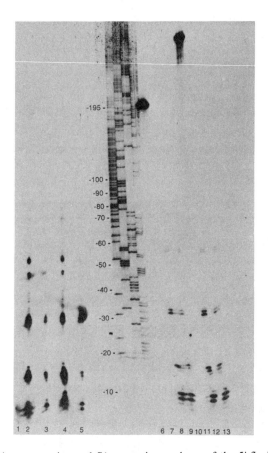

FIGURE 10. Primer extension and S1 protection analyses of the 5'-flanking region. The RNA was isolated from a 7-month-old fetal aorta (HFA). *Lanes 1–5* are primer extension and *lanes 6–13* are S1 protection. The sequencing reactions shown in the center of the figure used the same 24-nucleotide oligomer as primer as that used in the primer extension, and the ladder is in the order guanine, adenine, thymidine, and cytosine. The nucleotide positions are numbered starting with the first base before the ATG codon taken as −1. *Lane 1,* tRNA in primer extension reaction; *lane 2,* total HFA RNA; *lane 3,* poly(A+) HFA RNA; *lane 4,* total HFA RNA; *lane 5,* poly(A+) HFA RNA; *lane 6,* 219-nucleotide *Sma*I probe, undigested; *lane 7,* tRNA; *lane 8,* poly(A+) HFA RNA; *lane 9,* total HFA RNA; *lane 10,* 672-nucleotide *Pst*I probe, undigested; *lane 11,* tRNA; *lane 12,* poly(A+) HFA RNA; *lane 13,* total HFA RNA.

trolled by the steady-state level of elastin mRNA, and it has been presumed without direct experimental support that this steady-state level is predominantly controlled by transcription. However, even if this presumption proves to be correct, many problems remain. These include the identification of sequences within the 5'-flanking and first intron regions, which are physiologically important in controlling transcription initiation, and the significance, if any, of the multiple initiation sites. S1 mapping and primer extension analyses have been limited to

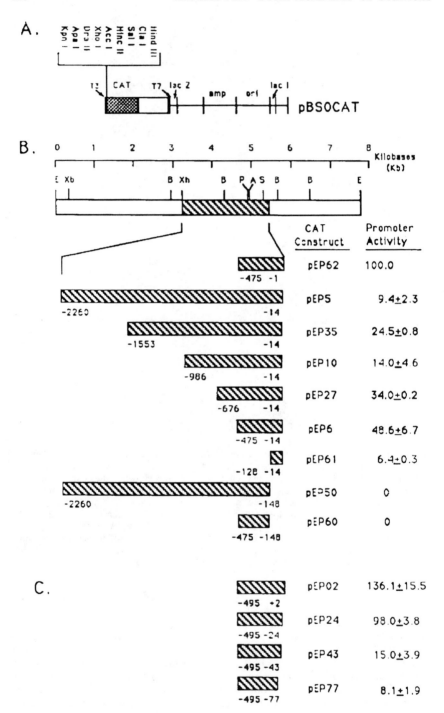

mRNA from fetal aorta, and it is possible that a different pattern of initiation may be found in elastin transcripts from other tissues. Elastin transcripts have been shown to undergo substantial alternative splicing, and it is presently unknown whether any relation exists between the position of transcript initiation and the pattern of alternative splicing.

Expression of Elastin Gene Promoter/CAT Reporter Gene Constructs in Cultured Cells

To test the ability of the putative elastin promoter to direct transcription, a series of elastin promoter/chloramphenicol acetyl transferase (CAT) reporter chimeric gene constructs was synthesized,[12,32] using segments extending from -1 to $-2,260$. Restriction endonuclease and exonuclease digestions were used to generate fragments containing deletions. All promoter fragments were subcloned into a promoterless plasmid construct pBSOCAT which contains the entire CAT gene, as well as small t intron and the polyadenylation signal[35] (FIG. 11A). To test the functional activity of the chimeric elastin/CAT gene constructs, they were utilized in transient transfections of freshly isolated neonatal rat aortic smooth muscle cell cultures, which actively express elastin.[36] Parallel transfection experiments were first performed with pEP62CAT (-475 to -1), together with pEP5CAT ($-2,260$ to -14), pEP6CAT (-475 to -14), and a promoterless construct pBSOCAT as a negative control (FIG. 11B). The CAT activity in cells transfected with pEP6CAT was ~50% of that noted with pEP62CAT, whereas the activity of pEP5CAT was only ~10% of that noted with pEP62CAT. Because the transfection efficiencies of all the constructs was comparable, as determined by the plasmid copy number measured in the transfected cell cultures, these observations suggested that the more 5' region ($-2,260$ to -475) contained down-regulatory elements for transcriptional activity.

To delineate potential regulatory segments more precisely, deletion clones were generated by exonuclease III digestions. The precise ends of selected deletion constructs were determined by nucleotide sequencing, and the constructs were transfected into the rat smooth muscle cells. Comparison of the promoter

---

FIGURE 11. Relative promoter activity of elastin/CAT gene constructs in transient transfections of rat aortic smooth muscle cell cultures. **(A)** The linearized promoterless plasmid construct pBSOCAT containing the entire CAT structural gene and small t intron together with a polyadenylation site. Upstream from the CAT gene is a multiple cloning site that allows insertion of DNA fragments of the elastin promoter region for construction of elastin promoter/CAT gene constructs. **(B)** The activity of different elastin promoter/CAT gene constructs. The construct pEP5CAT was subjected to 5' and 3' deletion analyses by exonuclease III digestion. The resulting clones, with a 3' end fixed at -14, contained variable 5' ends as shown in the figure. In addition, pEP5CAT and pEP6CAT were subjected to digestion by *Sac*II endonuclease to remove the region from -148 to -14, resulting in constructs pEP50CAT and pEP60CAT, respectively. **(C)** Relative promoter activity of 3' deletion constructs with variable 3' ends and the 5' end fixed at -495. The promoter activity on the right side is expressed in relation to pEP62CAT (100%) spanning from -475 to -1 and represents activity above that noted with pBSOCAT used for parallel transfection in each experiment. The numbers are means ± SD from three separate experiments, each point representing at least two independent cultures. The restriction sites indicated in **B** are: E = *Eco*RI; Xb = *Xba*I; B = *Bam*HI; Xh = *Xho*I; P = *Pvu*II; A = *Apa*I; S = *Sac*II.

activity of inserts with their 5' end residing between those of pEP5CAT (−2,260 to −14) and pEP6CAT (−475 to −14) revealed reproducible differences. The activity observed with pEP35CAT (−1,553 to −14) was consistently more than twofold greater than that with pEP5CAT, whereas the activity with pEP10CAT (−986 to −14) was always ~50–60% of that noted with pEP35CAT (FIG. 11B). These results suggest the presence of both up- and down-regulatory elements within the 5'-flanking region of the elastin gene. Further deletion analyses revealed that the smallest construct examined, pEP61CAT (−128 to −14), consistently showed activity that was ~60% that of pEP62CAT, and suggested that the basic elastin promoter elements necessary for expression were contained within the −128 to −1 region. To test this hypothesis, segment −147 to −14 was removed by SacII digestion, and the resulting constructs, pEP50CAT (−2,260 to −148) and pEP60CAT (−475 to −148), were then used in transfections of rat aortic smooth muscle cells. The deletion of the −147 to −14 segment abolished all CAT activity in these cells. Thus, the first 147 nucleotides 5' of the translation initiation site are of paramount importance in transcriptional control of elastin gene expression.

To further elucidate the role of distinct regions within the segment from −147 to −1, four different deletion constructs having a fixed 5' end and a variable 3' end were used in further transfection experiments (FIG. 11C). The most notable finding with these 3' deletion constructs was a marked reduction in promoter activity from pEP24CAT (−495 to −24) to pEP43CAT (−495 to −43), suggesting the importance of the 18 nucleotides between −43 and −24 in directing the elastin promoter activity. However, pEP43CAT (−495 to −43) and pEP77CAT (−495 to −77) consistently showed CAT activity that was clearly above that observed with pBSOCAT, suggesting that these constructs still contained elements allowing expression of minimal promoter activity. A consistent observation noted in parallel transfections was that pEP02CAT (−495 to +2) always yielded higher CAT activity than did pEP62CAT. Furthermore, somewhat surprisingly, the activity with pEP24CAT (−495 to −24) was about twice that of pEP6CAT (−475 to −14). These observations suggest that the 20 nucleotides between −495 and −475 contain an up-regulatory cis-element.

Because the experiments just described were performed with human elastin promoter/CAT gene constructs in rodent aortic smooth muscle cell cultures, it was of interest to compare the activities of these constructs in human cells. Therefore, cultures of human skin fibroblasts and human HT-1080 fibrosarcoma cells were subjected to parallel transfections with pEP5CAT and pEP6CAT. For comparison, mouse NIH-3T3 cells, which were shown to be highly efficient in transfection studies, were also included in these experiments. In all cell cultures examined, pEP6CAT had a higher activity than did pEP5CAT (data not shown). The activity obtained with pEP5CAT was only 27–43% of that detected with pEP6CAT, further supporting the finding that sequences within −2,260 to −475 contain negative, down-regulatory elements.

Therefore, comparison of the activity noted with different deletion constructs has allowed identification of several potential regulatory regions. (See FIG. 12 for model.) The basic promoter element (P) appears to reside within the sequence −128 to −1. Two regions appear to contain negative, down-regulatory elements (N1 and N2 at positions −2,260 to −1,554 and −986 to −476, respectively). At the same time, two distinct regions (E1 and E2) were shown to have up-regulatory activity. In particular, the region E2 extending from −495 to −129 showed a strong up-regulatory effect on the basic promoter activity. This positive regulatory activity may be explained, as least in part, by the presence of multiple SP-1 and AP2 binding sites within this region. It is unclear whether these positive regulatory regions are position-dependent, up-regulatory elements or whether they consist of position- and orientation-independent enhancer-like elements. However, collectively these observations suggest that the identified regulatory regions may contain several functional subregions.

Several interesting, potentially functional consensus sequences were identified within the elastin 5'-flanking region, including three glucocorticoid-responsive elements.[37] Such elements may explain previous reports indicating that elastin gene expression can be up-regulated by glucocorticoid steroids.[38] Particularly interesting was the observation that two different potential cAMP-responsive elements were found in the 5'-flanking region. First, two separate putative AP2 binding sites were found within the basic promoter element, and several additional AP2 binding sites were found further upstream. Such AP2 binding sequences have been shown to mediate cAMP and TPA modulation of gene expression.[39] Secondly, two distinct cAMP-responsive regulatory elements (CRE) were found to be present, and these may participate in cAMP-mediated regulation of elastin gene expression. Previous studies have shown that cGMP can stimulate elastin production of ligamentum nuchae fibroblasts, and that this stimulation can be inhibited with cAMP.[40] In addition, two consensus motifs resembling TPA-inducible elements were found, but the functional significance of these elements is

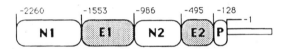

FIGURE 12. Schematic representation of functional domains of the 5'-flanking region of the human elastin gene. The basic promoter element (P) resides within region −128 to −24. Three separate up-regulatory, enhancer-like regions (E1–E3) were noted in functional CAT assays in transient transfections of cultured rat aortic smooth muscle cells. Similarly, two down-regulatory (negative) elements (N1 and N2) were delineated.

currently unknown. However, the elastin promoter/CAT gene constructs provide a means to study the transcriptional modulation of elastin gene expression by a variety of exogenous agents.

Most of the detailed analyses were performed in rat aortic smooth muscle cell cultures, which have previously been shown to be highly active in the expression of the endogenous elastin gene.[36] However, clearly detectable promoter activity also was noted in human HT-1080 cells, human skin fibroblasts, and mouse NIH-3T3 cell cultures. Human skin fibroblasts and fibrosarcoma HT-1080 cells (unpublished observations) have been shown to express the elastin gene, as demonstrated by the presence of 3.5-kb elastin mRNA. However, no such elastin mRNA has been seen in 3T3 cells. These observations suggest that all elements for tissue and developmental-specific expression of the elastin gene may not reside within the 2.2-kb 5'-flanking region.

SUMMARY

Recent isolation and characterization of cDNAs encompassing the full length of chicken, cow, and human elastin mRNA have led to the elucidation of the primary structure of the respective tropoelastins. Comparison of the tropoelastin from the different species has revealed that large segments of the sequence are conserved, but considerable variation also exists, ranging in extent from relatively small alterations, such as conservative amino acid substitutions, to large-scale deletions and insertions. Several distinct approaches have yielded compelling evidence of a single elastin gene per haploid genome. Analysis of the bovine and human elastin genes revealed that functionally distinct hydrophobic and cross-link domains of the protein are encoded in separate exons which alternate in the genes. The human gene contains 34 exons, the intron/exon ratio is unusually large (20 : 1), and the introns contain large amounts of repetitive sequences that may predispose to genetic instability. Comparison of the cDNA and genomic sequences has demonstrated that the primary transcript of both species is subject to considerable alternative splicing, which can account for the presence of multiple tropoelastin isoforms. It is likely that the conformation of elastin is, at least in part, that of a random coil, and therefore it might be expected that the stringency for conservation of the amino acid sequence would be less than that for other proteins with unique conformations. This suggests that functional elastin molecules that vary in their sequence and fitness may exist in the human population and be compatible with a normal life. Potentially though, these variations could have profound consequences on the properties of vital tissues found in the cardiovascular and pulmonary systems over the lifetime of the individual. Consequently, analysis of the structure of the elastin gene and its variation in what is regarded as the normal human population, rather that in those individuals with clearly heritable diseases, assumes greater importance.

The 5′-flanking region of the gene is G+C rich and contains several SP-1 and AP2 binding sites, as well as putative glucocorticoid, cAMP, and TPA responsive elements, but no consensus TATA box or functional CAAT box. Primer extension and S1 mapping of the elastin mRNA indicated that transcription was initiated at multiple sites. Transfection experiments using promoter elements/reporter gene constructs demonstrated that the basic promoter element was found within region −128 to −1. In addition, three distinct up-regulatory and two down-regulatory regions were delineated. Taken together, these findings suggest that the regulation of elastin gene expression is complex and takes place at several levels.

REFERENCES

1. PARTRIDGE, S. M. 1962. Elastin. Adv. Protein Chem. **17:** 227–297.
2. FRANZBLAU, C., F. M. SINEX, B. FARIS & R. LAMPIDIS. 1965. Identification of a new cross-linking amino acid in elastin. Biochem. Biophys. Res. Comm. **21:** 575–581.
3. PINNELL, S. R. & G. R. MARTIN. 1968. The cross-linking of collagen and elastin : enzymatic conversion of lysine in peptide linkage to amino-adipicsemialdehyde (allysine) by an extract from bone. Proc. Natl. Acad. Sci. USA **61:** 708–716.
4. SANDBERG, L. B., N. WEISSMAN & D. W. SMITH. 1969. The purification and partial characterization of a soluble elastin-like protein from copper-deficient aorta. Biochemistry **8:** 2940–2945.
5. FOSTER, J. A., R. SHAPIRO, P. VOYNOW, G. CROMBIE, B. FARIS & C. FRANZBLAU. 1975. Isolation of soluble elastin from lathyritic chicks. Comparison to tropoelastin from copper deficient pigs. Biochemistry **14:** 5343–5347.
6. GRAY, W. R., L. B. SANDBERG & J. A. FOSTER. 1973. Molecular model for elastin structure and function. Nature **246:** 461–466.
7. INDIK, Z., H. YEH, N. ORNSTEIN-GOLDSTEIN, P. SHEPPARD, N. ANDERSON, J. C. ROSENBLOOM, L. PELTANEN & J. ROSENBLOOM. 1987b. Alternative splicing of human elastin mRNA indicated by sequence analysis of cloned genomic and complementary DNA. Proc. Natl. Acad. Sci. USA **84:** 5680–5684.

8. RAJU, K. & R. A. ANWAR. 1987. Primary structures of bovine elastin a, b, and c deduced from the sequences of cDNA clones. J. Biol. Chem. **262:** 5755–5762.
9. YEH, H., N. ORNSTEIN-GOLDSTEIN, Z. INDIK, P. SHEPPARD, N. ANDERSON, J. C. ROSENBLOOM, G. CICILA, K. YOON & J. ROSENBLOOM. 1987. Sequence variation of bovine elastin mRNA due to alternative splicing. Collagen Rel. Res. **7:** 235–247.
10. BRESSAN, G. M., P. ARGOS & K. K. STANLEY. 1987. Repeating structures of chick tropoelastin revealed by complementary DNA cloning. Biochemistry **26:** 1497–1503.
11. CICILA, G., M. MAY, N. ORNSTEIN-GOLDSTEIN, Z. INDIK, S. MORROW, H. S. YEH, J. ROSENBLOOM, C. BOYD, J. ROSENBLOOM & K. YOON. 1985. Structure of the 3' portion of the bovine elastin gene. Biochemistry **24:** 3075–3080.
12. BASHIR, M. M., Z. INDIK, H. YEH, N. ORNSTEIN-GOLDSTEIN, J. C. ROSENBLOOM, W. ABRAMS, M. FAZIO, J. UITTO & J. N. ROSENBLOOM. 1989. Characterization of the complete human elastin gene: Delineation of unusual features in the 5'-flanking region. J. Biol. Chem. **264:** 8887–8891.
13. BURNETT, W., A. FINNIGAN-BUNICK, K. YOON & J. ROSENBLOOM. 1982. Analysis of elastin gene expression in the developing chick aorta using cloned elastin cDNA. J. Biol. Chem. **257:** 1569–1572.
14. YOON, K., M. MAY, N. GOLDSTEIN, Z. K. INDIK, L. OLIVER, C. BOYD & J. ROSENBLOOM. 1984. Characterization of a sheep elastin cDNA clone containing translated sequences. Biochem. Biophys. Res. Commun. **118:** 261–269.
15. GUBLER, U. & B. J. HOFFMAN. 1983. A simple and very efficient method for generating cDNA libraries. Gene **2:** 263–269.
16. SANDBERG, L. B. & J. M. DAVIDSON. 1984. Elastin and its gene. Peptide Protein Rev. **3:** 169–193.
17. CLEARLY, E. G. & M. A. GIBSON. 1984. Elastin-associated microfibrils and microfibrillar proteins. Int. Rev. Connect. Tissue Res. **10:** 97–209.
18. KYTE, J. & R. F. DOOLITTLE. 1982. A simple method for displaying the hydropathic character of a protein. J. Mol. Biol. **157:** 105–132.
19. URRY, D. W. & M. M. LONG. 1976. Conformations of the repeat peptides of elastin in solution. Crit. Rev. Biochem. **4:** 1–45.
20. BOEDTKER, H., F. FULLER & V. TATE. 1983. Structure of collagen genes. Int. Rev. Connect. Tissue Res. **10:** 1–63.
21. SCHMID, C. W. & W. R. JELLINEK. 1981. The Alu family of dispersed repetitive sequences. Science **216:** 1065–1069.
22. DEININGER, D. L., D. J. JOLLY, C. M. RUBIN, T. FRIEDMAN & C. W. SCHMID. 1981. Base sequence studies of 300 nucleotide renatured repeated human DNA clones. J. Mol. Biol. **151:** 17–33.
23. LEHRMAN, M. A., W. J. SCHNEIDER, T. C. SUDHOF, M. S. BROWN, J. L. GOLDSTEIN & D. W. RUSSELL. 1985. Mutation in LDL receptor: Alu : Alu recombination deletes exons encoding transmembrane and cytoplasmic domains. Science **227:** 140–146.
24. ORKIN, S. H. & H. MICHELSON. 1980. Partial deletion of the plobin structural gene in human-thalassaemia. Nature **286:** 538–540.
25. JAGADEESWARAN, P., D. TUAN, B. G. FORGET & S. M. WEISSMAN. 1982. A gene deletion ending at the midpoint of a repetitive DNA sequence in one form of hereditary persistence of fetal haemoglobin. Nature **296:** 469–472.
26. CALABRETTA, B., D. L. ROBBERSON, H. A. BERRERA-SALDANA, T. P. LAMBROU & G. F. SAUNDERS. 1982. Genome instability in a region of human DNA enriched in Alu repeat sequences. Nature **296:** 219–225.
27. PADGETT, R. A., P. J. GRABOWSKI, M. M. KONARSKA, S. SEILER & P. A. SHARP. 1985. Splicing of messenger RNA precursors. Ann. Rev. Biochem. **55:** 1119–1150.
28. RICH, C. B. & J. FOSTER. 1984. Isolation of tropoelastin a from lathyritic chick aorta. Biochem. J. **217:** 581–584.
29. DAVIDSON, J. M., K. SMITH, S. SHIBAHARA, P. TOLSTOSHEV & R. G. CRYSTAL. 1982. Regulation of elastin synthesis in developing sheep nuchal ligament by elastin mRNA levels. J. Biol. Chem. **257:** 747–754.
30. MECHAM, R. P., B. D. LEVY, S. L. MORRIS & D. S. WRENN. 1985. Glucocorticoids stimulate elastin production in differentiated bovine fibroblasts but do not induce elastin synthesis in undifferentiated cells. J. Biol. Chem. **259:** 12414–12418.

31. CHIPMAN, S. D., B. FARIS, L. M. BARONE, C. A. PRATT & C. FRANZBLAU. 1985. Processing of soluble elastin in cultured neonatal rat smooth muscle cells. J. Biol. Chem. **260:** 12780–12785.

32. KAHARI, V.-M., M. J. FAZIO, Y. Q. CHEN, M. M. BASHIR, J. ROSENBLOOM & J. UITTO. 1990. Deletion analyses of 5'-flanking region of the human elastin gene: Delineation of functional promoter and regulatory cis-elements. J. Biol. Chem. **265:** 9485–9490.

33. WANG, A. H.-J., G. J. QUIGLEY, F. J. KOLPAK, J. L. CRAWFORD, J. H. VANBOOM, G. VANDERMAREL & A. RICH. 1979. Molecular structure of a left handed double helical DNA fragment at atomic resolution. Nature **282:** 680–686.

34. GARDINER-GARDEN, M. & M. FROMMER. 1987. CpG islands in vertebrate genomes. J. Mol. Biol. **196:** 261–282.

35. FAZIO, M., V.-M. KAHARI, M. B. BASHIR, B. SAITTA, J. ROSENBLOOM & J. UITTO. 1990. Regulation of elastin gene expression: Evidence for functional promoter activity in the 5'-flanking region of the human gene. J. Invest. Dermatol. **94:** 191–196.

36. OAKES, B. W., A. C. BATTY, C. I. HANDLEY & L. B. SANDBERG. 1982. The synthesis of elastin, collagen, and glycosaminoglycans by high density primary cultures of neonatal rat aortic smooth muscle. An ultrastructural and biochemical study. Eur. J. Cell. Biol. **27:** 34–36.

37. BEATO, M. 1989. Gene regulation by steroid hormones. Cell **56:** 335–344.

38. EICHNER, R. & J. ROSENBLOOM. 1979. Collagen and elastin synthesis in the developing chick aorta. Arch. Biochem. Biophys. **198:** 414–423.

39. ROESLER, W. J., G. R. VANDERBARK & R. W. HANSON. 1988. Cyclic AMP and the induction of eukaryotic gene transcription. J. Biol. Chem. **263:** 9063–9066.

40. MECHAM, R. P., B. D. LEVY, S. L. MORRIS, J. G. MADARAS & D. S. WRENN. 1985. Increased cyclic GMP levels lead to a stimulation of elastin production in ligament fibroblasts that is reversed by cyclic AMP. J. Biol. Chem. **260:** 3255–3258.

Elastin Synthesis and Fiber Assembly[a]

ROBERT P. MECHAM

Respiratory and Critical Care Division
Department of Medicine, and
Department of Cell Biology and Physiology
Jewish Hospital at Washington University Medical Center
St. Louis, Missouri 63110

Elastin is found in most connective tissues in conjunction with other structural elements such as collagen and proteoglycans. When isolated in relatively pure form, fibers of elastin behave as rubber-like elastomers capable of being stretched to several times their length. The amino acid composition of elastin is unique, with one third of the amino acid residues being glycine and one ninth proline. In contrast to collagen, elastin contains very little hydroxyproline, no hydroxylysine, and a preponderance of the nonpolar amino acids alanine, valine, leucine, and isoleucine.[1,2] In addition to its unusual physical properties, insoluble elastin is extremely stable under normal physiologic conditions, perhaps lasting for the life of the organism.[3]

Elastin exhibits a rapid extensibility to two or three times its resting length under tension with an equally rapid recovery to the original size when the tension is released. Keys to the stability of this process are covalent cross-links that impart restriction to the molecule, so that while stretching, one chain is unable to slip past another. Several laboratories have shown that the cross-linkages in elastin originate through oxidative deamination of the ε-amino group of specific lysine residues and the subsequent condensation of the aldehyde residue that is formed with amino groups or other aldehyde moieties.[4–6] Several excellent reviews address the chemical and physical properties of elastin[1,2,7,8] as well as the complex process of cross-link formation.[1,9,10]

The elastic fiber is a complex structure that consists of several components. Early in development, the elastic fiber consists of 10–12-nm fibers, termed microfibrils, that define fiber location and morphology.[11–13] Over time, tropoelastin accumulates within the bed of microfibrils where the enzyme lysyl oxidase covalently couples one tropoelastin molecule with another to form the functional, polymeric protein that is known as elastin. Biosynthesis of elastic fibers thus presents complex problems for the connective tissue cell. Elastin must be synthesized intracellularly in soluble form and posttranslationally modified, packaged, and secreted into the extracellular space for insolubilization and fiber assembly. During development, the cell must adjust both the temporal, qualitative, and quantitative pattern of its synthetic activities to coordinate the amount, type, and spatial organization of elastic fibers and other matrix components, so as to maintain the phenotypic composition typical of a particular connective tissue. The cell must also remain responsive to physiologic and pathologic stresses that necessitate modification of the matrix. It is now evident that this elastogenic pathway is an integrated system regulated at many points by the cell, which can respond to

[a] The original work described in this report was funded by a grant from the National Institutes of Health (HL-26499).

common biologic signals (such as hormones or macromolecules from the extracellular matrix), thus permitting timing and control of elastogenesis by external factors.

TROPOELASTIN BIOSYNTHESIS AND SECRETION

Some controversy exists as to whether tropoelastin is the initial gene product in elastin biosynthesis. Several studies[14-16] reported the existence of high molecular weight species of soluble elastins (90–140,000 daltons), claimed to be primary gene products and to represent "pro" forms of tropoelastin analogous to the procollagen species. However, cell-free translation studies of elastin-rich mRNA on membrane-bound polysomes[17] and in reticulocyte lysate[18-20] have failed to substantiate these claims, and it now appears certain that pre-tropoelastin (containing the signal peptide) is first synthesized as a species of approximately 70,000 daltons. Nucleotide sequencing of cloned tropoelastin cDNAs confirms this conclusion.

Tropoelastin apparently undergoes little intracellular posttranslational modification. Although some hydroxylation of proline residues occurs in the endoplasmic reticulum, the functional significance of the presence of hydroxyproline in elastin is uncertain. In contrast to collagen, posttranslational hydroxylation of proline residues appears to be unimportant in elastin fibrillogenesis. Lack of proline hydroxylation does not affect the secretion of tropoelastin monomers,[21-23] the oxidation of lysyl residues,[24,25] or the incorporation of tropoelastin into elastic fibers.[26] Incorporation of proline analogs that disrupt triple-helical regions of collagen also does not affect tropoelastin synthesis.[22] It has been suggested that tropoelastin may be "accidentally" hydroxylated by collagen prolyl hydroxylase as a consequence of the concomitant synthesis of procollagen by the cell.[27] Evidence exists, however, that overhydroxylation can compromise the ability of tropoelastin to form mature fibers.[28]

At present it must be concluded that detailed analysis of the intracellular secretory pathway for elastin is lacking. Of particular interest is the question of whether tropoelastin is co-secreted with other components of the elastic fiber, such as microfibrillar proteins or lysyl oxidase, or whether it may be accompanied by a component of its own receptor (see below) that acts as a "molecular chaperone" to direct transport of tropoelastin to appropriate sites on the cell surface and to mediate fiber assembly.[29,30]

ELASTIC FIBER ASSEMBLY

The Microfibrillar Component

Ultrastructural studies have clearly demonstrated the intimate relation between the elastin-secreting cell and the developing elastic fiber. Elastic fibers first appear in fetal development as aggregates of 10–12-nm microfibrils arranged in parallel array, often occupying infoldings of the cell membrane (FIG. 1). Elastin then is deposited as small clumps of amorphous material within these bundles of microfibrils, which subsequently coalesce to form true elastic fibers (FIG. 2). The relative proportion of microfibrils to elastin declines with increasing age of the animal, adult elastic fibers having only a very sparse peripheral mantle of microfi-

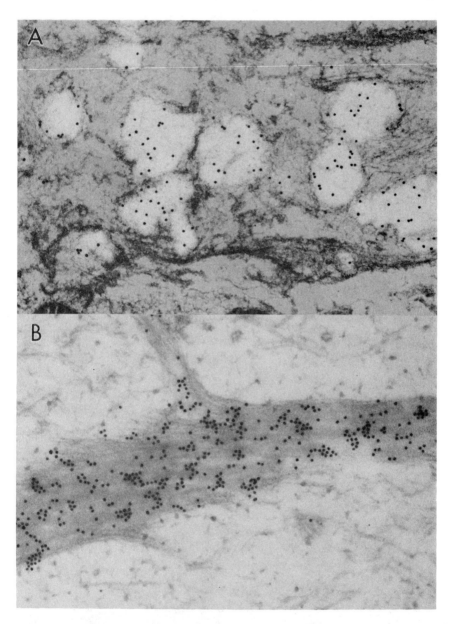

FIGURE 1. Ultrastructure of developing elastic fibers. Early in development, elastic fibers consist of 10–12 nm microfibrils **(A).** Tropoelastin associated with the microfibrils is only evident when visualized with gold particles coated with antibodies to elastin. As development proceeds, amorphous elastin becomes the predominant component of the fiber **(B)** with the microfibrils located around the periphery of the elastin globules. The gold particles in panels **A** and **B** are coated with a monoclonal antibody to elastin.

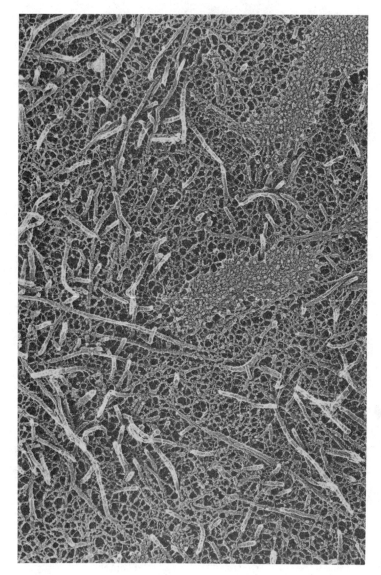

FIGURE 2. Quick-freeze, deep-etch images of elastic cartilage provides a three-dimensional view of the substructure of elastic fibers. (From Mecham and Heuser[73] with permission.)

brillar material. This sequence of elastogenesis is consistent in a variety of tissues from different species. Morphologically identical microfibrils have also been demonstrated in tissues that are devoid of elastin. These include the ciliary zonule[31,32] and a wide variety of tissues that contain oxytalan fibers.[33]

Isolation and characterization of microfibrillar components have proved a resistant problem. Early preparations were poorly characterized and highly heterogeneous, although the technique used for extraction (chaotropic and enzymatic solubilization of cellular and matrix components followed by extraction with a denaturing buffer under reducing conditions) has served as a model for later investigators (extensively reviewed by Cleary and Gibson[34]). Recently, three molecules have been identified that are likely components of microfibrils. One is a 31,000-dalton glycoprotein termed microfibril-associated glycoprotein (MAGP),[35,36] the second is a 35-kD protein with amine oxidase activity,[37] and the third is fibrillin, a 350,000-dalton glycoprotein derived from the medium of human fibroblast cell cultures.[38] Antibodies to all three of these molecules localize at the ultrastructural level to microfibrils in both elastic and nonelastic tissues, although it is not yet clear if these proteins are structurally related. It is of considerable interest that all three antibodies localize to morphologically similar microfibrils in nonelastic tissues, suggesting the possibility that such microfibrils may play a role in maintaining the physical and functional integrity of the tissues in which they are found, independently of the presence of elastin.

Despite advances in understanding the structural complexity of microfibrils, the mechanism(s) by which they participate in fibrillogenesis remains speculative. The observation that microfibrillar aggregates take the form and orientation of presumptive elastic fibers led Ross et al.[39] to suggest that they may direct the morphogenesis of elastic fibers by acting as a "scaffold" on which elastin is deposited. In this model, microfibrils serve as a "registration site" to align tropoelastin molecules in precise register, so that cross-linking regions are juxtaposed prior to oxidation by lysyl oxidase. Ultrastructural appearances of elastic fibers in experimental copper deficiency, in which superabundant microfibrils are found to surround sparse amounts of abnormally electron-dense elastin,[40] are also consistent with this hypothesis, as is the recent demonstration by Kagan et al.[41] of microfibrillar localization of anti-lysyl oxidase antibodies.

Incorporation of Tropoelastin into Elastic Fibers

Tropoelastin appears to be accreted to the surface of elastic fibers directly in a monomeric form, and there is no evidence that the molecule undergoes cleavage before cross-linking into the developing fiber. Newly synthesized and secreted tropoelastin apparently is rapidly incorporated into growing elastic fibers[42,43] which probably takes place very close to the cell membrane.[44–46] The observation that tropoelastin is able to undergo reversible, temperature-dependent, non-covalent self-aggregation (coacervation) under physiologic conditions[47,48] has been postulated as an initial step in fiber formation.

In its native form, insoluble elastin is a polymer of tropoelastin chains randomly cross-linked into a highly extensible, three-dimensional network by an unusual system of compounds derived from cyclization of the side-chains of paired lysine residues.[1,6] Cross-linking of both elastin and collagen is initiated by the action of lysyl oxidase, a copper-dependent metalloenzyme that catalyzes the oxidative deamination of lysine to allysine (α-aminoadipic acid δ-semialdehyde).

This appears to be the only enzymatic step involved in elastin cross-linking, subsequent formation of elastin cross-links (including the tetrafunctional amino acid isomers desmosine and isodesmosine) probably occurring as a series of spontaneous condensation reactions.[9,10]

The Elastin Receptor

Organization of a functional elastin fiber is a complex process involving interactions between tropoelastin and glycoprotein microfibrils. How this interaction occurs is not known, but it has long been presumed that tropoelastin is secreted from the cell and somehow finds its way through the extracellular matrix to the growing elastic fiber where it interacts with microfibrils and becomes oriented in the proper alignment for cross-linking. Repeated experiments over a period of years, however, have failed to provide convincing evidence to support this view. New studies suggest that the assembly process is not random but is mediated at the cell surface by the elastin receptor.[29,30,49] The elastin receptor is a 67-kD, bifunctional protein that contains a protein binding site that recognizes a hydrophobic sequence in elastin (VGVAPG)[50,51] and a carbohydrate binding site that associates with galactoside sugars (FIG. 3). Tropoelastin binds with high affinity to the receptor at the protein binding site. Binding of carbohydrate at the sugar binding site, however, lowers the affinity of the receptor for tropoelastin. Thus, with multiple binding sites, the 67-kD protein is ideally suited to direct the specific association between tropoelastin and highly glycosylated microfibrillar components. In the extracellular space, binding of glycoprotein microfibrils to the receptor at the sugar-binding site could serve to lower the affinity of receptor-tropoelastin binding, thereby facilitating transfer of the tropoelastin monomer to an assembly site on microfibrils.

Immunolocalization studies with elastin-producing cells demonstrate that tropoelastin and 67-kD receptor subunit colocalize intracellularly.[52] That the two proteins are associated in the secretory pathway is suggested by the observation that monoclonal antibody BA-4, specific for the receptor binding sequence on tropoelastin, only recognizes intracellular elastin in permeabilized cells that are treated with lactose. Because the effect of lactose is to dissociate the tropoelastin-receptor complex, it is inferred from these studies that tropoelastin may be bound to, and transported with, its receptor during periods of active elastin synthesis. The existence of a transporter protein for elastin is especially intriguing in light of questions concerning the intracellular segregation and secretion of tropoelastin. Lacking glycosylation, tropoelastin is unable to utilize sugar residues as intracellular trafficking signals, and even though the carboxy-terminus of tropoelastin has a domain similar to transmembrane domains on membrane-associated proteins, tropoelastin is translocated completely into the endoplasmic reticulum and does not remain associated with an intracellular membranous compartment.[53] These findings raise the interesting possibility that the receptor functions as a "molecular chaperone" by providing trafficking signals to direct the proper movement of the receptor-tropoelastin complex to the site of elastin fiber formation on the cell surface where the receptor mediates the interaction between tropoelastin and microfibrils. Although 67-kD protein has been localized immunologically to extracellular elastin fibers, it is not known whether it forms part of the final structure.

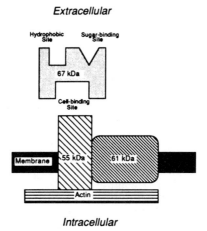

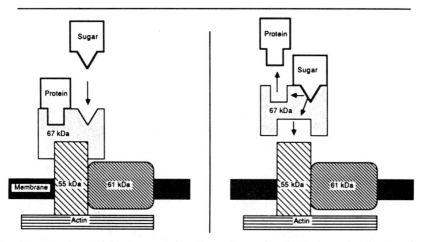

FIGURE 3. Model of the 67-kD elastin receptor. **(Upper panel)** The 67-kD elastin receptor contains a protein binding site that recognizes a hydrophobic sequence in elastin, a carbohydrate binding site that interacts with galactoside sugars, and a membrane binding site that associates with a 55-kD transmembrane protein. **(Lower panel)** Tropoelastin binds with high affinity to the receptor at the protein binding site. Binding of carbohydrate at the sugar binding site, however, lowers the affinity at both the protein and cell binding sites, resulting in the release of bound protein and dissociation of the receptor from its transmembrane anchor.

DEVELOPMENTAL REGULATION OF ELASTIN PRODUCTION IN LUNG

Elastin is unique among connective tissue proteins in that expression of the elastin gene is temporally regulated in mammals to a brief period of intense elastogenesis occurring in late fetal life and extending into the early postnatal period. Studies of the composition of developing connective tissues have indicated deposition of insoluble elastin at an exponential rate during this period in the extracellular matrix of a variety of tissues.[54–57] These studies further imply that elastin deposition is essentially complete in the first decade of life, consistent with the observed long physiologic half-life of insoluble elastin. Studies of elastin biosynthesis *in vitro* have also shown a clear correlation between donor age and rates of elastin synthesis. Maximal elastogenesis corresponded closely in each case to the previously observed perinatal window, suggesting a stable phenotypic change in cell activity occurring at this time.

Elastogenesis in the lung has been extensively studied under the light and electron microscope.[58–62] In human lung, elastic fibers are first detected at 3 months of gestation in the walls of the trachea and main stem bronchi, the pleura, and pulmonary artery. At about 6 months the lung loses its glandular appearance, and vascularization of the mesenchymal tissue begins. In the area of alveoli, development of elastic fibers appears in association with the origin of alveolar septal primordial cells[61,63,64] and with smooth muscle cells at bifurcations in the late glandular stage.[61] A large increase in elastogenesis occurs during the last developmental trimester as the respiratory portion of the lung differentiates. At the beginning of this period, elastic fiber bundles are confined essentially to areas immediately surrounding the mouths of the alveoli but expand into the alveolar wall during the neonatal and adolescent phase of life.[58,65]

The cells responsible for elastin synthesis in the lung are diverse, depending on anatomic location: auricular chondroblasts produce the elastin found in cartilagenous tissues such as trachea and main stem bronchi, pleural mesothelial cells deposit elastin in the pleura,[66,67] and smooth muscle cells produce and organize elastin in the lung vasculature.[68,69] Cells responsible for elastin production in the lung parenchyma have been more difficult to identify, however, because this area of the lung is comprised of multiple cell types and the phenotype of the interstitial cell changes during development. As the alveolar septae begin to form, myofibroblasts are found in close association with capillary structures and are considered the source of elastin at these sites.[60,61,64,70] In rat lung, immature interstitial cells differentiate into lipid-filled and non-lipid-filled fibroblasts, the latter being alveolar septal myofibroblasts.[71,72] It is likely that little if any elastin is made by type I or type II pneumocytes, although this has not been thoroughly investigated.

ACKNOWLEDGMENT

I thank Dr. Alek Hinek for providing the electron micrographs used in FIGURE 1.

REFERENCES

1. PARTRIDGE, S. M. 1962. Adv. Prot. Chem. **17:** 227–302.
2. FRANZBLAU, C. 1971. Comprehensive Biochem. **26c:** 659–712.

3. LEFEVRE, M. & R. B. RUCKER. 1980. Biochim. Biophys. Acta **630**: 519–529.
4. PARTRIDGE, S. M., D. F. ELSDEN & J. THOMAS. 1963. Nature **197**: 1297–1298.
5. PARTRIDGE, S. M., D. F. ELSDEN & J. THOMAS. 1964. Biochem. J. **93**: 30c–33c.
6. FRANZBLAU, C., F. M. SINEX, B. FARIS & R. LAMPIDIS. 1965. Biochem. Biophys. Res. Commun. **21**: 575–581.
7. GOSLINE, J. M. 1976. Int. Rev. Connect. Tissue Res. **7**: 211–249.
8. GOSLINE, J. M. & J. ROSENBLOOM. 1984. *In* Extracellular Matrix Biochemistry. K. A. Piez & A. H. Reddi, eds.: 191–228. Elsevier. New York.
9. PAZ, M. A., D. A. KEITH & P. M. GALLOP. 1982. Methods Enzymol. **82**: 571–587.
10. EYRE, E. R., M. A. PAZ & P. M. GALLOP. 1984. Ann. Rev. Biochem. **53**: 717–748.
11. FAHRENBACH, W. H., L. B. SANDBERG & E. G. CLEARY. 1966. Anat. Rec. **155**: 563–576.
12. GREENLEE, T. K. J., R. ROSS & J. L. HARTMAN. 1966. J. Cell Biol. **30**: 59–71.
13. ROSS, R. & P. BORNSTEIN. 1969. J. Cell Biol. **40**: 366–381.
14. FOSTER, J. A., R. P. MECHAM, C. B. RICH, M. F. CRONIN, A. LEVINE, M. IMBERMAN & L. L. SALCEDO. 1978. J. Biol. Chem. **253**: 2797–2803.
15. HENG-KHOO, C. S., R. B. RUCKER & K. W. BUCKINGHAM. 1979. Biochem. J. **177**: 559–567.
16. RUCKER, R. B., J. MURRA, W. RIEMANN, K. BUCKINGHAM, K. TAN & G. S. KHOO. 1977. Biochem. Biophys. Res. Commun. **75**: 358–365.
17. RYHANEN, L., P. H. GRAVES, G. M. BRESSAN & D. J. PROCKOP. 1978. Arch. Biochem. Biophys. **185**: 344–351.
18. BURNETT, W., R. EICHNER & J. ROSENBLOOM. 1980. Biochemistry **19**: 1106–1111.
19. DAVIDSON, J. M., B. LESLIE, T. WOLT, R. G. CRYSTAL & L. B. SANDBERG. 1982. Arch. Biochem. Biophys. **218**: 31–37.
20. PARKS, W. C., H. SECRIST, L. C. WU & R. P. MECHAM. 1988. J. Biol. Chem. **263**: 4416–4423.
21. UITTO, J., H. P. HOFFMAN & D. J. PROCKOP. 1976. Arch. Biochem. Biophys. **173**: 187–200.
22. ROSENBLOOM, J. & A. CYWINSKY. 1976. FEBS Lett. **65**: 246–250.
23. KAO, W. W. Y., G. W. BRESSAN & D. J. PROCKOP. 1982. Connect. Tiss. Res. **10**: 263–274.
24. NARAYANAN, A. S., R. C. PAGE & F. KUZAN. 1977. Adv. Exp. Med. Biol. **79**: 491–508.
25. NARAYANAN, A. S., R. C. PAGE, F. KUZAN & C. G. COOPER. 1978. Biochem. J. **173**: 857–862.
26. ABRAHAM, P. A., D. W. SMITH & W. H. CARNES. 1974. Biochem. Biophys. Res. Commun. **58**: 597–604.
27. ROSENBLOOM, J. 1982. Connect. Tiss. Res. **10**: 73–91.
28. BARONE, L. M., B. FARIS, S. D. CHIPMAN, P. TOSELLI, B. W. OAKES & C. FRANZBLAU. 1985. Biochim. Biophys. Acta **840**: 245–254.
29. HINEK, A., D. S. WRENN, R. P. MECHAM & S. H. BARONDES. 1988. Science **239**: 1539–1541.
30. MECHAM, R. P., A. HINEK, R. ENTWISTLE, D. S. WRENN, G. L. GRIFFIN & R. M. SENIOR. 1989. Biochemistry **28**: 3716–3722.
31. RAVIOLA, G. 1971. J. Biol. Chem. **262**: 5755–5762.
32. STREETEN, B. W. & P. A. LICARI. 1983. Invest. Opthal. Visual Sci. **21**: 130–135.
33. FULLMER, H. M., J. H. SHEETZ & A. J. NARKATES. 1974. J. Oral Pathol. **3**: 291–316.
34. CLEARY, E. G. & M. A. GIBSON. 1983. Int. Rev. Connect. Tiss. Res. **10**: 97–209.
35. PROSSER, I. W., M. A. GIBSON & E. G. CLEARY. 1984. Aust. J. Exp. Biol. Med. Sci. **62**: 287–294.
36. GIBSON, M. A., J. L. HUGHES, J. C. FANNING & E. G. CLEARY. 1986. J. Biol. Chem. **261**: 11429–11436.
37. SERAFINI-FRACASSINI, A., G. VENTRELLA, M. J. FIELD, J. HINNIE, N. I. ONYEZILI & R. GRIFFITHS. 1981. Biochemistry **20**: 5424–5429.
38. SAKAI, L. Y., D. R. KEENE & E. ENGVALL. 1986. J. Cell Biol. **103**: 2499–2509.
39. ROSS, R., P. J. FIALKOW & K. ALTMAN. 1977. Adv. Exp. Med. Biol. **79**: 7–17.
40. WAISMAN, J. & W. H. CARNES. 1967. Am. J. Pathol. **51**: 117–135.

41. KAGAN, H. M., C. A. VACCARO, R. E. BRONSON, S. S. TANG & J. S. BRODY. 1986. J. Cell Biol. **103:** 1121–1128.
42. NARAYANAN, A. S. & R. C. PAGE. 1976. J. Biol. Chem. **251:** 1125–1130.
43. BRESSAN, G. M. & D. J. PROCKOP. 1977. Biochemistry **16:** 1406–1412.
44. ALBERT, E. N. 1972. Am. J. Pathol. **69:** 89–102.
45. KUHN, C. I., S. Y. YU, M. CHRAPLYVY, H. E. LINDER & R. M. SENIOR. 1976. Lab. Invest. **34:** 372–380.
46. SERAFINI-FRACASSINI, A. 1984. *In* Ultrastructure of the Connective Tissue Matrix. A. Ruggeri & P. M. Motta, eds.: 140–150. Martinus Nijhoff Publishers. Boston.
47. SMITH, D. W., D. M. BROWN & W. H. CARNES. 1972. J. Biol. Chem. **247:** 2427–2432.
48. WHITING, A. H., B. C. SYKES & S. M. PARTRIDGE. 1974. Biochem. J. **179:** 35–45.
49. MECHAM, R. P., A. HINEK, G. L. GRIFFIN, R. M. SENIOR & L. LIOTTA. 1989. J. Biol. Chem. **264:** 16652–16657.
50. SENIOR, R. M., G. L. GRIFFIN, R. P. MECHAM, D. S. WRENN, K. U. PRASAD & D. W. URRY. 1984. J. Cell Biol. **99:** 870–874.
51. WRENN, D. S., G. L. GRIFFIN, R. M. SENIOR & R. P. MECHAM. 1986. Biochemistry **25:** 5172–5176.
52. HINEK, A. & R. P. MECHAM. 1990. *In* Elastin: Chemical and Biological Aspects. T. Tamburro & J. M. Davidson, eds.: 369–381. Galatina. Potenza.
53. GROSSO, L. & R. P. MECHAM. 1988. Biochem. Biophys. Res. Commun. **151:** 822–826.
54. DUBICK, M., R. B. RUCKER, J. A. LAST, L. O. LOLLINI & C. E. CROSS. 1981. Biochim. Biophys. Acta **672:** 303–306.
55. CLEARY, E. G., L. B. SANDBERG & D. S. JACKSON. 1983. J. Cell Biol. **33:** 469–479.
56. BARRINEAU, L. L., C. B. RICH & J. A. FOSTER. 1981. Connect. Tiss. Res. **8:** 189–491.
57. DAVIDSON, J. M., S. SHIBAHARA, K. SMITH & R. CRYSTAL. 1981. Connect. Tiss. Res. **8:** 209–212.
58. LOOSLI, C. G. & E. L. POTTER. 1959. Am. Rev. Respir. Dis. **80:** 5–23.
59. JONES, A. W. & A. J. BARSON. 1971. J. Anat. **110:** 1–15.
60. COLLETT, A. J. & G. DES BIENS. 1974. Anat. Rec. **179:** 343–360.
61. FUKUDA, Y., V. J. FERRANS & R. G. CRYSTAL. 1983. Am. J. Anat. **167:** 405–439.
62. AMY, R. W. M., D. BOWES, P. H. BURRI, J. HAINES & W. M. THURLBECK. 1977. J. Anat. **24:** 131–151.
63. FUKUDA, Y., V. J. FERRANS & R. G. CRYSTAL. 1984. Am. J. Anat. **170:** 597–629.
64. NOGUCHI, A., R. REDDY, J. D. KURSAR, W. C. PARKS & R. P. MECHAM. 1989. Exp. Lung Res. **15:** 537–552.
65. EMERY, J. L. 1970. Respiration **27** (suppl): 41–50.
66. RENNARD, S. I., M. C. JAURAND, J. BRIGNON, O. KAWANAMI, V. J. FERRANS, J. DAVIDSON & R. G. CRYSTAL. 1984. Am. Rev. Respir. Dis. **130:** 267–274.
67. CANTOR, J. O., M. WILLHITE, B. A. BRAY, S. KELLER, I. MANDL & G. M. TURINO. 1986. Proc. Soc. Exp. Biol. Med. **181:** 387–391.
68. HAUST, M. D., R. H. MORE, S. A. BENSCOME & J. U. BALIS. 1965. Exp. Mol. Pathol. **4:** 508–524.
69. MECHAM, R. P., L. A. WHITEHOUSE, D. S. WRENN, W. C. PARKS, G. L. GRIFFIN, R. M. SENIOR, E. C. CROUCH, K. R. STENMARK & N. F. VOELKEL. 1987. Science **237:** 423–426.
70. CAMPAGNONE, R., J. REGAN, C. B. RICH, M. MILLER, D. R. KEENE, L. SAKAI & J. A. FOSTER. 1987. Lab. Invest. **56:** 224–230.
71. VACCARO, C. & J. S. BRODY. 1978. Anat. Rec. **192:** 457–480.
72. BRODY, J. S. & C. VACCARO. 1979. Fed. Proc. **38:** 215–223.
73. MECHAM, R. P. & J. HEUSER. 1990. Connect. Tiss. Res. **24:** 83–93.

Measurement of Elastin-Derived Peptides

G. WEINBAUM,[a] U. KUCICH,[a] P. KIMBEL,[a] S. METTE,[a]
S. AKERS,[b] AND J. ROSENBLOOM[c]

[a]*Department of Medicine*
The Graduate Hospital
Philadelphia, Pennsylvania 19146

[b]*University of Medicine and Dentistry of New Jersey*
Robert Wood Johnson Medical School
Camden, New Jersey 08103

[c]*University of Pennsylvania School of Dental Medicine*
Philadelphia, Pennsylvania 19104

HISTORICAL PERSPECTIVE

With the development of the "protease-protease inhibitor" hypothesis described throughout this monograph, and the identification and characterization of elastases from neutrophils and macrophages, cells that are routinely increased in numbers in the lungs of smokers,[1] a number of investigators have focused on the elastase : elastin interaction as the essential event in the production of pulmonary emphysema. Because elastin degradation appears to be the *sine qua non* of emphysema, it was natural to attempt quantitation of the hydrolysis products of elastin as a potential early index of emphysema induction.

Emphasis was initially directed to measurement of desmosine and isodesmosine, because these unique amino acids are produced from lysines only during the cross-linking of tropoelastin precursor molecules into the mature, extracellular, insoluble elastin[2] and are not found in any other lung protein. In addition, desmosines are very stable and do not appear to be catabolized in the body, a characteristic suggesting that they could be satisfactory indicators of increased elastin turnover in the emphysematous patient. Indeed, in 1978, Goldstein and Starcher[3] used a model of elastase-induced emphysema to show that urinary desmosines increased rapidly after intratracheal instillation of pancreatic elastase and that the amount of urinary desmosine could account for more than 50% of the elastin lost from the lungs. Furthermore, the desmosines excreted in the urine were components of polypeptides, whose sizes approximated 9–27 kD. The desmosine-containing polypeptides reached an excretion maximum 1 or 2 days after administration of intratracheal elastase, and little urinary desmosine was measured by the third day. Control animals had no detectable urinary desmosines.

This early study did not attempt to relate urinary desmosine excretion to the amount of elastase instilled or to alterations in pulmonary architecture or function. These relationships were examined later by Janoff *et al.*[4] using a radioimmunoassay of urinary desmosine to monitor lung injury after endobronchial elastase instillation in sheep. The increase in desmosine excretion measured by these investigators correlated well with the dose of elastase, alveolar damage (as shown by an increase in mean linear intercept), and altered pulmonary function (as indicated by a decrease in lung perfusion and a decrease in ventilation). Unfortunately, as will be described herein, the promise shown by the urinary desmosine assay in models of experimental emphysema has not been fulfilled in human studies.

Marfan's syndrome, a heritable defect in connective tissue matrix in humans, is apparently characterized by extensive loss of elastin-rich structures in blood vessel walls with concomitant aneurysm formation, implying excessive elastin degradation in this disease. Analyses of peptidyl-desmosine in urine specimens of these patients yielded consistently lower values than those obtained from healthy control volunteers.[5] It is possible that these patients have an attenuated conversion of lysyl-derived precursor molecules into desmosines, and thus urinary desmosine levels may not *per se* be indicative of either rapid elastin destruction or high elastin turnover.

On the other hand, urinary desmosine measurement in normal nonsmokers, when compared to the values obtained from heavy smokers with a variety of pulmonary disorders, seemed to suggest that smoking subjects with chronic obstructive pulmonary disease or lung infection had elevated excretion of desmosine.[6] The relatively small number of subjects examined (23 controls and 18 patients) in this preliminary study precluded making any firm predictions, and a more extensive and controlled study was performed by Davies *et al.*[7] Analyses of 157 subjects, including 52 nonsmokers, 46 current smokers, and 59 exsmokers, did not show increased desmosine excretion in subjects with definite spirometric abnormalities (FEV_1 observed/predicted <70%) as compared to individuals with normal pulmonary function. Because the high background excretion levels of desmosine observed in normal nonsmokers may mask small changes that may be observed in smokers, a second study was performed using patients with alpha$_1$-proteinase inhibitor (α_1-PI) deficiency, with the expectation that lung deterioration would occur more rapidly in those individuals with markedly skewed elastase : elastase inhibitor balance and that urinary desmosine excretion should be significantly elevated.[8] As stated in that article, ". . .no significant differences in the urinary desmosine measurements could be detected between normal adults, adults with interstitial lung disease, adults with α_1-PI deficiency plus emphysema, and asymptomatic adults with homozygous α_1-PI deficiency (p >0.1, all comparisons)."

Therefore, because desmosine has an uncertain future as an indicator of early emphysema, it has been necessary to explore other potential markers of early lung destruction associated with the elastin matrix breakdown observed in emphysema. One such approach has been the development of an assay for the detection of urinary valyl-proline, inasmuch as that dipeptide appears to be a relatively specific degradation product of elastin.[9] The assay has been used in a small number of humans and in an animal model. Although preliminary data concerning val-pro analyses were encouraging, no further published work is available to evaluate the utility of this interesting approach. However, measurement of other polypeptide digestion products of lung elastin have shown provocative results in studies involving animal models as well as humans.

IMMUNOLOGIC ANALYSES OF ELASTIN-DERIVED PEPTIDES FOR DETECTION OF EMPHYSEMA

Early work at a number of laboratories, including ours,[10–12] established methods for the preparation of antibodies against insoluble elastin and peptide derivatives of elastin. Such antibodies have been very useful in the ultrastructural immunolocalization of extracellular mature elastin and intracellular newly synthesized elastin[13,14] as well as the radioimmunologic identification of tropoelastin, the biologic precursor of the mature, cross-linked amorphous component of the elastic fiber.[15] In addition, such antibodies have been used for monitoring elastin destruction occurring concomitantly with emphysema development in animal models.[16,17]

Studies in an Animal Model

A dose-dependent production of differing severities of experimental emphysema could be induced using aerosol administration of elastase. We developed methods to detect elastin degradation products in such a model.[16] Because the antibodies produced against the lung elastin-derived peptide antigens were of the nonprecipitating type, we used a hemagglutination inhibition assay to quantitate any appearance of circulating elastin-derived peptides after elastase aerosol exposure. Four different doses of elastase (10, 25, 50, or 100 mg) were aerosolized in a single administration in separate experiments. Controls received saline aerosol. Blood samples were taken periodically after aerosol exposure, and the sera were analyzed for elastin-derived peptides. Exposure to the lowest dose of elastase gave no detectable elastin-derived peptides nor any histologic evidence of emphysema in the lungs of such treated animals. The two intermediate doses of elastase caused a transient increase in immunologically detectable elastin-derived peptides which remained elevated for the first 12 days after elastase administration. Both these doses of elastase caused a very mild, patchy distribution of emphysematous changes. At the largest dose of elastase administered, the rise in circulating elastin-derived peptides was most pronounced and remained elevated throughout the entire study, that is, 42 days, after a single exposure to elastase. Interestingly, this dose induced more severe and increased areas of emphysematous changes when compared with the histologic changes in the lungs of animals exposed to the other doses.

Although a correlation appeared to exist between elastase dosage and serum concentrations of elastin-derived peptides in this study,[16] no attempt was made to establish a relation between the concentration of elastin peptides in the circulation and the severity of the emphysema produced, as measured either histologically or by pulmonary function analyses. However, the apparent success of these studies suggested that more sensitive assays should be developed and analyses of patient populations should be initiated.

Human Studies

Two laboratories developed methods for detecting circulating elastin-derived peptides in human subjects, one using a solid-phase radioimmunoassay[18] while we chose to develop an indirect enzyme-linked immunosorbent assay (ELISA).[19] A sensitive ELISA procedure has the advantages that radioactive antigen is not needed and that the endpoint of the reaction can be monitored by a standard spectrophotometer, usually found in most clinical laboratories.

As we had previously shown the significant species immunospecificity of elastin,[12] the antibodies used in the experimental model just described could not be used for human studies, and polyclonal antibody specific for human lung elastin-derived peptides was prepared. The ELISA technique initially used in our laboratory could detect as little as 15 ng of elastin peptide compared to 400 ng detected by the hemagglutination assay mentioned earlier. Such an ELISA procedure was used to analyze sera from normal nonsmoking subjects, smokers with normal pulmonary function as determined by spirometry, and patients with well-documented chronic obstructive pulmonary disease (COPD) having a mean $FEV_1/FVC\%$ of 50 ± 16.[19] Preliminary ELISA results for the three different populations showed that both the smoking group and the COPD group had significantly higher ($p < 0.001$) concentrations of elastin-derived peptides than did the control group. This first attempt to use immunologic analysis of elastin-derived peptides to distinguish individuals with normal lungs from those either who presently have emphysema or who are at risk of developing it, involved small numbers of subjects, and only a single sample was analyzed for each subject. So that we could expand that initial study, further modification of the original ELISA assay was undertaken.

To improve the use, sensitivity, reproducibility, and rapidity of the ELISA, we adapted a peroxidase-antiperoxidase method, originally used in immunoelectron microscopy, to an assay carried out in microtiter wells. The color that was produced could be read spectrophotometrically by an automatic plate reader.[20] Such a technical change lowered the limit of detection of elastin-derived peptides to 2 ng.

A series of new studies was performed using this modified ELISA, and these investigations duplicated the earlier observations that circulating elastin-derived peptides are elevated in patients with COPD and in some normal smokers.[20] To minimize variability and improve reproducibility it was found that plasma, collected in the presence of a mixture of protease inhibitors, gave the most reproducible results for multiple analyses and gave smaller standard errors when a population of normal nonsmokers was evaluated. Elastin-derived peptide levels were initially monitored on single random plasma samples, but when plasma samples were repetitively sampled from a group of subjects over a period as long as 3 months, it was observed that the values obtained for each individual subject remained relatively constant over the time interval that was monitored. It therefore appears that random sampling of a given individual is representative of the average circulating concentration of elastin-derived peptides. If the subject is a smoker, then the sample value appears to be independent of the time of completion of smoking the last cigarette.

In addition, other studies in our laboratory have established that a correlation is not apparent between the amount of circulating elastin-derived peptides and the age or sex of the subject. However, analyses of plasma from patients with other lung diseases (such as pneumonia, adult respiratory distress syndrome, upper respiratory infections, and asthma) and other diseases in which connective tissue damage may be present (ie, inflammatory arthritis and atherosclerotic vascular disease) show that only patients with COPD have a significant increase in plasma elastin antigens.[21] We also find such elastin-derived peptides in bronchoalveolar lavage fluids from patients with COPD. However, it is extremely difficult to evaluate the meaning of that kind of observation because quantitation of the recovered antigen is fraught with problems, especially with regard to the identification of an internal marker to indicate the dilution factor characteristic of the specific lavage protocol. In preliminary studies, we found a correlation between increased plasma elastin antigen and the degree of impairment of pulmonary function as determined by spirometric measurements. It is reassuring that Schriver et al.[22] have reproduced our observations of increased elastin-derived peptides in plasma and lavage fluids of patients with COPD, and that McLennan *et al.*[23] showed a highly positive association between circulating elastin-derived peptides and the exponential analysis of static pressure-volume curves as an index of pulmonary distensibility and hence pulmonary function. Although the data, to date, suggest that analysis of circulating elastin-derived peptides may be useful in the early diagnosis of emphysema, a number of questions remain to be answered in current or future studies, for example: What are the characteristics of the endogenously generated elastin antigens? Are they tissue specific? Is there continuous antigen processing after the elastin fragments are released into the circulation from the amorphous elastin component of the elastic fiber? Can the assay be made still more selective and specific?

FUTURE DIRECTIONS

The questions raised in the previous section are currently being addressed in a number of laboratories. Our laboratory has taken advantage of the fact that the human elastin gene has been cloned,[24] and now the amino acid sequence of that matrix protein is known. This knowledge has allowed the chemical synthesis of selected polypeptide domains within the elastin molecule and the preparation of monospecific antibodies to those domains.[25] This approach has permitted the following preliminary observations to be made: circulating elastin antigens are reactive with antibody against the carboxy-terminus polypeptide (amino acids 667–684) but do not react with antibody against the tyrosine-rich domain (amino acids 200–221) of elastin; both of these domain-specific antibodies react with the mature amorphous elastin within the alveolar interstitium; certain antigens that are reactive in plasma are reactive in urine. However, the carboxy-terminus antibody, which can be used to measure plasma elastin-derived peptides, cannot measure urinary elastin antigens.[26] The measurement of urinary elastin peptides is reminiscent of the desmosine studies mentioned earlier.[3] Our recent studies suggest that there is significant processing of elastin antigens, a possibility that is corroborated by molecular sizing of the elastin antigens found in plasma.[27] In addition, the observation that excretion of elastin-derived peptides is increased in the urine of patients with COPD provides a ready source of starting material for the isolation and characterization of the endogenously produced elastin degradation products. Complete purification of the *in vivo* generated elastin-derived peptides and preparation of either monospecific or monoclonal antibodies to those urinary antigens or to the unique low molecular weight plasma antigens may aid us in developing more sensitive or selective assays for the early diagnosis of emphysema using such biochemical markers.

REFERENCES

1. REYNOLDS, H. Y. 1987. Bronchoalveolar lavage. Am. Rev. Respir. Dis. **135:** 250–263.
2. PARTRIDGE, S. M., D. F. ELSDON, J. THOMAS, A. DORFMAN, A. TELSER & T.-L. HO. 1966. Incorporation of labeled lysine into the desmosine cross-bridges in elastin. Nature (London) **209:** 399–400.
3. GOLDSTEIN, R. A. &. B. C. STARCHER. 1978. Urinary excretion of elastin peptides containing desmosine after intratracheal injection of elastase in hamsters. J. Clin. Invest. **61:** 1286–1290.
4. JANOFF, A., A. D. CHANANA, D. D. JOEL, H. SUSSKIND, P. LAURENT, S. Y. YU & R. DEARING. 1983. Evaluation of the urinary desmosine radioimmunoassay as a moni-

tor of lung injury after endobronchial elastase instillation in sheep. Am. Rev. Respir. Dis. **128:** 545–551.

5. GUNJA-SMITH, Z. & R. J. BOUCEK. 1981. Desmosines in human urine. Biochem. J. **193:** 915–918.

6. HAREL, S., A. JANOFF, S. Y. YU, A. HUREWITZ & E. H. BERGOFSKY. 1980. Desmosine radioimmunoassay for measuring elastin degradation *in vivo*. Am. Rev. Respir. Dis. **122:** 769–773.

7. DAVIES, S. F., K. P. OFFORD, M. G. BROWN, H. CAMPE & D. NIEWOEHNER. 1983. Urine desmosine is unrelated to cigarette smoking or to spirometric function. Am. Rev. Respir. Dis. **128:** 473–475.

8. PELHAM, F., M. WEWERS, R. CRYSTAL, A. S. BUIST & A. JANOFF. 1985. Urinary excretion of desmosine (elastin cross-links) in subjects with PiZZ alpha-1-antitrypsin deficiency, a phenotype associated with hereditary predisposition to pulmonary emphysema. Am. Rev. Respir. Dis. **132:** 821–823.

9. SOSKEL, N. J. & L. B. SANDBERG. 1983. Detection of urinary valyl-proline as an indicator of elastin degradation. Liq. Chromat. **1:** 434–436.

10. DAMIANO, V. V., A. TSANG, P. CHRISTNER, J. ROSENBLOOM & G. WEINBAUM. 1979. Immunologic localization of elastin by electron microscopy. Am. J. Pathol. **96:** 439–456.

11. DARNULE, T. V., V. LIKHITE, G. M. TURINO & I. MANDL. 1977. Immune response to peptides produced by enzymatic digestion of microfibrils and elastin of human lung parenchyma. Conn. Tiss. Res. **5:** 67–73.

12. KUCICH, U., P. CHRISTNER, J. ROSENBLOOM & G. WEINBAUM. 1981. An analysis of the organ and species immunospecificity of elastin. Conn. Tiss. Res. **8:** 121–126.

13. DAMIANO, V. V., A. TSANG, U. KUCICH, G. WEINBAUM & J. ROSENBLOOM. 1981. Immunoelectron microscopic studies on cells synthesizing elastin. Conn. Tiss. Res. **8:** 185–188.

14. DAMIANO, V. V., A-L. TSANG, G. WEINBAUM, P. CHRISTNER & J. ROSENBLOOM. 1984. Secretion of elastin in embryonic chick aorta as visualized by immunoelectron microscopy. Collagen Rel. Res. **4:** 153–164.

15. CHRISTNER, P., M. DIXON, A. CYWINSKI & J. ROSENBLOOM. 1976. Radioimmunological identification of tropoelastin. Biochem. J. **157:** 525–528.

16. KUCICH, U., P. CHRISTNER, G. WEINBAUM & J. ROSENBLOOM. 1980. Immunologic identification of elastin-derived peptides in the serums of dogs with experimental emphysema. Am. Rev. Respir. Dis. **122:** 461–465.

17. DARNULE, T. V., M. OSMAN, A. T. DARNULE, I. MANDL & G. M. TURINO. 1980. Immunologic detection of lung elastin peptides in the serum of rats with elastase induced emphysema (abstr). Am. Rev. Respir. Dis. **121:** A331.

18. DARNULE, T. V., M. MCKEE, A. T. DARNULE, G. M. TURINO & I. MANDL. 1982. Solid phase radioimmunoassay for estimation of elastin peptides in human serum. Anal. Biochem. **122:** 302–307.

19. KUCICH, U., P. CHRISTNER, M. LIPPMANN, A. FEIN, A. GOLDBERG, P. KIMBEL, G. WEINBAUM & J. ROSENBLOOM. 1983. Immunologic measurement of elastin-derived peptides in human serum. Am. Rev. Respir. Dis. **127:** S28–S30.

20. KUCICH, U., P. CHRISTNER, M. LIPPMANN, P. KIMBEL, G. WILLIAMS, J. ROSENBLOOM & G. WEINBAUM. 1985. Utilization of a peroxidase antiperoxidase complex in an enzyme-linked immunosorbent assay of elastin-derived peptides in human plasma. Am. Rev. Respir. Dis. **131: 709–713.**

21. AKERS, S., U. KUCICH, G. WEINBAUM, M. GLASS, J. ROSENBLOOM, M. SWARTZ, G. ROSEN & P. KIMBEL. 1987. Sensitivity and specificity of plasma elastin-derived peptides in chronic obstructive pulmonary disease (abstr.). Am. Rev. Respir. Dis. **135:** A147.

22. SCHRIVER, E. E., G. R. BERNARD, B. B. SWINDELL, J. S. SUTCLIFFE & J. M. DAVIDSON. 1989. Elastin fragment levels in human plasma, urine and bronchoalveolar lavage fluid (abstr.). Chest **96:** 153S.

23. MCLENNAN, G., T. DILLON, R. WALSH, M. ECKERT, M. CREW, R. SCICCHITANO, A. THORNTON & R. ANTIC. 1990. Correlation of plasma elastin derived peptides with pulmonary function tests (abstr.). Am. Rev. Respir. Dis. **141:** A113.

24. INDIK, Z., H. YEH, N. ORNSTEIN-GOLDSTEIN, P. SHEPPARD, N. ANDERSON, J. C. ROSENBLOOM, L. PELTONEN & J. ROSENBLOOM. 1987. Alternative splicing of human elastin mRNA indicated by sequence analysis of cloned genomic and complementary DNA. Proc. Natl. Acad. Sci. USA **84:** 5680–5684.
25. KUCICH, U., W. R. ABRAMS, S. AKERS, P. KIMBEL, G. WEINBAUM & J. ROSENBLOOM. 1988. Determination of elastin peptide levels in human plasma using elastin domain-specific antibodies (abstr.). Am. Rev. Respir. Dis. **137:** A372.
26. KUCICH, U., J. HAMILTON, S. AKERS, P. KIMBEL, J. ROSENBLOOM & G. WEINBAUM. 1990. Urine from emphysema patients contains elevated levels of elastin-derived peptides (abstr.). Am. Rev. Respir. Dis. **141:** A232.
27. KUCICH, U., J. ROSENBLOOM, P. KIMBEL, G. WEINBAUM & W. R. ABRAMS. 1991. Size distribution of human lung elastin-derived peptide antigens generated *in vitro* and *in vivo*. Am. Rev. Respir. Dis., in press.

Development and Application of Assays for Elastase-Specific Fibrinogen Fragments[a]

JEFFREY I. WEITZ[b]

Department of Medicine
McMaster University and
Hamilton Civic Hospitals Research Centre
Hamilton, Ontario

Human neutrophil elastase (HNE) degrades a variety of macromolecular substrates other than elastin. One of these substrates, fibrinogen, is cleaved by HNE into progressively smaller unclottable fragments.[1,2] We hypothesized that if we could identify a specific site of HNE proteolysis on the fibrinogen molecule, isolate and characterize the released fragment, and develop an assay for this peptide, we would have a useful marker of unopposed HNE action on fibrinogen. We reasoned that this test would provide important information given that neutrophils and fibrinogen are integral components of the inflammatory response. Accordingly, we set out to explore the interaction between HNE and fibrinogen.

IDENTIFICATION OF AN ELASTASE CLEAVAGE SITE ON FIBRINOGEN

Our initial studies confirmed previous reports that crude leukocyte extracts contain neutral proteinases capable of degrading fibrinogen into unclottable fragments[3,4] and cleaving the amino-terminal region of the fibrinogen Aα-chain.[4] We extended these observations by demonstrating that HNE is the enzyme responsible for both the fibrinogenolytic activity of the extracts and Aα-chain cleavage. Furthermore, we isolated and characterized the peptide released from the amino-terminal region of the Aα-chain.[2]

Whereas thrombin hydrolyzes the Aα16 (Arg)-Aα17 (Gly) bond on the Aα-chain of fibrinogen,[5] thereby releasing fibrinopeptide A (FPA or Aα1-16), HNE cleaves the Aα21 (Val)-Aα22 (Glu) bond and releases the FPA-containing fragment Aα1-21 (FIG. 1). This peptide appears to be a specific product of HNE action on fibrinogen because it is not released by thrombin, plasmin, trypsin, chymotrypsin, porcine pancreas elastase,[2] cathepsin G, or the metalloelastase isolated from *Pseudomonas aeruginosa* (unpublished observations). In addition, the Aα21-Aα22 bond is a very early site of HNE proteolysis, so that the Aα1-21 fragment is a major product of HNE attack on fibrinogen.[6] Finally, only free HNE releases

[a] This work was supported by grants from the Medical Research Council of Canada and the Ontario Heart and Stroke Foundation. Dr. Weitz is a Scholar of the Heart and Stroke Foundation of Ontario.

[b] Address for correspondence: Dr. Jeffrey Weitz, Henderson General Hospital, 711 Concession Street, Hamilton, Ontario, L8V 1C3, Canada.

the Aα1-21 fragment. HNE complexed to either of its physiologic inhibitors, α$_1$-proteinase inhibitor or α$_2$-macroglobulin, will not cleave the Aα21-Aα22 bond.[2] On the basis of these considerations, therefore, we set out to develop assays to quantify the HNE-derived fibrinopeptide Aα1-21.

DEVELOPMENT OF ASSAYS FOR ELASTASE-SPECIFIC FIBRINOPEPTIDES

Although the original assay to quantify Aα1-21 was indirect, more recently we developed direct assays for this fragment. To accomplish this, it proved necessary to determine the biologic fate of the HNE-derived peptide. Both the indirect and the direct assays will be briefly described.

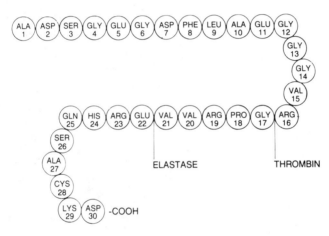

FIGURE 1. Sites of thrombin and neutrophil elastase cleavage on the amino-terminal region of the Aα-chain of fibrinogen. Thrombin cleaves the Aα16-Aα17 bond, thereby releasing Aα1-16 (fibrinopeptide A), whereas elastase hydrolyzes the Aα21-Aα22 bond and releases the fibrinopeptide A-containing fragment Aα1-21. Blood carboxypeptidases then degrade Aα1-21 first to Aα1-20 and then to Aα1-19.

Indirect Assay for Aα1-21

The original assay for the quantification of Aα1-21 was indirect and used a specific antiserum against FPA.[2] This antiserum, designated R2, has its antigenic determinant at the carboxy-terminal region of the FPA molecule.[7-11] Because this epitope is inaccessible in larger FPA-containing fragments, the antibody reacts poorly with Aα1-21.[8,9,11] When Aα1-21-containing samples are incubated *in vitro* with thrombin, however, the FPA portion of the peptide is released. This results in an approximately 1,000-fold increase in immunoreactivity, which we designated thrombin-increasable FPA immunoreactivity or TIFPA.[2]

Although the TIFPA assay provides excellent recovery of synthetic Aα1-21 added to blood or plasma,[2] the drawback of this technique is that it will measure any FPA-containing fragment. Accordingly, we set out to develop a direct assay for the Aα1-21 fragment.

Direct Assays for Aα1-21

To develop a specific assay for Aα1-21, we produced antibodies directed against the carboxy-terminus of the peptide. An Aα14-21 homolog with an amino-terminal cysteine residue at position Aα13 was synthesized. With a heterobifunctional cross-linking agent, the amino-terminal residue of the peptide was coupled to keyhole limpet hemocyanin, thereby leaving the carboxy-terminal free to act as a hapten. Antibodies against this immunogen were then raised in sheep. One high titer antibody (designated S-847) was specific for Aα1-21 and did not cross-react with FPA or with FPA-containing fragments larger than Aα1-21. The Val (Aα21) residue is a critical determinant of the antigenic site of this antibody because it does not recognize FPA-containing fragments shorter than Aα1-21 (ie, Aα1-20, Aα1-19, or Aα1-18).

With antiserum S-847, a radioimmunoassay specific for Aα1-21 was developed. When the levels of Aα1-21 measured with this assay were compared with TIFPA values in plasma samples collected from 30 subjects with a variety of clinical disorders, however, the levels of Aα1-21 were considerably lower than those of TIFPA. Two lines of evidence indicate that this discrepancy is due to proteolysis of Aα1-21 by blood carboxypeptidases. First, the incubation of Aα1-21 in blood results in rapid and progressive loss of immunoreactivity when the peptide is quantified with antiserum S-847 (FIG. 2). The addition of o-phenanthroline totally inhibits this loss of immunoreactivity, indicating that peptidase-mediated proteolysis is responsible for this phenomenon. Second, incubation of Aα1-21 with carboxypeptidases A, B, or Y in buffer results in a rapid loss of immunoreactivity similar to that occurring in blood. HPLC analysis of the time course of Aα1-21 proteolysis indicates that these carboxypeptidases progressively degrade the peptide first to Aα1-20 and then to Aα1-19.

Carboxypeptidase-mediated proteolysis of other fibrinopeptides has been well described. Both the thrombin-derived fragment fibrinopeptide B and Bβ1-42, a product of plasmin action on fibrinogen, are rapidly degraded by blood carboxypeptidases and circulate as desarginine FPB and Bβ1-41, respectively.[12,13] It is important to consider this phenomenon when developing specific assays for these fibrinogen fragments inasmuch as antibodies that have their epitopes at the carboxypeptidase-sensitive portion of the peptides are not useful for measuring circulating peptide levels.[13,14]

In an attempt to circumvent the loss of Aα1-21 immunoreactivity that occurs as a result of carboxy-terminal proteolysis, antibodies were raised in rabbits against synthetic Aα1-20 coupled to bovine serum albumin with glutaraldehyde. One high titer antibody (designated antiserum R20) cross-reacts completely with Aα1-21, Aα1-20, and Aα1-19, but does not react with FPA-containing fragments larger than Aα1-21. Although Aα1-18 also is recognized, R20 does not cross-react with FPA or its homologs, indicating that its antigenic determinant is at the carboxy-terminal region of the Aα1-20 fragment.

Antiserum R20 was used to develop a specific radioimmunoassay to quantify Aα1-21 and its carboxypeptidase-derived degradation products which we have designated elastase-specific fibrinopeptides or ESF. In patient samples, peptide levels measured with this antibody are not significantly different from the TIFPA values. This is not surprising because both assays measure intact Aα1-21 and its degradation products.

In Vivo *Proteolysis of Aα1-21*

Two lines of evidence indicate that carboxypeptidase-mediated proteolysis of Aα1-21 also occurs *in vivo*. First, the results of Aα1-21 clearance studies performed in primates demonstrate that the peptide is rapidly degraded first to Aα1-20 and then to Aα1-19. Second, HPLC analysis of patient plasma samples indicates that the bulk of ESF or TIFPA immunoreactivity coelutes with Aα1-19.

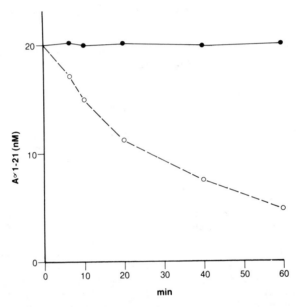

FIGURE 2. Stability of Aα1-21 immunoreactivity in whole blood when assayed with an antibody specific for the intact peptide. Synthetic Aα1-21 was incubated in whole blood for 60 minutes at 37°C in the absence (*dashed line*) or presence (*solid line*) of *o*-phenanthroline. At the times indicated, aliquots were taken, fibrinogen was precipitated with ethanol, and the ethanol supernatants were then assayed for Aα1-21 using antiserum S-847.

Clearance Studies

Marmosets were injected with synthetic Aα1-21, and the disappearance of the peptide was then monitored immunochemically using the TIFPA, ESF, and Aα1-21 assays. More rapid disappearance of immunoreactivity occurs when the peptide is measured with the specific Aα1-21 assay than when it is quantified with either the ESF or the TIFPA assay. HPLC analysis of the samples indicates that this is the result of rapid degradation of the peptide first to Aα1-20 and then to Aα1-19.[15] Because unlike the ESF and TIFPA assays, only intact Aα1-21 is measured in the Aα1-21 assay, these findings explain the more rapid disappearance of immunoreactivity when the peptide is quantified with the specific assay.

HPLC Analysis of Patient Samples

To further characterize the peptides identified by the ESF and TIFPA assays, patient samples were subjected to HPLC analysis using a gradient that separates Aα1-21 from its carboxypeptidase-derived degradation products. The fractions were then collected and assayed for ESF and TIFPA. Over 95% of the initial immunoreactivity is recovered, and approximately 85% coelutes with Aα1-19. Less than 3% coelutes with intact Aα1-21, whereas the remainder reflects Aα1-20.[15] Like Aα1-21 proteolysis by carboxypeptidases in a buffer system therefore, Aα1-19 appears to be the major degradation product of Aα1-21 *in vivo*.

APPLICATION OF ASSAYS FOR ELASTASE-SPECIFIC FIBRINOGEN FRAGMENTS

Results of the studies just described clearly indicate that quantification of HNE activity in blood requires assays that recognize intact Aα1-21 and its carboxypeptidase-derived degradation products. As both the TIFPA assay and the assay that uses antiserum R20 will measure these peptides, these tests were used to study HNE activity *in vivo* and *in vitro*.

HNE Activity and Proteinase Inhibitor Phenotype

Our initial studies demonstrated that the mean plasma level of TIFPA was significantly higher in six patients with congenital α_1-proteinase inhibitor deficiency than in 12 healthy controls (7.9 and 0.2 nM, respectively; p <0.0001), consistent with increased systemic HNE activity.[2] There are two possible explanations for these findings. The elevated TIFPA levels may reflect normal HNE activity that is unchecked in individuals lacking the major regulator of the enzyme. Alternatively, the increased HNE activity may be secondary to the inflammatory component of the pulmonary and/or hepatic disease that all of these patients had. To distinguish between these two potential mechanisms, we set out to determine whether a correlation exists between plasma ESF levels and proteinase inhibitor phenotype.

A total of 128 subjects were investigated. Forty-nine subjects had congenital deficiency of α_1-proteinase inhibitor of the PiZ type, whereas the remaining 76 were their first-degree relatives. Of these, 56 were heterozygotes of the PiMZ phenotype, and the remaining 23 had the normal PiMM phenotype. Using plasma ESF levels (as measured with either the TIFPA assay or the assay that uses antiserum R20 to measure Aα1-21 and its degradation products) as an index of HNE activity, there was a biologic gradient in peptide levels depending on proteinase inhibitor phenotype.[15] Thus, subjects with a normal phenotype had low levels of circulating peptide, heterozygotes had intermediate values, and those with homozygous deficiency had the highest levels.

These findings indicate that systemic HNE activity correlates with α_1-proteinase inhibitor concentration because heterozygotes have an approximately 50% reduction in inhibitor concentration, whereas those with homozygous deficiency have less than 30% α_1-proteinase inhibitor. Furthermore, because heterozygotes have no clinical evidence of lung or liver disease, these data suggest that plasma levels of the elastase-specific fibrinopeptides reflect normal HNE activity that is unchecked in subjects with deficiency of the major inhibitor of the enzyme. Finally, the findings support the concept that plasma levels of the elastase-specific fibrinopeptides may be useful in monitoring the activity of HNE-mediated diseases.

Effect of Smoking on Elastase Activity

Cigarette smoking is a major risk factor in the development of obstructive pulmonary diseases.[16] For many years, it has been hypothesized that cigarette smoking leads to lung disease by increasing elastase activity in the airways.[17] However, previous attempts to show increased enzyme activity in cigarette smokers have been inconclusive,[18–20] possibly because the assays used to measure elastase activity lacked specificity. Using plasma levels of TIFPA as an index of unopposed HNE activity, we set out to determine whether cigarette smoking results in an increase in enzyme activity.

Plasma levels of TIFPA in healthy cigarette smokers were compared with those in nonsmokers. The mean TIFPA level in 10 cigarette smokers was fivefold higher (95% confidence interval [CI] 3.0–9.6) than that in 20 healthy nonsmokers (2.0 nM compared with 0.4 nM; $p < 0.0001$), even after a 12-hour abstinence from smoking. To evaluate the acute effect of smoking on enzyme activity, a second group of 10 smokers was studied. After refraining from smoking for 12 hours, each individual smoked three cigarettes. The mean TIFPA level in the second group of smokers was not different from that in the first group of smokers (1.8 and 2.0 nM, respectively), but it was fivefold higher (95% CI 2.6–8.7) than that in nonsmokers (1.8 and 0.4 nM, respectively; $p < 0.0001$). After smoking three cigarettes, there was a twofold elevation (95% CI 1.6–3.5) in the mean TIFPA concentration (from 1.8–4.1 nM; $p = 0.002$), and peptide levels did not return to baseline for at least 1 hour after the subjects stopped smoking.[21]

These data indicate that cigarette smoking is associated with an increase in systemic HNE activity. There are many potential mechanisms by which smoking could increase enzyme activity. Previous studies have shown that cigarette smoke recruits neutrophils into the airways,[22,23] increases HNE release from neutrophils,[24] and inactivates α_1-proteinase inhibitor.[25–27] This latter effect occurs because cigarette smoke directly oxidizes the inhibitor, thereby rendering it considerably less effective as a regulator of HNE.[28,29] The oxidized form of the inhibitor has been recovered from the lung lavage fluid of rats exposed to cigarette smoke[28,29] and is found in the bronchoalveolar lavage fluid of cigarette smokers.[19,30,31]

These findings represent the first demonstration that cigarette smoking is associated with a systemic increase in HNE activity and suggest a possible mechanism by which cigarette smoking could lead to chronic lung disease. Furthermore, they raise the possibility that other systemic complications of smoking (such as atherosclerotic disease) may be the result of unopposed HNE activity.

Neutrophil Interactions with Fibrinogen

Although an increase in TIFPA levels in patients lacking α_1-proteinase inhibitor was an expected finding, the presence of circulating peptide in individuals with normal plasma concentrations of antiproteinases was puzzling given the rapidity of the interaction between HNE and α_1-proteinase inhibitor. *In vitro* studies of proteolysis by neutrophils, however, suggested a possible explanation. These studies demonstrated that enzymes released from stimulated neutrophils can degrade a variety of susceptible macromolecular substrates in the presence of antiproteinases,[32-35] thereby raising the possibility that cell-associated proteinases are more resistant to inhibition than are the free enzymes. To examine this hypothesis, we studied neutrophil-mediated fibrinogenolysis in the presence and absence of antiproteinases.[36]

Neutrophils were stimulated to migrate across ^{125}I-fibrinogen-coated nitrocellulose filters in response to 10^{-7} M formyl-met-leu-phe (FMLP), and the extent of fibrinogenolysis was determined by measuring release of Aα1-21 and ^{125}I-labeled fibrinogen degradation products. The fibrinogenolytic activity of migrating neutrophils was then compared with that of free HNE present in neutrophil lysates (TABLE 1) or secreted by neutrophils stimulated with FMLP (TABLE 2). The

TABLE 1. Effect of Inhibitors on Aα1-21 Release from Fibrinogen-Coated Filters Produced by PMN Lysates

Inhibitor[a]	ESF Release at 60 Minutes of Incubation (pmol)
None	550.0 ± 18.6
Z-Gly-Leu-PheCH$_2$Cl (0.1 mM)	561.3 ± 20.2
Plasma (1%)	0.9 ± 0.1
Serum (1%)	1.1 ± 0.2
α_1-Proteinase inhibitor (0.03 mM)	0.1 ± 0.01
Soybean trypsin inhibitor (0.3 mM)	2.1 ± 0.3
MeO-Suc-Ala$_2$-Pro-ValCH$_2$Cl (0.01 mM)	0.1 ± 0.01

[a] Inhibitors at the concentrations indicated were added to lysates from 10^6 PMN and incubated with fibrinogen-coated filters for 60 minutes at 37°C, and Aα1-21 was then assayed. 0.03 mM α_1-proteinase inhibitor (1.35 mg/100 ml) is within the physiologic range of this inhibitor in normal human plasma. The results shown are the means $\pm$ SD from two separate experiments.

fibrinogenolytic activity of soluble HNE was completely inhibited by 1% plasma or serum and by macromolecular antiproteinases (α_1-proteinase inhibitor and soybean trypsin inhibitor). In contrast, even in the presence of undiluted plasma or serum, the activity of the migrating neutrophils (TABLE 3) was incompletely blocked (81–85%). Furthermore, concentrations of α_1-proteinase inhibitor and soybean trypsin inhibitor that totally inhibited free HNE activity incompletely blocked (88–89% inhibition) the fibrinogenolytic activity of migrating neutrophils, indicating that FMLP-stimulated neutrophils demonstrate significant fibrinogenolytic activity in the presence of antiproteinases as small as 20,000 daltons. A specific low molecular weight (500 daltons) HNE inhibitor (MeO-Suc-Ala$_2$-Pro-ValCH$_2$Cl), however, totally blocked neutrophil-mediated fibrinogenolysis without affecting intracellular HNE activity, HNE secretion from neutrophils, or neutrophil migrating in response to FMLP.

These findings suggest that chemoattractant-stimulated neutrophils form zones of close contact with fibrinogen, thus preventing access of macromolecular

TABLE 2. Effect of Inhibitors on Aα1-21 Release by Supernatants of FMLP-Stimulated PMN[a]

Inhibitor	Aα1-21 Release at 60 Minutes of Incubation (pmol)
None	151.0 ± 10.6
Z-Gly-Leu-PheCH$_2$Cl (0.1 mM)	162.3 ± 15.2
Plasma (1%)	0.1 ± 0.01
Serum (1%)	1.5 ± 0.9
α$_1$-Proteinase inhibitor (0.03 mM)	0.1 ± 0.01
Soybean trypsin inhibitor (0.3 mM)	0.9 ± 0.1
MeO-Suc-Ala$_2$-Pro-ValCH$_2$Cl (0.01 mM)	0.1 ± 0.01

[a] The results shown are the means ± SD from two separate experiments.

antiproteinases to HNE released at the cell-substrate interface. Formation of a protected compartment between the neutrophils and fibrinogen is similar to that compartment that forms between macrophages and substrates coated with IgG or the complement protein C3.[37]

The receptor on neutrophils that mediates the interaction between FMLP or phorbol dibutyrate-stimulated neutrophils and fibrinogen-coated surfaces is complement receptor 3 (CR3;CD11b/CD18), a molecule previously identified as a receptor for the complement fragment C3bi.[38] Thus, monoclonal antibodies against CR3 that inhibit the binding of C3bi also block the binding of neutrophils to fibrinogen-coated surfaces. In addition, these antibodies inhibit the formation of a protected compartment.[38] The region of fibrinogen recognized by CR3 lies at the carboxy-terminus of the gamma-chain of fibrinogen, because peptides based on this sequence effectively inhibit the binding of stimulated neutrophils to fibrinogen-coated surfaces and block the binding of C3bi to CR3.[38] This KXXGD-containing gamma-chain sequence (His-His-Leu-Gly-Gly-Ala-Lys-Glu-Ala-Gly-Asp-Val) is the same region that is recognized by the glycoprotein IIb/IIIa (GPIIb/IIIa) receptor on platelets.[39] Sequence analysis shows strong structural similarity between the gamma-chain region on fibrinogen and other known ligands of CR3 such

TABLE 3. Effect of Inhibitors on Aα1-21 Release by FMLP-Stimulated PMN

Inhibitor[a]	Aα1-21 Release at 60 Minutes of Incubation (pmol)	Inhibition (%)
None	161.3 ± 18.1	—
Plasma (100%)	24.6 ± 8.4	85 ± 3
Serum (100%)	30.0 ± 2.0	81 ± 1
α$_1$-Proteinase inhibitor (0.03 mM)	19.3 ± 10.2	88 ± 4
Soybean trypsin inhibitor (0.3 mM)	17.2 ± 5.8	89 ± 1
MeO-Suc-Ala$_2$-Pro-ValCH$_2$Cl (0.01 mM)	0.5 ± 0.2	>99

[a] Inhibitors at the concentrations indicated were present in both upper and lower chambers of chemotaxis apparatus. 10^6 PMNs were used in each assay. Note that plasma and plasma-derived serum were used in 100-fold excess of the amounts required to inhibit 99.5% of all the elastase activity secreted by FMLP-stimulated PMNs (TABLE 2), or present in lysates of 10^6 PMNs (TABLE 1). The results shown are the means ± SD from three separate experiments.

as the 21-residue sequence in complement protein C3[40] and the gp63 of *Leishmania*,[41] both of which contain the sequence Arg-Gly-Asp.

Altieri *et al.*[42,43] recently demonstrated that CR3 on monocytes also can function as a receptor for fibrinogen and for coagulation factor X. The binding of both factor X and fibrinogen to activated monocytes is blocked by the monoclonal antibodies 7E3 and OKMI, which are directed against GPIIb/IIIa on platelets and CD11b, respectively. In contrast, these antibodies do not inhibit fibrinogen binding to activated neutrophils.[38] Furthermore, whereas the KXXGD-containing peptide derived from the carboxyl terminal of the gamma-chain of fibrinogen inhibits the binding of activated neutrophils to fibrinogen-coated surfaces and prevents the formation of a protected compartment, this peptide does not affect fibrinogen binding to monocytes.[42] These differences suggest that the binding site for fibrinogen on activated neutrophils may be different from that on monocytes.

The findings of these *in vitro* studies may provide an important link to understanding why measurable levels of the HNE-derived peptide are found in plasma of normal individuals,[2,15,21] and why elevated levels are found in cigarette smokers.[21] Each day, approximately 10^9 neutrophils emigrate from the vascular compartment to the extravascular space.[44] Neutrophils move across endothelial cell monolayers only when stimulated by a chemotactic gradient.[45] The close contact of stimulated neutrophils with intravascular and extravascular fibrinogen can lead to protected compartment formation and subsequent proteolysis despite normal concentrations of antiproteinases. Through this mechanism, a low level of fibrinogenolysis may be an inescapable consequence of neutrophil migration in response to many environmental or microbial stimuli.

Cigarette smoking may amplify neutrophil emigration from the vasculature into the lungs. The peripheral leukocyte count is elevated in cigarette smokers,[46] cigarette smoking delays neutrophil transit through the pulmonary circulation,[47] and increased numbers of neutrophils are recovered in the bronchial lavage fluid of smokers.[48] The mechanism for neutrophil attraction into the lungs is unknown, but macrophage-derived chemoattractants,[49,50] nicotine,[51] complement components,[52] and elastin fragments with chemoattractant properties[53] have been implicated. The elevated levels of elastase-specific fibrinopeptides in cigarette smokers, therefore, may reflect the increase in neutrophil traffic through the lungs. In addition to fibrinogen proteolysis, enzymes released from neutrophils may degrade components of the pulmonary interstitium, thereby explaining why cigarette smoking is a major risk factor in the development of chronic lung disease.

Other Assays for Elastase-Derived Fibrinopeptides

In this volume, Mumford and coworkers describe their assays for elastase-derived fibrinopeptides. Although casual inspection suggests that their findings are in conflict with those reported herein, more careful analysis indicates agreement in several areas. First, like Goretzki *et al.*,[6] these investigators have confirmed our finding that the Aα21-Aα22 bond is an early and specific site of HNE cleavage of fibrinogen. Second, using an antibody specific for Aα1–21 that does not cross-react with smaller fragments, Mumford and coworkers find little or no immunoreactivity in patient samples. These observations are not unexpected, given that Aα1-21 is rapidly degraded by blood carboxypeptidases, and we obtained similar results when we attempted to measure the elastase-specific peptides with an antibody that only recognizes intact Aα1-21.

The disagreement occurs when Mumford *et al.* attempted to develop and apply an assay similar to the TIFPA assay described herein. In contrast to our results, using their assay these authors found little or no immunoreactivity in samples collected from healthy volunteers, smokers, or subjects with α_1-proteinase inhibitor deficiency. There is no support for their suggestion that the TIFPA we are measuring is due to the carry-over of trace amounts of fibrinogen after ethanol precipitation, which could lead to an increase in FPA immunoreactivity after *in vitro* thrombin treatment that is not due to the presence of Aα1-21 or its degradation products. Four lines of evidence indicate that fibrinogen carry-over is not responsible for our observations. First, we have demonstrated that the TIFPA levels that we measure are unaffected by solid-phase extraction of the ethanol supernatants using C18 cartridges, a technique that ensures total removal of fibrinogen. Second, we have shown a good correlation between the results obtained with the TIFPA assay and those measured using antiserum R20 (which recognizes Aα1-21 and its carboxypeptidase-derived fragments). As R20 does not cross-react to any great extent with fibrinogen, these findings do not support the concept that the TIFPA is a result of trace amounts of contaminating fibrinogen in the ethanol supernatants. Third, on HPLC analysis of patient samples (using reverse-phase C18 columns), over 95% of the TIFPA immunoreactivity is recovered, and more than 85% of this coelutes with Aα1-19. In addition, we have shown that peptide levels correlate with proteinase inhibitor phenotype and smoking status, thereby indicating a biologic relevance of our findings. Fourth, assay of ethanol supernatants for thrombin-increasable fibrinopeptide B (TIFPB) was used as an indirect test for measuring plasma levels of the fibrinopeptide B-containing fibrinogen fragment Bβ1-42.[54] With the generation of a specific antibody against Bβ1-42 that does not react with the carboxypeptidase-sensitive portion of the peptide, an assay for Bβ1-42 was developed which gave results similar to those obtained using the TIFPB assay.[13] These findings again support the concept that carry-over of fibrinogen is insignificant after the ethanol precipitation techniques used in our studies.

A more likely explanation for the discrepancy between our results and those of Mumford *et al.* is that there are differences in the assay techniques. In experiments that we conducted with their antiserum against FPA, we detected two potential problems. First, their assay is approximately 10-fold less sensitive than ours. Second, and probably more important, we noted a significant matrix effect when Aα1-21 standard is added to plasma that has been processed according to their protocol. This results in a marked decrease in the sensitivity of their assay. These problems seriously limit the utility of their assay and may well explain their inability to confirm our findings.

CONCLUSIONS

These studies indicate that the HNE-derived fibrinopeptide Aα1-21 is rapidly degraded by blood carboxypeptidases primarily to Aα1-19. We have developed both a direct and an indirect assay to measure Aα1-21 and its carboxypeptidase-derived degradation products. Using these assays, we found a correlation between circulating peptide levels and proteinase inhibitor phenotype. In addition, higher peptide values are found in cigarette smokers than in healthy nonsmokers. These data support the concept that the levels of the elastase-specific fibrinopeptides can be used *in vitro* or *in vivo* as an index of HNE activity.

REFERENCES

1. GRAMSE, M., C. BINGENHEIMER, W. SCHMIDT, R. EGBRING & K. HAVEMAN. 1978. Degradation products of fibrinogen by elastase-like neutral protease from human granulocytes. Characterization and effects on blood coagulation in vitro. J. Clin. Invest. **61**: 1027–1033.
2. WEITZ, J. I., S. L. LANDMAN, K. A. CROWLEY, S. BIRKEN & F. J. MORGAN. 1986. Development of an assay for in vivo human neutrophil elastase activity: Increased elastase activity in patients with α_1-proteinase inhibitor deficiency. J. Clin. Invest. **78**: 155–162.
3. PLOW, E. F. & T. S. EDGINGTON. 1975. An alternative pathway of fibrinolysis. I. The cleavage of fibrinogen by leukocyte proteases at physiological pH. J. Clin. Invest. **56**: 30–38.
4. BILEZIKIAN, S. B. & H. L. NOSSEL. 1977. Unique pattern of fibrinogen cleavage by human leukocyte proteases. Blood **50**: 21–28.
5. BLOMBACK, B., B. HESSEL, D. HOGGE & L. THERIKILDSEN. 1978. A two-step fibrinogen-fibrin transition in blood coagulation. Nature (Lond.) **275**: 501–505.
6. GORETZKI, L., E. MULLER & A. HENSCHEN. 1987. Cleavage pathway and specificity of leucocyte elastase as compared to plasmin during fibrinogen degradation (abstr.) Thromb. Haemostasis **58**: 1098.
7. NOSSEL, H. L., I. YUDELMAN, R. E. CANFIELD, V. P. BUTLER, JR., K. SPANONDIS, G. D. WILNER & G. D. QURESHI. 1974. Measurement of fibrinopeptide A in human blood. J. Clin. Invest. **54**: 43–53.
8. CANFIELD, R. E., J. DEAN, H. L. NOSSEL, V. P. BUTLER, JR. & G. D. WILNER. 1976. Reactivity of fibrinogen and fibrinopeptide A containing fragments with antisera to fibrinopeptide A. Biochemistry **15**: 1203–1209.
9. WILNER, G. D., H. L. NOSSEL, R. E. CANFIELD & V. P. BUTLER, JR. 1976. Immunochemical studies of human fibrinopeptide A using synthetic peptide homologues. Biochemistry **15**: 1209–1213.
10. NOSSEL, H. L., V. P. BUTLER, JR., G. D. WILNER, R. E. CANFIELD & E. J. HARFENIST. 1976. Specificity of antisera to human fibrinopeptide A used in clinical fibrinopeptide A assays. Thromb. Haemostasis **35**: 101–109.
11. WEITZ, J. I., M. K. CRUICKSHANK, B. THONG, B. LESLIE, M. N. LEVINE, J. GINSBERG & T. ECKHARDT. 1988. Human tissue-type plasminogen activator releases fibrinopeptides A and B from fibrinogen. J. Clin. Invest. **82**: 1700–1707.
12. LAGAMMA, K. S. & H. L. NOSSEL. 1978. The stability of fibrinopeptide B immunoreactivity in blood. Thromb. Haemostasis **35**: 101–109.
13. WEITZ, J. I., J. A. KOEHN, R. E. CANFIELD, S. L. LANDMAN & R. FRIEDMAN. 1986. Development of a radioimmunoassay for the fibrinogen-derived fragment Bβ1-42. Blood **67**: 1014–1022.
14. ECKHARDT, T., H. L. NOSSEL, A. HURLET-JENSEN, K. S. LAGAMMA, J. OWEN & M. AUERBACK. 1981. Measurement of desarginine fibrinopeptide B in human blood. J. Clin. Invest. **67**: 809–816.
15. WEITZ, J. I., E. K. SILVERMAN, B. THONG & E. J. CAMPBELL. Plasma levels of elastase specific fibrinopeptides correlate with proteinase inhibitor phenotype: Evidence for increased elastase activity in subjects with homozygous and heterozygous deficiency of α_1-proteinase inhibitor. Manuscript submitted.
16. NATIONAL HEART, LUNG, AND BLOOD INSTITUTE. 1984. Chronic Obstructive Lung Disease: A summary of the health consequences of smoking: A report of the Surgeon General. Rockville, Maryland. Department of Health and Human Services.
17. JANOFF, A. 1985. Elastases and emphysema: Current assessment of the protease-antiprotease hypothesis. Am. Rev. Respir. Dis. **132**: 417–433.
18. GADEK, J., G. A. FELLS & R. G. CRYSTAL. 1979. Cigarette smoking induces functional antiprotease deficiency in the lower respiratory tract of humans. Science **206**: 1315–1316.
19. CARP, H., F. MILLER, J. HOIDAL & A. JANOFF. 1982. Potential mechanism of emphysema: Alpha-1-proteinase inhibitor purified from lungs of cigarette smokers contains oxidized methionine and has decreased elastase inhibitory capacity. Proc. Natl. Acad. Sci. USA **79**: 2041–2045.

20. STONE, P., J. D. CALORE, S. E. McGOWAN, J. BERNARDO, G. L. SNIDER & C. FRANZBLAU. 1983. Functional alpha-1-protease inhibitor in the lower respiratory tract of cigarette smokers is not decreased. Science 221: 1187–1189.
21. WEITZ, J. I., K. A. CROWLEY, S. L. LANDMAN, B. I. LIPMAN & J. YU. 1987. Increased neutrophil elastase activity in cigarette smokers. Ann. Intern. Med. 107: 680–682.
22. KILBURN, K. H. & W. McKENZIE. 1975. Leukocyte recruitment to airways by cigarette smoke and particle phase in contrast to cytotoxicity of vapor. Science 189: 634–637.
23. HUNNINGHAKE, G. W. & R. G. CRYSTAL. 1983. Cigarette smoking and lung destruction: Accumulation of neutrophils in the lungs of cigarette smokers. Am. Rev. Respir. Dis. 128: 833–838.
24. BLUE, M. L. & A. JANOFF. 1978. Possible mechanisms of emphysema in cigarette smokers: Release of elastase from human polymorphonuclear leukocytes by cigarette smoke condensate in vitro. Am. Rev. Respir. Dis. 117: 317–325.
25. CARP, H. & A. JANOFF. 1978. Possible mechanisms of emphysema in smokers: In vitro suppression of serum elastase-inhibitory capacity by fresh cigarette smoke and its prevention by antioxidants. Am. Rev. Respir. Dis. 118: 617–621.
26. COHEN, A. B. & H. L. JAMES. 1982. Reduction of the elastase inhibitory capacity of alpha-1-antitrypsin by peroxides in cigarette smoke: An analysis of brands and filters. Am. Rev. Respir. Dis. 126: 25–30.
27. PRYOR, W. A., M. M. DOOLEY & D. F. CHURCH. 1984. Inactivation of human alpha-1-proteinase inhibitor by gas-phase cigarette smoke. Biochem. Biophys. Res. Commun. 122: 676–681.
28. BEATTY, K., J. MATHESON & J. TRAVIS. 1984. Kinetic and chemical evidence for the inability of oxidized alpha-1-proteinase inhibitor to protect lung elastin from elastolytic degradation. Hoppe Seylers Z. Physiol. Chem. 365: 731–736.
29. BEATTY, K., J. BIETH & J. TRAVIS. 1980. Kinetics of association of serine proteinases with native and oxidized alpha-1-proteinase inhibitor and alpha-1-antichymotrypsin. J. Biol. Chem. 25: 3931–3934.
30. JANOFF, A., H. CARP, D. K. LEE & R. T. DREW. 1979. Cigarette smoke inhalation decreases alpha-1-antitrypsin activity in rat lung. Science 206: 1313–1314.
31. ABBOUD, R., T. FERA, A. RICHTER, M. Z. TABONA & S. Z. JOHAL. 1985. Acute effect of smoking on the functional activity of alpha-1-protease inhibitor in bronchoalveolar lavage fluid. Am. Rev. Respir. Dis. 131: 79–85.
32. JOHNSON, K. J. & J. VARANI. 1981. Substrate hydrolysis by immune complex-activated neutrophils: Effect of physical presentation of complexes and protease inhibitors. J. Immunol. 172: 1875–1879.
33. CAMPBELL, E. J., R. M. SENIOR, J. A. McDONALD & D. L. COX. 1982. Proteolysis by neutrophils. Relative importance of cell-substrate contact and oxidative inactivation of proteinase inhibitors in vitro. J. Clin. Invest. 70: 845–852.
34. WEISS, S. J., J. T. CURNUTTE & S. REGIANI. 1986. Neutrophil-mediated solubilization of the subendothelial matrix: Oxidative and nonoxidative mechanisms of proteolysis used by normal and chronic granulomatous disease phagocytes. J. Immunol. 136: 636–641.
35. CHAPMAN, H. A., JR. & O. L. STONE. 1984. Comparison of live human neutrophil and alveolar macrophage elastolytic activity in vitro. J. Clin. Invest. 74: 1693–1700.
36. WEITZ, J. I., A. J. HUANG, S. L. LANDMAN, S. C. NICHOLSON & S. C. SILVERSTEIN. 1987. Elastase-mediated fibrinogenolysis by chemoattractant-stimulated neutrophils occurs in the presence of physiologic concentrations of antiproteinases. J. Exp. Med. 166: 1836–1850.
37. WRIGHT, S. D. & S. C. SILVERSTEIN. 1984. Phagocytosing macrophages exclude proteins from the zones of contact with targets. Nature (Lond.) 309: 359–361.
38. WRIGHT, S. D., J. I. WEITZ, A. J. HUANG, S. M. LEVIN, S. C. SILVERSTEIN & J. D. LOIKE. 1988. Complement receptor type three (CDIIb/CD18) of human polymorphonuclear leukocytes recognizes fibrinogen. Proc. Natl. Acad. Sci. USA 85: 7734–7738.
39. KLOCZEWIAK, M., S. TIMMONS, T. J. LUKAS & J. HAWIGER. 1984. Platelet receptor recognition site on human fibrinogen: Synthesis and structure-function relationship

of peptides corresponding to the carboxy-terminal segment of the gamma chain. Biochemistry **23:** 1767–1774.

40. WRIGHT, S. D., P. A. REDDY, M. T. C. JONG & B. W. ERICKSON. 1987. C3bi receptor (complement receptor type 3) recognizes a region of complement protein C3 containing the sequence arg-gly-asp. Proc. Natl. Acad. Sci. USA **84:** 1965–1968.

41. RUSSELL, D. G. & S. D. WRIGHT. 1988. Complement receptor type 3 (CR3) binds to Arg-Gly-Asp-containing region of the major surface glycoprotein, gp63, of Leishmania promastigotes. J. Exp. Med. **168:** 279–292.

42. ALTIERI, D. C., R. BADER, P. M. MANNUCCI & T. S. EDGINGTON. 1988. Oligospecificity of the cellular adhesion receptor MAC-1 encompasses an inducible recognition specifity for fibrinogen. J. Cell. Biol. **107:** 1893–1900.

43. ALTIERI, D. C. & T. S. EDGINGTON. 1988. The saturable high affinity association of factor X to ADP-stimulated monocytes defines a novel function of the MAC-1 receptor. J. Biol. Chem. **263:** 7007–7015.

44. CRONKITE, E. P. 1979. Kinetics of granulopoiesis. Clin. Haematol. **8:** 351–370.

45. FURIE, M. B., B. L. NAPRSTEK & S. C. SILVERSTEIN. 1987. Migration of neutrophils across monolayers of cultured microvascular endothelial cells: An in vitro model of leukocyte extravasation. J. Cell. Sci. **88:** 161–175.

46. GALDSTON, M., E. L. MELNICK, R. M. GOLDRING, V. LEVYSKA, C. A. CURASI & A. L. DAVIS. 1977. Interaction of neutrophil elastase, serum trypsin inhibitory activity, and smoking history as risk factors for chronic obstructive pulmonary disease in patients with MM, MZ, and ZZ phenotypes for alpha-1-antitrypsin. Am. Rev. Respir. Dis. **116:** 837–846.

47. MACNEE, W., B. WIGGS, A. BELZBERG & J. C. HOGG. 1989. The effect of cigarette smoking on neutrophil kinetics in human lungs. N. Engl. J. Med. **321:** 924–928.

48. HUNNINGHAKE, G. W. & R. G. CRYSTAL. 1983. Cigarette smoking and lung destruction. Accumulation of neutrophils in the lungs of cigarette smokers. Am. Rev. Respir. Dis. **128:** 833–838.

49. HUNNINGHAKE, G., J. GADEK & R. CRYSTAL. 1980. Human alveolar macrophage neutrophil chemotactic factor: Stimuli and partial characterization. J. Clin. Invest. **66:** 473–483.

50. MERRILL, W. W., G. P. NAEGEL, R. A. MATHAY & H. Y. REYNOLDS. 1980. Alveolar macrophage-derived chemotactic factor. Kinetics of in vitro production and partial characterization. J. Clin. Invest. **65:** 268–276.

51. TOTTI, N., K. T. MCCUSKER, E. J. CAMPBELL, G. L. GRIFFEN & R. M. SENIOR. 1984. Nicotine is chemotactic for neutrophils and enhances neutrophil responsiveness to chemotactic peptides. Science (Washington, DC) **233:** 169–171.

52. KAZMIEROWSKI, J. A., J. I. GALLIN & H. Y. REYNOLDS. 1977. Mechanism for the inflammatory response in primate lungs. J. Clin. Invest. **5:** 273–281.

53. SENIOR, R. M., G. L. GRIFFEN & R. P. MECHAM. 1980. Chemotactic activity of elastin-derived peptides. J. Clin. Invest. **66:** 859–862.

54. NOSSEL, H. L., J. WASSER, K. L. KAPLAN, K. S. LAGAMMA, I. YUDELMAN & R. E. CANFIELD. 1979. Sequence of fibrinogen proteolysis and platelet release after intrauterine infusion of hypertonic saline. J. Clin. Invest. **64:** 1371–1378.

Direct Assay of Aα(1–21), a PMN Elastase-Specific Cleavage Product of Fibrinogen, in the Chimpanzee

RICHARD A. MUMFORD,[a] HOLLIS WILLIAMS,[a]
JENNIFER MAO,[a] MARY ELLEN DAHLGREN,[a] DALE
FRANKENFIELD,[b] THOMAS NOLAN,[b] LINDA
SCHAFFER,[d] JAMES B. DOHERTY,[c] DANIEL
FLETCHER,[a] KAREN HAND,[a] ROBERT BONNEY,[a,e]
JOHN L. HUMES,[a] STEPHEN PACHOLOK,[a]
WILLIAM HANLON,[a] AND PHILIP DAVIES[a]

[a]Department of Immunology & Inflammation,
[b]Laboratory Animal Resources Department,
[c]Department of Medicinal Chemistry, and
[d]Department of Pharmacology
Merck Sharp & Dohme Research Laboratories
Rahway, New Jersey 07065 and
West Point, Pennsylvania 19486

Polymorphonuclear leukocyte (PMN) elastase is a 218 amino acid single-chain glycoprotein[1,2] serine proteinase that can degrade a wide range of natural substrates. The enzyme, which has specificity for sites after small, uncharged amino acid residues, is present at high concentrations in the azurophilic primary lysosome of PMN. The only other source of the enzyme is the primary lysosome of blood monocytes.[3] PMN elastase, which is optimally active around neutral pH, cleaves a wide range of protein substrates in addition to elastin. These include connective tissue components such as proteoglycan, collagen (including type IV), and soluble proteins such as fibronectin, fibrinogen, and precursors of inflammatory mediators of the complement and kinin systems.

Extracellularly released PMN elastase has been implicated in a number of disease states including emphysema, chronic bronchitis, cystic fibrosis, and bronchiectasis, and increased levels of the enzyme complexed with its natural inhibitor alpha$_1$-proteinase inhibitor (α_1-PI) is found in plasma from patients with a wide variety of acute inflammatory diseases.[4–22]

It is widely assumed that PMN elastase functions primarily as an intracellular enzyme; however, a large number of cell biologic studies indicate that the enzyme can be released extracellularly by inflammatory stimuli. The evaluation of the extracellular activity of PMN elastase in both physiologic and pathologic states has been hampered by the lack of a specific assay to detect its activity *in vivo*. The direct measurement of the enzymatic activity of extracellular PMN elastase is difficult because the released enzyme rapidly binds to substrate or to plasma proteinase inhibitors. Detection of enzyme by immunoassay does not distinguish

[e] Present address: Bristol-Myers Co., 100 Forest Avenue, Buffalo, NY 14213.

between free enzyme and enzyme bound to inhibitor, and determinations of enzyme-inhibitor levels has the disadvantage that they detect only the inactivated enzyme.

As direct detection of enzymatically active PMN elastase in physiologic fluids has not proven feasible, alternate assay procedures dependent on the detection of its cleavage products have been pursued. Two of these methods have used PMN elastase-cleaved products of elastin as markers. One method measures urinary desmosine, a major cross-linking amino acid of elastin, by radioimmunoassay.[23-34] A similar result has been reported using ELISA measurements of elastase-generated elastin-derived peptides in serum of patients with lung disease and normal controls.[35-37] These methods suffer a lack of clear elevation in emphysema in the case of desmosine, and both methods lack specificity. This is particularly true with elastin peptides in which the lack of a well-defined immunogen is a critical drawback. Such drawbacks can be overcome by identification and assay of specific cleavage products of macromolecular natural substrates of elastase.

Fibrinopeptide A is a specific cleavage product from fibrinogen generated by thrombin from the N-terminus of the Aα chain. This peptide contains the first 16 residues of the Aα chain of fibrinogen [Aα(1–16)]. Nossel and his colleagues[38,39] developed a specific radioimmunoassay for this peptide which was used to demonstrate elevated levels of Aα(1–16) in disseminated intravascular coagulation, a disease associated with activation of blood coagulation. Several investigators[40-43] have demonstrated that fibrinogen is a substrate for PMN elastase. Weitz, working in Nossel's laboratory, demonstrated that elastase cleaves fibrinogen at the Val^{21}-Glu^{22} position on the Aα chain. He reasoned that the antibody developed by Nossel to measure Aα(1–16) could be used to measure levels of the larger PMN elastase-generated fibrinopeptide Aα(1–21) by a differential assay of Aα(1–16) levels before and after sample treatment with thrombin. On the basis of this differential approach, using a specific antiserum for Aα(1–16), he and his colleagues showed elevated amounts of thrombin-increasable fibrinopeptide A (TIFPA) in PiZZ individuals[44] and individuals before and after smoking.[45] They attributed the source of TIFPA to Aα(1–21), shown to be generated from fibrinogen by elastase in a stoichiometric way. However, no direct evidence was presented in these two reports to document that the clinical samples actually contained Aα(1–21).

DIRECT ASSAY OF Aα(1–21), A PMN ELASTASE-SPECIFIC CLEAVAGE PRODUCT OF FIBRINOGEN

In view of these findings we set out to characterize in detail the cleavage sites in the human fibrinogen molecule generated by PMN elastase. In addition to confirming the Val^{21}-Glu^{22} of the Aα chain of fibrinogen as a primary site of elastase cleavage, eight additional elastase cleavage sites were characterized.[46] The primary sites of cleavage were determined to be Aα Val^{21}-Glu^{22}, Aα Val^{450}-Ile^{451}, Aα Val^{464}-Thr^{465}, Aα Met^{476}-Asp^{477}, Aα Thr^{568}-Ser^{569}, γ-Thr^{305}-Ser^{306}, γ-Val^{347}-Tyr^{348}, and γ-Ala^{357}-Ser^{358}.

The Aα(1–21) cleavage peptide was chosen as an immunogen because it is the product of a single elastase cleavage in the fibrinogen molecule. Six of the other characterized peptide fragments required two cleavages to be released because of the disulfide network of the protein. H$_2$N-Cys-Nle-Aα (Gly^{17}-Pro^{18}-Arg^{19}-Val^{20}-Val^{21}-OH) was coupled to BSA with sulfo-MBS to give a product of 10 mol of immunogen peptide per mole of BSA. A rabbit polyclonal antisera, R-20, was found to have a sensitivity limit of 0.04 pmol/ml in a competitive RIA using ^{125}I-Tyr Aα(1–21) as probe.

This polyclonal antiserum does not cross-react with Aα(1–22), Aα(1–20), Aα(1–19), Aα(1–18), Aα(1–17) or FPA, Aα(1–16). Peptides shorter than Aα(1–21) which possess the carboxyl terminal Val^{21} residue [Aα(12–21) through Aα(16–21)] all cross-react with R-20. Amidation of the carboxyl terminal Val residue at position 21 resulted in a 2,000-fold loss of recognition by R-20. These results demonstrate the necessity for a free carboxyl group of the Val^{21} residue for recognition by this antiserum. Also, this observation rules out reaction of R-20 with an Aα(17–21) fragment within a peptide sequence, that is, within intact fibrinogen or within a fibrinogen-derived peptide cleaved at a site distal to the 21–22 bond.

The utility of this RIA has been demonstrated in a number of *in vitro* and *in vivo* situations.[46] Isolated PMN (10^6) incubated in the presence of fibrinogen release elastase in response to PMA to produce Aα(1–21) in a time- and concentration-dependent manner. A cephalosporin inhibitor[47] of PMN elastase, L-658,758 (0.5 μM), inhibits this activity. Additionally, intravenous injection of human leukocyte elastase (HLE) (2.5 mg) into marmosets produces a transient rise in Aα(1–21) to 70 nM at 1 minute, which is inhibited by prior injection of 10 mg/kg of L-658,758. Human blood incubated with A23187 (150 μM) produces Aα(1–21) in a time- and concentration-dependent manner [3.2 ± 1.2 nM Aα(1–21) for 17 normal individuals]. Aα(1–21) is undetectable in blood not stimulated with A23187. Blood from four PiZZ individuals produced considerably greater amounts of Aα(1–21) than did blood from normal subjects on stimulation with A23187 [range = 8–35 nM Aα(1–21), $n = 4$]. Blood from PiZZ individuals was also evaluated in this assay before and 1 day following α_1-PI augmentation therapy. Two PiZZ individuals were previously untreated with α_1-PI and three PiZZ individuals had received α_1-PI 30 days previously. Before receiving α_1-PI these individuals produced elevated amounts of Aα(1–21) peptide in the blood assay [range = 8–70 nM Aα(1–21)]. One day following α_1-PI reconstitution both groups produced amounts of Aα(1–21) in the range of 1–5 nM, similar to that seen in normal individuals.[48]

The experiments described in this manuscript represent a novel model system for evaluating the functional activity of a bioavailable PMN elastase inhibitor in the chimpanzee (*Pan troglodytes*). These experiments could be used clinically to demonstrate the presence of a functional elastase inhibitor in blood following administration in man.

MATERIALS AND METHODS

Production of Aα(1–21) in Chimpanzee Blood by the Calcium Ionophore A23187

To replicate 2-ml aliquots of freshly drawn heparinized (50 U/ml) chimpanzee blood, the calcium ionophore A23187 was added at concentrations up to 150 μM. Aliquots of blood containing 2 nM of synthetic Aα(1–21) peptide were also included as controls to assess recovery through the sample preparation protocol. All blood aliquots were incubated at 37°C for 25 minutes in the assay. Following the 37°C incubation, the tubes were placed on ice for 5 minutes, then centrifuged at 12,000 × g in a Beckman® microfuge (model-12) to prepare plasma. One milliliter of plasma was removed from each sample and transferred into a 12 × 75 mm glass tube, and 3 ml of chilled ethanol was added to precipitate protein. The sample was allowed to incubate on ice for 30 minutes prior to centrifugation. The ethanol supernatants were evaporated to dryness in a Savant Speed-Vac® (model SVC 100) concentrator and subsequently reconstituted to original plasma volume with water.

Radioimmunoassay for Aα(1–21)

Radioiodination of Probe Peptides

Radioiodination of the probe peptide Tyr Aα(1–21) was accomplished by reaction with chloramine T[49]. Tyr Aα(1–21) was dissolved in water at a concentration of 220 μg/ml. A 50-μl aliquot of this solution (containing 11 μg) was added to 10 μl of 0.5 M potassium phosphate, pH 7.4, and then combined with 2 mCi of Na^{125}I and 10 μl (0.1 mg/ml) of freshly prepared chloramine T in water. The mixture was allowed to react for 30 seconds and the reaction stopped with 10 μl of 1 mg/ml NaI plus 1 mg/ml sodium thiosulfate. The radioiodinated probe was purified by HPLC using Supelcosil C-8 (0.4 × 2.4 cm). The iodinated probe was eluted with a 30-minute gradient (2% per minute) of 90% eluant A, 10% eluant B to 30% eluant A, 70% eluant B at a flow rate of 1 ml per minute. Eluant A consisted of 0.1% trifluoroacetic acid in water, and eluant B consisted of 0.1% trifluoroacetic acid in acetonitrile. This purification procedure resolves labeled ^{125}I-Tyr Aα(1–21) from unlabeled Tyr Aα(1–21) peptide and other reaction byproducts. The specific activity of ^{125}I-Tyr Aα(1–21) peptide was determined by integration of the peak area from the HPLC with standard Tyr Aα(1–21) vs ^{125}I cpm associated with the peak material collected by HPLC. The specific activity routinely averages 2,000 Ci/mmol.

Immunoassay Protocol for the Determination of Peptide Concentration in Unknown Samples

The radioimmunoassay is conducted in a total volume of 300 μl. Assay buffer consists of Dulbecco's calcium- and magnesium-free phosphate-buffered saline solution (PBS) (purchased as a 10-fold concentrate from GIBCO Laboratories, Cat. #310–4200), supplemented with a 0.1% gelatin (DIFCO Cat. #0143–01), 0.01% thimerasol antiseptic, and 1.0 mM EDTA. Volumes up to 100 μl of an unknown solution can be assayed. Standard solutions are prepared in assay buffer at concentrations in the range of 1 × 10^{-10} M to 2 × 10^{-8} M which represents amounts of peptide ranging from 0.01–2.0 pmol in the 100-μl aliquot used in the assay. The assay includes antibody-free controls to measure nonspecific binding, and tubes containing antibody and probe without standard are used to determine the control level of peptide binding. To 100 μl of buffer, standard solution or sample is added 100 μl of rabbit antiserum #20, diluted 1 : 1000 in assay buffer. This is followed by the addition of 100 μl of radioactive probe, ^{125}I-Tyr Aα(1–21) fibrinogen peptide, diluted in assay buffer to yield approximately 15,000–20,000 cpm per 100 μl of aliquot.

The assay is incubated overnight at 4°C. Then non-antibody-bound radioactive material is separated from antibody-bound counts by adsorption onto dextran-coated activated charcoal. After a brief incubation period in an ice/water slurry, the charcoal was sedimented by centrifugation, and the supernatant was decanted and counted. Each assay also includes charcoal-free controls, to which 1 ml of PBS is added for determination of total counts. The antibody-free blank and the zero-peptide antibody control tubes are counted to determine the 0% bound and 100% bound values, respectively. The assay standards are then counted and related to the antibody control to determine the percent bound, and a standard curve is generated. Only those values falling between 80 and 20% displacement on the RIA standard curve are used to calculate data.

Effect of L-658,758 on Ex Vivo *Production of Aα(1–21) in Chimpanzee Blood*

Chimpanzees were immobilized with ketamine HCI at a dose rate of 3–5 mg/kg intramuscularly. Light anesthesia was maintained with Isoflurane at 1–1.5% in 100% oxygen. They were weighed and fitted with saphenous vein catheters. Heparinized blood samples were removed from each animal at intervals over a 60-minute period. One animal then received an intravenous injection of 10 mg/kg of L-658,758 and the other received an equivalent volume of normal saline solution. Additional heparinized blood samples were taken from both animals at intervals over the next 60 minutes. Samples were incubated and processed for radioimmunoassay measurement of Aα(1–21) as previously described.

HPLC Analysis of Chimpanzee Blood Samples for L–658,758

At various time intervals following the intravenous administration of 10 mg/kg of L-658,758 to chimpanzees, a 100-μl aliquot of heparinized blood was extracted with 5 ml of acetonitrile containing 0.2% trifluoroacetic acid. The tubes were vortexed vigorously for 15 seconds and centrifuged at 2,000 rpm for 15 minutes. The supernatant was transferred to a second set of glass tubes (16 × 150 mm) and evaporated to dryness under a stream of nitrogen. The samples were resuspended in 300 μl of HPLC buffer H_2O:ACN:TFA-78:22:0.2 v/v/v. HPLC was performed with a DuPont-Zorbax® C-18 column (0.4 × 2.5 cm) at ambient temperature equipped with a Beckman® 160 UV spectrophotometer. A 50-μl aliquot of sample was injected for HPLC determination of drug content, identified by the retention time of an authentic standard. The amount of drug present was quantitated by integration of the peak area using proprietary algorithms run on a Waters® 730 data module.

RESULTS AND DISCUSSION

FIGURE 1 shows a displacement curve for the measurement of Aα(1–21) peptide using rabbit polyclonal antisera R-20 in assay buffer. The limits of detection for the measurement of Aα(1–21) is approximately 0.04 pmol in the assay. The 50% displacement value is 0.12 pmol. Only those values falling between 80 and 20% displacement on the RIA standard curve are used to calculate data. This figure also shows a standard displacement curve in ethanol-precipitated chimpanzee plasma, demonstrating that no interfering material is present. A standard curve from each chimpanzee's plasma is used to calculate data. No endogenous Aα(1–21) peptide was detected in chimpanzee processed plasma.

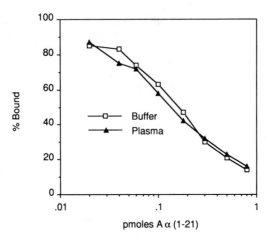

FIGURE 1. RIA displacement curve of Aα(1–21). Increasing concentrations of synthetic Aα(1–21) were titrated in either buffer (- □ -) or chimpanzee plasma (- • -).

FIGURE 2 demonstrates the time-dependent production of Aα(1–21) peptide in chimpanzee blood stimulated with the calcium ionophore A23187. The maximal amount of Aα(1–21) peptide is produced by 20 minutes. No endogenous Aα(1–21) peptide is detected in blood alone or in controls incubated for 60 minutes at 37°C. Additionally, other stimuli such as zymosan, FMLP, or PMA in the presence of cytochalasin B also give similar results (data not shown). These findings demonstrate the cleavage of an endogenous substrate by elastase released from PMN in chimpanzee blood in the presence of large concentrations of the natural inhibitors α_1-PI and α_2-macroglobulin.

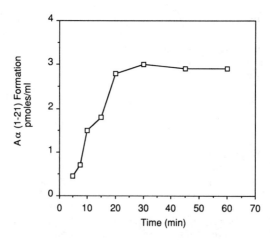

FIGURE 2. Time-dependent production of Aα(1–21) in chimpanzee whole blood stimulated with A23187.

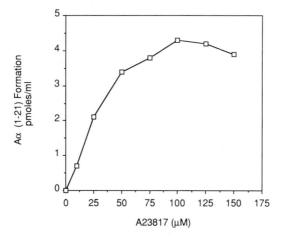

FIGURE 3. Calcium ionophore A23187-dependent formation of Aα(1–21) in chimpanzee blood.

FIGURE 3 displays the concentration dependence of calcium ionophore A23187 stimulated formation of Aα(1–21) in chimpanzee whole blood incubated at 37°C for 25 minutes. Maximal Aα(1–21) peptide formation was achieved with A23187 concentrations of approximately 100 μM. This concentration of A23187 (100 μM) does not cause a release of LDH, demonstrating a noncytotoxic effect of ionophore in the milieu of whole blood (data now shown).

FIGURE 4 demonstrates the inhibition of Aα(1–21) formation in the calcium ionophore blood assay by the PMN elastase inhibitor L-658,758 in a concentration-dependent fashion with an IC_{50} of 35 μg/ml. L-658,758 (FIG. 5) has been

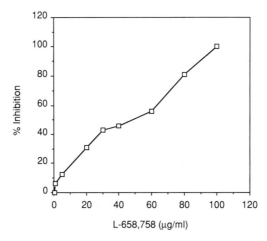

FIGURE 4. Inhibition of Aα(1–21) formation in chimpanzee whole blood by the PMN elastase cephalosporin inhibitor L-658,758.

shown to be a time-dependent cephalosporin inhibitor of elastase.[47] The inhibition of the formation of Aα(1–21) in blood allows the demonstration of the potency of synthetic elastase inhibitors in a biologic environment. With this background data we determined whether L-658,758, administered intravenously to a chimpanzee at 10 mg/kg, would inhibit the formation of Aα(1–21) in blood removed from the animal and stimulated *in vitro* with A23187 (100 μm). FIGURE 6 shows that in a series of pre-bleeds, -60 minutes to 1 minute before L-658,758 administration, the amount of Aα(1–21) produced was constant at approximately 3 nM. It is also shown that the blood samples drawn from the chimpanzee immediately after infusion of L-658,758 produced markedly lower levels of Aα(1–21) in response to A23187 stimulation. Over the course of the next 30–40 minutes the amount of ionophore-stimulated Aα(1–21) peptide production in freshly drawn samples gradually returned to the pretreatment level as the plasma level of drug decreased. In contrast, no consistent change was observed in the untreated animal (data not shown). Thus, the elastase inhibitory activity of L-658,758 can be demonstrated in circulating blood in a chimpanzee after intravenous infusion.

The findings are consistent with other observations of Aα(1–21) formation in human blood stimulated with A23187.[46] Human blood stimulated with the calcium ionophore A23187, zymosan, FMLP, or PMA in the presence of cytochalasin B also produce Aα(1–21) in a concentration- and time-dependent fashion. These experiments will be published elsewhere. In summary, the results presented in this publication document that neutrophils present in chimpanzee blood can be stimulated with the calcium ionophore A23187 to release elastase from the azurophilic granule and hydrolyze endogenous fibrinogen to produce Aα(1–21). This system allows the evaluation of elastase inhibitors in a biologic environment. The experiments presented here could be extended to allow the evaluation of elastase inhibitors in human blood following administration in man.

ACKNOWLEDGMENT

The assistance of Lorena Bennett in the preparation of this manuscript is gratefully acknowledged.

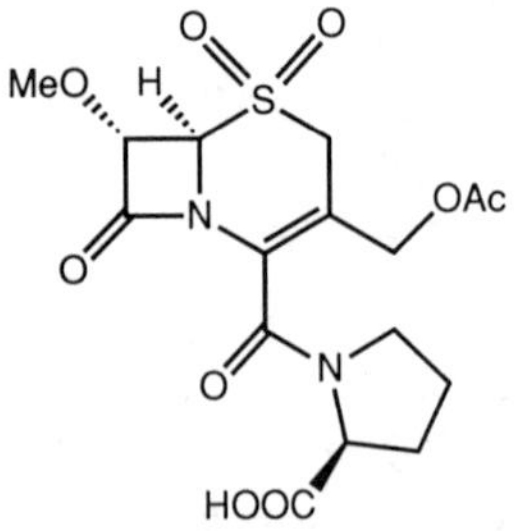

FIGURE 5. Structure of L-658,758.

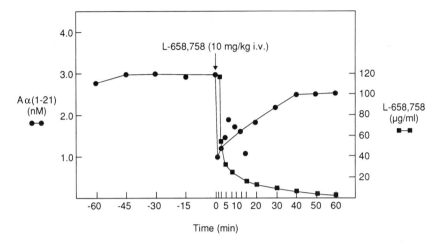

FIGURE 6. Inhibition of production of Aα(1–21) in chimpanzee blood by prior iv infusion of L-658,758. Blood was drawn at several time intervals from a chimpanzee 1 hour before and up to 1 hour after an iv bolus of L-658,758 (10 mg/kg) was given. At the indicated times, blood was stimulated *ex vivo* with A23187 (150 μM), incubated at 37°C for 25 minutes, and then plasma was prepared and samples processed for assay of Aα(1–21) levels according to Materials and Methods.

REFERENCES

1. SINHA, S., W. WATOREK, S. KARR, J. GILES, W. BODE & J. TRAVIS. 1987. Primary structure of human neutrophil elastase. Proc. Natl. Acad. Sci. USA **84:** 2228–2232.
2. TAKAHASHI, H., T. NUKIWA, K. YOSHIMURA C. D. QUICK, D. J. STATES, M. D. HOLMES, J. WHANG-PENG, T. KNUTSEN & R. G. CRYSTAL. 1988. Structure of the human neutrophil elastase gene. J. Biol. Chem. **263:** 14739–14747.
3. CAMPBELL, E. J., E. K. SILVERMAN & M. A. CAMPBELL. 1989. Elastase and cathepsin G of human monocytes. J. Immunol. **143:** 2961–2968.
4. STOCKLEY, R. A. & D. BURNETT. 1980. Serum derived protease inhibitors and leukocyte elastase in sputum and the effect of infection. Bul. Eur. Physiopathol. Resp. **16** (Suppl.): 261–271.
5. NEUMANN, S., N. HEINRICH, G. GUNZA & H. LANG. 1981. Enzyme linked immunoassay for elastase from leucocytes in human plasma. J. Clin. Chem. Clin. Biochem. **19:** 232–237.
6. KLEESICK, K., S. NEUMANN & H. GREILING. 1982. Determination of the elastase α_1 proteinase inhibitor complex-elastase activity and proteinase inhibitors in the synovial fluid. Fresenius 2. Anal. Chem. **311:** 434–435.
7. COHEN, G., K. FEHR & F. J. WAGENHAUSER. 1983. Leukocyte elastase and free collagenase activity in synovial effusions: Relation to numbers of polymorphonuclear leukocytes. Rheumatol. Int. **3:** 89–95.
8. JOCHUM, M., K. H. DUSWALD, E. HILLER & H. FRITZ. 1983. Plasma levels of human granulocytic elastase α_1-proteinase inhibitor complex (E-α_1 PI) in patients with septicemia and acute leukemia. *In* Selected Topics in Clinical Enzymology. D. M. Goldberg & H. Werner, eds.: 85–100. de Gruyter. Berlin. New York.
9. HÖRL, W. H., M. JOCHUM, A. HEIDLAND & H. FRITZ. 1983. Release of granulocyte proteinase during hemodialysis. Am. J. Nerphrol. **3:** 213–217.

10. MERRIT, T. A., C. G. COCHRANE, K. HOLCOMB, B. BOHL, M. HALLMAN, D. STRAYER, D. K. EDWARDS & L. GLUCK. 1983. Elastase and α_1-proteinase inhibitor activity in tracheal aspirates during respiratory distress syndrome. J. Clin. Invest. **72:** 656–666.
11. HILLER, E. & M. JOCHUM. 1984. Plasma levels of human granulocytic elastase-α_1 P proteinase inhibitor complex (E-α_1 PI) in leukemia. Blut **48:** 1–7.
12. NEUMANN, S., G. GUNZER, H. HEANRICH & H. LANG. 1984. PMN elastase assay: Enzyme immunoassay for human polymorphonuclear elastase complexed with α_1 proteinase inhibitor. J. Clin. Chem. Clin. Biochem. **22:** 693–697.
13. STOCKLEY, R. A. & C. OFFORD. 1984. The effect of leukocyte elastase on the immunoelectrophoretic behavior of alpha$_1$-antitrypsin. Clin. Sci. **66:** 217–224.
14. STOCKLEY, R. A., H. M. MORRISON, S. SMITH & T. TETLEY. 1984. Low molecular mass bronchial proteinase inhibitor and α_1-proteinase inhibitor in sputum and bronchoalveolar lavage. Z. Biol. Chem. Hoppe-Seyler. **363:** 587–595.
15. STOCKLEY, R. A. & S. C. OFFORD. 1984. Qualitative studies of lung lavage α_1-proteinase inhibitor. Z. Biol. Chem. Hoppe-Seyler **365:** 503–510.
16. VIRGA, G. D., R. K. MALLYA, M. B. PEPYS & H. P. SCHNEBLI. 1984. Quantitation of human leukocyte elastase, cathepsin G, alpha$_2$-macroglobulin and alpha$_1$-proteinase inhibitor in osteoarthrosis and rheumatoid arthritis synovial fluids. Adv. Exp. Biol. **167:** 345–353.
17. COOKE, A. L., F. C. HORY & D. A. ISENBERG. 1984. Elastase activity in serum and synovial fluid of patients with connective tissue disorders. J. Rheumatol. **11:** 666–671.
18. SCHNEBLI, H. P., P. CHRISTEN, M. JOCHUM, R. K. MALLYA & M. B. PEPYS. 1984. Plasma levels of inhibitor-bound leukocyte elastase in rheumatoid arthritis patients. Adv. Exp. Med. Biol. **167:** 355–362.
19. JOCHUM, M., A. PELLETIER, C. BOUDIER, G. PAULI & J. G. BIETH. 1985. The concentration of leukocyte elastase-α_1-proteinase inhibitor complex in bronchoalveolar lavage fluids from healthy subjects. Am. Rev. Respir. Dis. **132:** 913–914.
20. ADEYEMI, E. O., S. NEUMANN, V. S. CHADWICK, H. J. F. HODGSON & M. PEPYS. 1985. Circulating human leukocyte elastase levels in patients with inflammatory bowel disease. Gut **26:** 306–311.
21. KLEESIEK, K., R. REINARDS, D. BRACKERTZ, S. NEUMANN, H. LANG & H. GREILING. 1986. Granulocyte elastase as a new biochemical marker in the diagnosis of chronic joint diseases. Rheumatol. Int. **6:** 161–169.
22. ADEYEMI, E. O., R. G. HULL, V. S. CHADWICK, G. R. V. HUGHES & H. J. F. HODGSON. 1986. Circulating human leucocyte elastase in rheumatoid arthritis. Rheumatol. Int. **6:** 57–60.
23. STARCHER, B. 1977. Determination of the elastin content in tissues by measuring desmosine and isodesmosine. Anal. Biochem. **79:** 11–15.
24. GOLDSTEIN, R. & B. STARCHER. 1978. Urinary excretion of elastin peptides containing desmosine after intratracheal injections of elastases in hamsters. J. Clin. Invest. **61:** 1286–1290.
25. STARCHER, B. C. & R. A. GOLDSTEIN. 1979. Studies on the absorption of desmosines and isodesmosines. J. Lab. Clin. Med. **89:** 848–852.
26. KING, G. S., V. S. MOHAN & B. C. STARCHER. 1980. Radioimmunoassay for desmosine. Conn. Tis. Res. **7:** 263–267.
27. HAREL, S., A. JANOFF, S. YU & A. HUREWITZ. 1980. Desmosine radioimmunoassay for measuring elastin degradation *in vivo*. Am. Rev. Respir. Dis. **122:** 769–773.
28. YU, S. Y., A. JANOFF, S. HAREL & V. H. AYVAZIAN. 1981. Urinary excretion of desmosine in patients with severe burns. Metabolism **30:** 497–501.
29. DAVIES, S., K. OFFORD, M. BROWN, H. CAMPE & D. NIEWOEHNER. 1983. Urinary desmosine is unrelated to cigarette smoking or to spirometric function. Am. Rev. Respir. Dis. **128:** 473–475.
30. JANOFF, A., A. CHANANA, D. JOEL, H. SUSSKIND, P. LAURENT, S. YU & R. DEARING. 1983. Evaluation of the urinary desmosine radioimmunoassay as a monitor of lung injury after endotracheal elastase instillation in sheep. Am. Rev. Respir. Dis. **128:** 545–551.

31. PELHAM, R., M. WEWER, R. CRYSTAL, A. BUIST & A. JANOFF. 1985. Urinary excretion of desmosine (elastin cross-links) in subjects with PiZZ alpha-1-antitrypsin deficiency, a phenotype with hereditary predisposition to pulmonary emphysema. Am. Rev. Respir. Dis. **132:** 821–823.
32. BRUCE, M., K. WEDIG, N. JENTOFT, R. MARTIN, P. CHENG, T. BOAT, & A. FANAROFF. 1985. Altered urinary excretion of elastin cross-links in premature infants who develop bronchopulmonary displasia. Am. Rev. Respir. Dis. **131:** 568–572.
33. ROGNIECKI, J., J. STODKOWSKA & U. RUTA. 1986. The urinary content of desmosine in experimental and clinical studies as a test for some lung injuries. Eur. J. Respir. Dis. **69:** 225–232.
34. IDELL, S., R. S. THRALL, R. MAUNDER, T. R. MARTIN, J. MCLARTY, M. SCOTT & B. C. STARCHER. 1989. Bronchoalveolar lavage desmosine in bleomycin-induced lung injury in marmosets and patients with adult respiratory distress syndrome. Exp. Lung Res. **15:** 739–753.
35. KUCICH, V., P. CHRISTNER, G. WEINBAUM & J. ROSENBLOOM. 1980. Immunologic identification of elastin-derived peptides in the serum of dogs with experimental emphysema. Am. Rev. Respir. Dis. **122:** 461–465.
36. KUCICH, U., W. R. ABRAMS, P. CHRISTNER, J. ROSENBLOOM, P. KIMBEL & G. WEINBAUM. 1984. Molecular weight distribution of elastin peptides in plasmas from human non-smokers, smokers and emphysema patients. Am. Rev. Respir. Dis. **129:** 503–507.
37. KUCICH, U., P. CHRISTNER, M. LIPPMANN, P. KIMBEL, G. WILLIAMS, J. ROSENBLOOM & G. WEINBAUM. 1985. Utilization of a peroxidase-antiperoxidase complex in an enzyme-linked immunosorbent assay of elastin-derived peptides in human plasma. Am. Rev. Respir. Dis. **131:** 709–713.
38. NOSSEL, H. L., V. P. BUTLER, JR., G. D. WILNER, R. E. CANFIELD & E. J. HARFNIST. 1976. Specificity of antisera to human fibrinopeptide A used in clinical fibrinopeptide A assays. Thromb. Haemost. **35:** 109–115.
39. NOSSEL, H. L., M. TI, K. L. KAPLAN, K. SPANONDIS, T. SOLAND & V. P. BUTLER, JR. 1976. The generation of fibrinopeptide A in clinical blood samples. J. Clin. Invest. **58:** 1136–1144.
40. GRAMSE, M., C. BINGENHEIMER, W. SCHMIDT, R. EGBRING & K. HAVEMANN. 1978. Degradation products of fibrinogen by elastase-like neutral protease from granulocytes: Characterization and effects on blood coagulation *in vitro*. J. Clin. Invest. **61:** 1027–1033.
41. PLOW, E. F. 1982. Leukocyte elastase release during blood coagulation. J. Clin. Invest. **69:** 564–572.
42. PLOW, E. F., M. GRAUSE & K. HAVERMANN. 1983. Immunochemical discrimination of leukocyte elastase from plasmic degradation products of fibrinogen. J. Lab. Clin. Med. **102:** 858–869.
43. WALLIN, R. & T. SALDEEN. 1987. A specific radioimmunoassay for determination of peptides derived from elastase degradation of human fibrinogen. Acta Univ. Upsaliensis **102:** 1–26.
44. WEITZ, J. I., S. L. LANDMAN, K. A. CROWLEY, S. BIRKEN & F. J. MORGAN. 1986. Development of an assay for *in vivo* neutrophil elastase activity. Increased elastase activity in patients with alpha-1-proteinase inhibitor deficiency. J. Clin. Invest. **78:** 155–163.
45. WEITZ, J. I., K. A. CROWLEY, S. L. LANDMAN, B. I. LIPMAN & J. YU. 1987. Increased neutrophil elastase activity in cigarette smokers. Ann. Int. Med. **107:** 680–682.
46. MUMFORD, R. A., O. F. ANDERSEN, J. BOGER, S. BONDY, R. J. BONNEY, D. BOULTON, P. DAVIES, J. DOHERTY, D. FLETCHER, K. HAND, J. MAO, H. WILLIAMS & M. E. DAHLGREN. 1989. Development of a direct, sensitive and highly specific RIA for a primary human leukocyte elastase (HLE) generated fibrinogen cleavage product, Aα(1–21). Am. Rev. Respir. Dis. **139:** A574.
47. DOHERTY, J. B., B. M. ASHE, L. W. ARGENBRIGHT, P. L. BARKER, R. J. BONNEY, G. O. CHANDLER, M. E. DAHLGREN, C. P. DORN JR., P. E. FINKE, R. A. FIRE-

STONE, D. FLETCHER, W. K. HAGMANN, R. MUMFORD, L. O'GRADY, A. L. MAY-COCK, J. M. PISANO, S. K. SHAH, K. R. THOMPSON & M. ZIMMERMAN. 1986. Cephalosporin antibiotics can be modified to inhibit human leukocyte elastase. Nature **322:** 192–194.

48. MUMFORD, R. A., H. WILLIAMS, J. MAO, M. E. DAHLGREN, R. HUBBARD, R. CRYSTAL & P. DAVIES. 1990. Production of Aα(1–21), the human PMN elastase generated cleavage product of fibrinogen, in PiZZ individual's blood stimulated by calcium ionophore (A23187) before and after reconstitution therapy with α_1 trypsin. Am. Rev. Respir. Dis. **141:** A111.

49. HUNTER, W. M. & F. C. GREENWOOD. 1962. Preparation of iodine-131 labelled human growth hormone of high specific activity. Nature **194:** 495–496.

Quantifying Emphysema by CT Scanning

Clinicopathologic Correlates

WILLIAM MacNEE,[a,b] GRAHAM GOULD,[a]
AND DAVID LAMB[c]

[a]Unit of Respiratory Medicine
Department of Medicine (RIE) and
[c]Department of Pathology
University of Edinburgh
Edinburgh, Scotland, UK

Emphysema has been defined as "a condition of the lungs characterized by abnormal, permanent enlargement of airspaces distal to the terminal bronchioles, accompanied by destruction of their walls, without obvious fibrosis."[1] However, the "normal" size of distal airspaces is not well defined.[2-4] Therefore, any method that claims to diagnose, or indeed quantitate, emphysema in life must correlate with the pathologic features that define emphysema.

The clinical features of emphysema may only be recognized late in the course of the disease and are imprecise in the diagnosis of this condition. Consequently, in 48 patients who died during the National Institute of Health trial of intermittent positive pressure breathing (IPPB), no correlation was noted between the clinical variables of cough, wheeze, or dyspnea and the alveolar surface area, calculated from measurements of the mean linear intercept in postmortem lungs.[5] The radiologic signs of emphysema on the plain chest radiograph, such as hyperinflation and reduced vascularity, are no more precise than the clinical features in diagnosing this condition.[6] Thurlbeck and Simon,[7] in a large series of patients, found a relatively poor correlation between the radiographic diagnosis of emphysema and a picture-grading technique to assess emphysema subsequently in autopsy lungs. Less than half the patients had pathologically proven emphysema detected by plain chest radiography, and almost one third the patients with the most severe grades of emphysema were misdiagnosed on the chest film.[7]

The hope that tests of respiratory function would be sensitive and specific enough to detect emphysema in life has also not been realized. Many studies have tried to relate impaired respiratory function to pathologic measurements of emphysema.[5,6,8-16] The conflicting results of these studies may partly be explained by the different material, such as postmortem or surgically resected lungs, obtained for analysis. A major disadvantage of postmortem studies is the variable time delay between measurements of respiratory function and pathologic assessment. This is avoided by studying patients undergoing lung resection. However, surgical studies have the disadvantage that a selected group of patients are studied whose respiratory function falls within a fairly narrow range and is usually well preserved. A review of published structure-function studies reveals that the physiologic variables that most consistently correlate with the severity of emphysema are the diffusing capacity for carbon monoxide (DLCO)[5,6,8-16] and, to a lesser

[b] Address for correspondence: Dr. W. MacNee, Unit of Respiratory Medicine, City Hospital, Greenbank Drive, Edinburgh EH10 5SB, Scotland, UK.

extent, measurements derived from the pressure/volume curve of the lung.[8,17] However, even these measurements correlate relatively poorly with the severity of emphysema.[5,6] They are also insensitive in mild emphysema[8] and, in the case of the DLCO, nonspecific in the diagnosis of emphysema.[18]

PATHOLOGIC ASSESSMENT OF EMPHYSEMA

Many previous clinicopathologic studies have concentrated on emphysema that is visible to the naked eye, that is, the "holes" of macroscopic emphysema. Macroscopic emphysema is usually assessed on sagittal lung slices from resected or autopsy lungs. The most frequently used technique is a visual comparison with a set of standardized pictures of lung sections, showing a wide range of emphysema "scores."[19] This picture-grading score is semiquantitative, but apparently is reproducible in expert hands.[4] However, this technique measures only those lesions of emphysema that are confluent and >1 mm in diameter; it does not assess microscopic emphysema. It has been estimated that three quarters of the alveolar surface area of the lungs must be lost before macroscopic emphysema develops.[20] Furthermore, the picture-grading technique measures emphysema on a nonlinear scale, which may explain the relatively poor correlations between lung function and emphysema score when linear regression analysis is used.[2,8] It also ignores the contribution to lung function of diffuse microscopic emphysema. Diffuse airspace enlargement can be measured pathologically by the mean linear intercept.[6,21] These previous studies used lungs embedded in paraffin which, when sectioned, require a correction factor to account for shrinkage and distortion.[21] Moreover, the calculation of the alveolar surface area from the mean linear intercept involves a number of mathematical assumptions. However, this measurement of distal airspace size has been shown by Thurlbeck et al.[6] and others[5,22] to correlate with measurements of lung function, particularly the DLCO; however, the values of the DLCO are widely scattered, particularly in subjects with little or no macroscopic emphysema.

AIRSPACE WALL PER UNIT VOLUME

To satisfy the definition of emphysema,[1] it is important to measure both micro- and macroscopic emphysema which may make differing contributions to lung function. We therefore devised a morphometric technique to measure the size of distal airspaces in lungs resected for peripheral lung tumors.[23]

Immediately after resection the lungs were fixed in inflation with formol saline solution for 24 hours and then cut into 1-cm sagittal slices. The lateral two parasagittal slices were overlaid with a transparent sheet carrying a 2 × 2 cm grid. Six 2 × 2 cm blocks were cut from each slice using a series of random numbers to provide the coordinates for each block. Shrinkage and distortion of the blocks were avoided by embedding in glycol methacrylate. Sections 3μ thick were cut from each block. Random fields were selected from these 12 sections, using numbered coordinates derived from an England finder. Each field was magnified 63 times and examined with a computer-assisted image analysis system that enhanced and edited the image (FIGS. 1 and 2) before direct measurements of the length of the airspace perimeter in a 1 mm² field were made. This length of airspace wall (mm/mm^2) was then converted to the surface area of the airspace walls per unit volume of lung (AWUV) in mm^2/mm^3, using the formula:[24]

$$AWUV = \text{Distal Airspace Perimeter} \times 4/\pi$$

Between 20 and 35 microscopic fields were examined in each specimen to obtain a stable running mean,[25] which did not vary by more than 3% in five consecutive fields. With a hand lens, macroscopic emphysema was also assessed in each resected lung as the percentage of the area of the mid-sagittal slice involved by macroscopic emphysema (airspaces >1 mm in diameter). These lesions were traced on a polythene sheet placed over the lung slice. The combined areas of the lesions were measured using a digitizing tablet and a microprocessor.

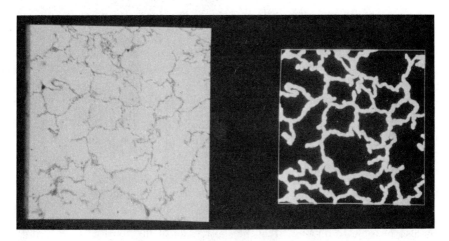

FIGURE 1. (**Left**) Microscopic field of lung tissue used to measure AWUV before editing and enhancing the image. Note intraalveolar cells and debris. (**Right**) Binary image produced before measurement of alveolar perimeter length by computer. AWUV for this field = 24.5 mm^2/mm^3.

COMPUTERIZED TOMOGRAPHIC SCANNING

Computerized tomographic (CT) scanning was developed for clinical use in the 1970s by Hounsfield,[26] who originally envisaged the scanner as a measuring device, because it provides precise information on the density of tissues derived from the attenuation of X-rays. However, this application has not been widely utilized compared with its use in imaging. In fact, the earliest CT scanners were more reliable as densitometers than were many of the later devices, whose design has been modified to improve the quality of the image at the expense of a reduction in the accuracy of measurements of tissue density. The EMI CT 5005 used for most of our work produces X-ray attenuation values that bear a linear relation to true electron densities for substances whose physical density is in the range of

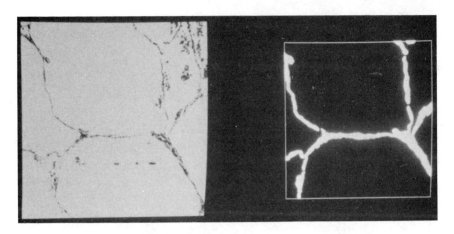

FIGURE 2. As in Figure 1. AWUV for this field = 7.3 mm^2/mm^3.

0.001 (air) to 1.0 g cm^2 (water).[27] The CT density of a tissue is measured on an arbitrary scale, where an EMI number of 0 is equivalent to water, −500 to air, and +500 to bone. Hounsfield numbers are exactly twice those of the EMI scale (FIG. 3). Each CT scan is composed of a large number of picture elements, or pixels, for which tissue density is calculated on the EMI scale. These data can be converted by a computer to a Grey scale to produce an image. With an appropriate density window setting, different structures can be highlighted. For example, density window setting in the EMI range of −500 to −200 highlights lung parenchyma, whereas a setting of −150 to +150 highlights the tissues in the mediastinum and chest wall. Early workers found that the CT density of normal lungs lay within the range of −450 to −200 EMI units when measured at arrested deep inspiration. It is also possible to obtain a frequency histogram of the density of every pixel within each lung slice or a cumulative histogram for the whole lung.

Our early work[28] showed a marked shift in the lung density histogram towards low densities in emphysematous patients than in normal subjects (FIG. 4). From such density histograms the mean CT density can be obtained. As the frequency

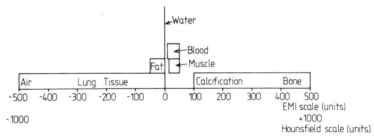

FIGURE 3. Computerized tomography x-ray absorption chart (EMI scale).

histogram of CT lung density is skewed towards higher densities because of the presence of large vessels and airway walls in the lung fields, the mean density obtained from this histogram may not be very sensitive to the change in density occurring in emphysema. Therefore, we chose to focus on the lowest fifth percentile of the density histogram, which is more sensitive to changes in low density produced by areas of emphysema.

POTENTIAL ERRORS IN CT DENSITOMETRY

Several factors may produce erroneous results when CT lung density is being measured. Most important among these is the effect of lung volume.[29-31] In nor-

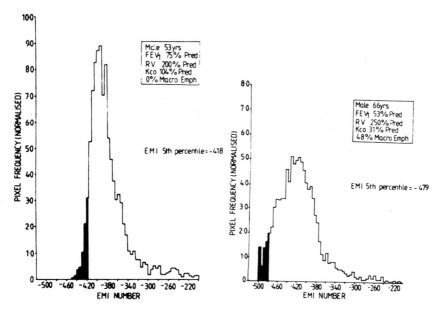

FIGURE 4. CT density histogram of **(left)** a subject with no emphysema and **(right)** a patient with severe emphysema. The *shaded area* represents the lowest 5% of the distribution.

mal subjects the CT density histogram shifts towards lower densities on expiration than on inspiration (FIG. 5). This effect is much less marked in patients with severe emphysema, in whom little change occurs particularly in the lowest fifth percentile of the density histogram (FIG. 5). Thus, to be reproducible, CT lung density should be measured at similar lung volumes. To achieve both patient comfort and the most reproducible results, we studied patients at arrested deep inspiration.

A standardized technique should also be employed to outline the lung fields from which the density histogram is obtained. We developed a density discriminatory program[32] that identifies the density gradient between lung parenchyma and the soft tissues of the chest wall. When measuring CT lung density, attention should also be paid to certain technical problems, such as movement artifact, which can be eliminated by the use of a short scan time. The partial volume effect, which occurs when a pixel (or voxel, which is the cross-sectional area of the pixel multiplied by the thickness of the scan slice) is occupied by tissues of differing densities, can be minimized by studying the periphery of the lung, where lung tissue is relatively homogenous. As the measurement of CT lung density varies between different scanners,[33] phantom studies must be performed if the results from two scanners, even from the same manufacturer, are to be compared.

CT LUNG DENSITY IN THE ASSESSMENT OF EMPHYSEMA

Using these techniques we set out to correlate CT lung density with morphometric measurements of the size of distal airspaces. We hypothesized that as emphysema developed, a decrease in alveolar surface area would occur, as alveolar walls are lost, associated with an increase in distal airspace size, which would result in decreased lung CT density (FIG. 6). We also considered that these sequelae would be associated with a decrease in lung function.

To correlate measurements of distal airspace size with CT lung density, we studied 28 patients undergoing lung resection for peripheral lung tumors.[23] Respiratory function and a limited CT scan (two slices) were performed 48 hours before the operation. The tumor was excluded from the analysis of CT lung density. A

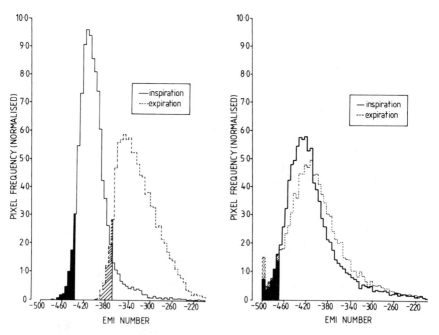

FIGURE 5. Change in CT density histogram with respiration in **(left)** a normal subject and **(right)** a patient with severe emphysema. *Solid line* = inspiration; *dotted line* = expiration. *Shaded area* represents lowest 5% of the distribution.

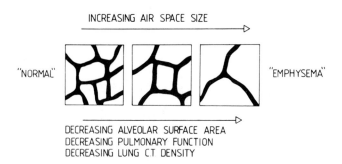

FIGURE 6. Diagrammatic representation of the possible sequence of associated events that occur during the development of emphysema.

direct comparison was made between CT lung density in the lateral two fifths of the lung fields (which excludes large airways and vessels and therefore is representative of the lung parenchyma) with morphometric measurements from the lateral two parasagittal slices, in the subsequently resected lung.[23] Lung CT density, as measured by the lowest fifth percentile of the cumulative density histogram, correlated best with the mean AWUV of the lowest five fields which were examined in the resected lung (FIG. 7). Furthermore, AWUV correlated significantly with volume-corrected diffusing capacity for carbon monoxide (DLCO/VA) (FIG. 8) and, to a lesser extent, with the total lung capacity (TLC) $r = -0.36$, $p < 0.05$). Interestingly, this correlation between CT lung density and distal airspace size was present in a surgical population with relatively mild or no macroscopic emphysema. Only five individuals had $>5\%$ of the surface area of the midsagittal lung slice involved by macroscopic emphysema. This implies that these relationships are independent of the presence of macroscopic emphysema (FIGS. 7 and 8). These results extend those of an earlier study in which we found that CT lung density was significantly lower in patients with than in those without emphysema.[28] However, we did not demonstrate a significant correlation between CT lung density and the number of lesions of macroscopic emphysema in that study.

Previous studies[34-36] including our own[23,28] have shown that visual interrogation of the CT scan can locate areas of macroscopic emphysema in postmortem or resected lungs. Improvement in this subjective method for localizing areas of emphysema was developed by Goddard and associates,[37] who scored each lung field from 1–4 depending on the degree of vascular destruction and the number of low density areas. They found only a weak correlation between mean lung density and their visual score for emphysema on the CT scan ($r = 0.34$, $p < 0.005$). However, the diagnosis of emphysema in this study was made on clinical grounds only.

A similar scoring system, employed by Bergin *et al.*[38] who studied patients undergoing lung resection, showed a correlation ($r = 0.88, p < 0.001$) between the visual score for emphysema on the CT scan and macroscopic emphysema, using the picture-grading system of Thurlbeck *et al.*[6] This relation was present despite the fact that the picture-grading score does not measure emphysema on a linear scale. Furthermore, the correlation is heavily weighted by one individual with marked emphysema.[38] Both the DLCO and the FEV_1/FVC were also shown to correlate with the pathologic emphysema score. A relation between CT and pathologic emphysema scores was confirmed in three further studies ($r = 0.82–0.91$)

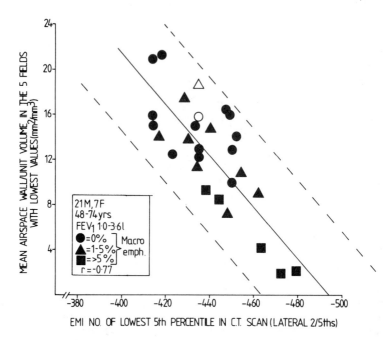

FIGURE 7. Relation between lung CT density (fifth percentile EMI No. of lateral two-fifths of lung field) and the mean alveolar wall/unit lung volume in the lowest five fields that were measured (AWUV/L5F) in 28 subjects. *Circles* = subjects with no macroscopic emphysema. *Triangles* = subjects with 1–5% macroscopic emphysema. *Squares* = subjects with greater than 5% macroscopic emphysema. *Open symbols* indicate nonsmokers. Regression line and 95% confidence limits are shown ($r = 0.84$, $p < 0.001$).

by the same group[39–41] and by other authors.[42] In these studies, CT scanning tended to underestimate the severity of the disease, particularly as most centriacinar lesions smaller than 5 mm were missed.

As a visual assessment of the extent of macroscopic emphysema by CT scanning is insensitive, subjective, and has intra- and interobserver variability,[38–41] Müller and colleagues described a more objective method of highlighting pixels within the lung fields in a predetermined low density range, a facility that is available on most scanners. They chose a density range between −910 and −1000 Hounsfield units, equivalent to −450 to −500 on the EMI scale, based on an earlier study by our group showing that patients with centriacinar emphysema had significantly more pixels within this range.[28] Such a "density mask" program with a predetermined density range is fairly arbitrary, and we were unable to show a linear relation between centriacinar emphysema and the number of pixels within this range, albeit in a small group of patients.[28] However, Müller et al.[43] were able to show a very good correlation between pathologic emphysema scores and both the CT visual assessment and "density mask" score, the latter being less time-consuming and less subjective than the visual scoring technique and able to be done by a technician rather than an experienced radiologist. It is important to stress that the use of this technique does not mean that airspace size is "normal" in areas of lung outside the highlighted range. Furthermore, although this technique may diagnose macroscopic emphysema, the degree of emphysema is still underestimated, because it failed to identify three cases of mild emphysema.[43]

Our study is the only one to quantify microscopic emphysema and correlate it with CT lung density.[23] However, in those patients without macroscopic emphysema, we observed a very wide range of values of AWUV, and the correlations between AWUV and DLCO/VA or CT lung density were independent of the presence of macroscopic emphysema. This suggests that as emphysema develops, enlargement of distal airspaces occurs generally throughout the lungs, leading to a decrease in diffusing capacity. These data also suggest that macroscopic emphysema is an exaggeration of the generalized increase in airspace size due to local factors, not clearly defined, perhaps related to local mechanical factors or local differences in proteolytic imbalance. This may explain why macroscopic centrilobular emphysema has an upper lobe predominance, whereas preliminary work from our group in 20 resected lungs confirms that measurements of AWUV do not show a regional variation throughout the lung.

Macroscopic and microscopic emphysema have been shown to correlate in several studies and the presence of microscopic emphysema also correlates with measurements of lung function.[22,44–47] Inasmuch as both macroscopic and microscopic emphysema coexist, it is difficult to separate the effect of either process on lung function. However, in a large series of surgically resected lungs, the relation between macroscopic emphysema and airflow limitation was poor and those between the DLCO/VA and the elastic recoil and macroscopic emphysema are significant, although relatively weak.[8] Clinicopathologic studies in resected lungs have the major disadvantage, as stated earlier, that the patients are highly selected and usually have relatively mild emphysema.[8]

Our demonstration that distal airspace size or AWUV correlates with CT lung density gave us the opportunity to measure distal airspace size, as assessed by CT

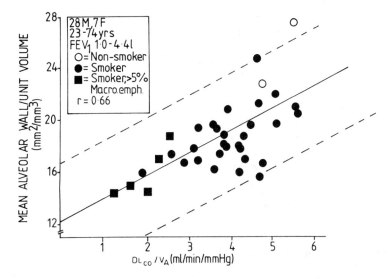

FIGURE 8. Relation between gas transfer (DLCO/VA) and alveolar wall/unit lung volume (AWUV/L5F) in 34 subjects. *Symbols* as in FIGURE 7 ($r = 0.84$, $p < 0.001$).

scanning, in a large group of normal subjects and patients with COPD, with a wide range of functional impairment. In 80 subjects (aged 23–82 years; FEV_1 8–116% predicted; DLCO/VA 15–139% predicted), correlations between respiratory function and CT lung density were most significant when density was measured as the lowest fifth percentile of the frequency histogram, rather than the mean CT density.[48] The lowest fifth percentile of the CT lung density did not correlate significantly with height, but correlated weakly with age ($r = 0.27$, $p < 0.05$) and with total lung capacity ($r = 0.47$, $p < 0.001$). CT lung density also correlated strongly with measurements of airflow limitation (FEV_1/FVC), the diffusing capacity (DLCO/VA),[48] and the shape parameter "K" of the pressure volume relationship.[49] These data support the concept that CT lung density reflects distal airspace enlargement in this group of patients with COPD, because the correlations with function are similar to those in our series of surgical patients in whom emphysema was confirmed pathologically.[23]

Lung CT density, however, is influenced by both macroscopic and microscopic emphysema. To determine whether macroscopic or microscopic emphysema is more important functionally, we recently undertook a study of 23 patients with bullous emphysema.[50] Areas of bullous lung were identified on the CT scan by highlighting confluent areas of low density whose pixels had EMI numbers

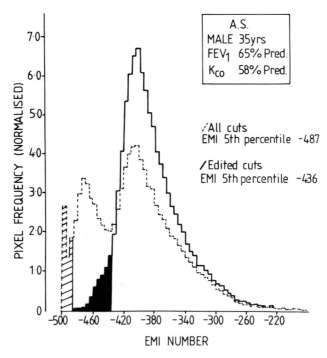

FIGURE 9. CT histograms of male aged 35 years; FEV_1 65% predicted; DLCO/VA 58% predicted. *Interrupted line* = whole lung histogram. *Solid line* = nonbullous lung only. *Shaded area* represents lowest 5% of the distribution. Note the bimodal distribution of the whole lung histogram due to the inclusion of the bulla and the absence of parenchymal emphysema.

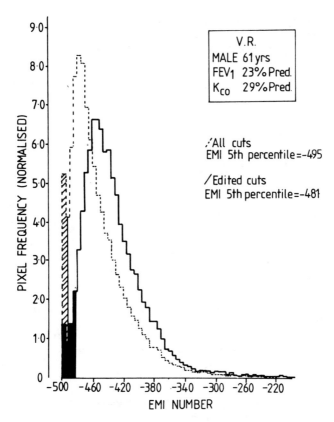

FIGURE 10. CT histograms of male aged 61 years; FEV_1 23% predicted; DLCO/VA 29% predicted. Legend as in FIGURE 9. Note the similarity of the whole lung histogram and the nonbullous lung histogram due to the presence of severe parenchymal emphysema.

smaller than -450. The area of each lung slice involved in the bullous process could then be calculated, and the mean and lowest fifth percentile from the CT lung density histogram from the nonbullous lung could also be measured. Two variations of the density histogram were observed. The first was a bimodal distribution in which a density histogram of the nonbullous lung remained relatively "normal" (FIG. 9). The other pattern resulted from a shift in CT density of the nonbullous lung towards low density values, indicating the presence of generalized emphysema (FIG. 10). Interestingly in this study significant correlations were noted between measurements of both airflow limitation and the diffusing capacity and CT lung density, but not between measurements of lung function and the percentage of the lung involved by bullous disease.[50] These data lend further support to our hypothesis that the amount of microscopic emphysema has an important effect on lung function. In addition, the similarity of the correlations between CT lung density and lung function in the nonbullous areas of the lungs in patients with bullous disease and those inpatients with generalized COPD[48] suggests that in most cases the bullae may be an exaggeration of a generalized emphysematous process. Measurements of CT lung density can therefore not only determine the extent of bullous disease, but also assess the severity of the emphysema in the intervening lung and may thus help in planning surgical intervention in such patients.

OTHER FACTORS AFFECTING CT LUNG DENSITY

We have considered two factors that may influence CT lung density. The first is local blood volume. Clearly loss of alveolar walls will be associated with a loss of pulmonary capillaries. To assess the contribution of local blood volume to the CT density of the periphery of the lung, we studied the effect of intravenous contrast on CT lung density. Preliminary, as yet unpublished data in normal subjects do not suggest that the injection of contrast has any significant effect on the CT density in the lung periphery in contrast to its effect on the density of large vessels. Thus, the contribution of capillary blood volume to CT lung density is small, as would be expected because the calculated capillary blood volume for the whole lung is only 75 ml.[51] Although a recent study demonstrated a reduction in capillary blood volume in patients with pathologically proven emphysema, this measurement did not prove to be a better discriminator of the pathologic emphysema score than the DLCO/VA.[40]

Secondly, we studied the effect of the hyperinflation of asthma on CT lung density. Preliminary results[52] indicate that in nonsmoking persons with chronic asthma, who had a degree of irreversibility to their airflow limitation, CT lung density in some cases is within the range that we consider to represent macroscopic emphysema. However, in this group, a reduction in lung volume after nebulized bronchodilators was not associated with any significant change in CT lung density. More recent data, in which CT lung density was measured in asthmatic subjects at the end of an exacerbation of their condition and again 6 weeks later when their condition was stable, indicate that an improvement in the degree of hyperinflation is mirrored by an increase in CT lung density.[53] These preliminary data suggest that chronic hyperinflation in persons with severe asthma or the hyperinflation of acute asthma may result in reduced CT lung density, which mimics the pattern of microscopic, but not macroscopic emphysema. However, because we do not have accurate measurements on the distribution or the size of distal airspaces in normal subjects, the distinction between chronic hyperinflation and microscopic emphysema remains unclear.

APPLICATIONS OF MEASURING CT LUNG DENSITY

Diagnosis and Quantification of Emphysema

We believe our data indicate that in chronic stable COPD, CT lung density, as measured by the density number defining the lowest fifth percentile of the density histogram from the lung periphery, reflects the size of distal airspaces and hence could be used to assess emphysema, even at an early stage. By defining a range of CT lung density, macroscopic emphysema and bullous emphysema can also be quantified. Provided acute asthma can be excluded, the technique should be specific for emphysema. However, more data are needed to support this view, and more information is required on the comparison of the frequency distributions of both CT lung density and AWUV to determine the difference between normality, microscopic emphysema, and chronic hyperinflation before microscopic emphysema can be truly quantified.

"Blue Bloaters" and "Pink Puffers"

Traditional teaching suggests that patients with COPD can be divided into two clinical patterns: "blue bloaters," who characteristically have hypoxemia, CO_2 retention, cor pulmonale, and secondary polycythemia, and "pink puffers" with mild hypoxemia, without CO_2 retention, cor pulmonale, or polycythemia. Previously it was considered that pink puffers had emphysema predominantly and blue bloaters largely had bronchitis, although these represented two ends of the spectrum in which most patients had a mixed clinical pattern. We recently used CT scanning, together with measurements of pulmonary hemodynamics, in patients with COPD and failed to demonstrate a significant correlation between pulmonary arterial pressure, hypoxemia, and measurements of CT density, indicating that the presence or absence of emphysema is not related to the clinical pattern of COPD.[54]

Progression of Emphysema

CT scanning clearly has the potential to assess the progression of emphysema. This technique could be a powerful tool with many applications. These include studies of the natural history of emphysema in smokers to determine if the accelerated fall in FEV_1, which occurs in these individuals, is due to worsening emphysema. CT scanning could also be used to identify susceptible smokers at an early stage and to assess the factors related to the pathogenesis of the disease that distinguish such individuals. The potential to study the effects of therapeutic interventions, such as α_1-antitrypsin replacement therapy in deficient patients, is another obvious application of this technique.

To use CT scanning to follow the progression of emphysema, it is essential to ensure that the technique is reproducible. We recently began to study this problem and have shown that CT lung density in normal subjects does not change over a period of 5 years, whereas in 70 patients with COPD studied over an interval of 30 ± 4 months, a decrease in DLCO/VA was paralleled by a decrease in CT lung density, indicating worsening emphysema.[55] Preliminary results in 36 patients with COPD followed up over 18–36 months did not show a difference in CT lung density in survivors or nonsurvivors,[56] although more data are needed to support the hypothesis that the presence of emphysema per se does not affect survival in patients with COPD.

REFERENCES

1. SNIDER, G. L., J. KLEINERMAN, W. M. THURLBECK & Z. H. BENGALI. 1985. The definition of emphysema. Report of a National Heart, Lung and Blood Institute, Division of Lung Diseases Workshop. Am. Rev. Respir. Dis. **132:** 182–185.
2. WEIBEL, E. R. 1963. Morphometry of the Human Lungs.:60. Springer Verlag. Berlin.
3. DUNNILL, M. S. 1962. Postnatal growth of the lung. Thorax **17:** 329–333.
4. THURLBECK, W. M. 1983. Overview of the pathology of pulmonary emphysema in the human. Clin. Chest Med. **4:** 337–350.

5. NAGAI, A., W. W. WEST & W. M. THURLBECK. 1985. The National Institutes of Health Intermittent Positive-Pressure Breathing Trial: Pathology studies. II. Correlation between morphologic findings, clinical findings and evidence of airflow obstruction. Am. Rev. Respir. Dis. 132: 946–953.

6. THURLBECK, W. M., J. A. HENDERSON, R. G. FRASER & D. V. BATES. 1970. Chronic obstructive lung disease. A comparison between clinical, roentgenologic, function and morphological criteria in chronic bronchitis, asthma and bronchiectasis. Medicine 49: 81–145.

7. THURLBECK, W. M. & G. SIMON. 1978. Radiographic appearance of the chest in emphysema. Am. J. Roentgenol. 130: 429–440.

8. PARÉ, P. D., L. A. BROOKS, J. BATES, L. M. LAWSON, K. G. NELEMS, J. L. WRIGHT & J. C. HOGG. 1982. Exponential analysis of the lung pressure-volume curve as a predictor of pulmonary emphysema. Am. Rev. Respir. Dis. 126: 54–61.

9. MORRISON, N. J., R. J. ABBOUD, F. RAMADAN, R. R. MILLER, N. N. GIBSON, K. G. EVANS, W. NELEMS & N. L. MULLER. 1989. Comparison of single breath carbon monoxide diffusing capacity and pressure/volume curves in detecting emphysema. Am. Rev. Respir. Dis. 139: 1179–1187.

10. BEREND, N., A. J. WOOLCOCK & G. E. MARLIN. 1979. Correlation between the function and structure of the lung in smokers. Am. Rev. Respir. Dis. 119: 695–705.

11. BURROWS, B., C. M. FLETCHER, B. E. HEARD, N. L. JONES & J. S. WOOTLIFF. 1966. Emphysematous and bronchial types of chronic airways obstruction. A clinicopathological study of patients in London and Chicago. Lancet i: 830–835.

12. SYMONDS, G., A. D. RENZETTI & M. M. MITCHELL. 1974. The diffusing capacity in pulmonary emphysema. Am. Rev. Respir. Dis. 109: 391–394.

13. JENKINS, D. E., S. D. GREENBERG, S. F. BOUSHY, H. I. SCHWEPPE & R. M. O'NEAL. 1965. Correlation of morphologic emphysema with pulmonary function parameters. Trans. Assoc. Physician 72: 218–230.

14. SWEET, H. C., J. P. WYATT & P. W. KINSELLA. 1960. Correlation of lung macrosections with pulmonary function in emphysema. Am. J. Med. 24: 277–281.

15. PARK, S. S., M. JANIS, C. S. SHIM & M. H. WILLIAMS. 1970. Relationship of bronchitis and emphysema to altered pulmonary function. Am. Rev. Respir. Dis. 102: 927–936.

16. WATANABE, S., M. MITCHELL & A. D. RENZETTI. 1965. Correlation of structure and function in chronic pulmonary emphysema. Am. Rev. Respir. Dis. 92: 221–227.

17. GREAVES, I. A. & H. J. H. COLEBATCH. 1980. Elastic behaviour and structure of normal and emphysematous lungs post mortem. Am. Rev. Respir. Dis. 121: 127–136.

18. WEIBEL, E. R. 1973. Morphological basis of the alveolar-capillary gas exchange. J. Clin. Invest. 53: 419–495.

19. THURLBECK, W. M., M. S. DUNNILL, W. HARTUNG, B. E. HEARD, A. G. HEPPELSTON & R. C. RYDER. 1970. A comparison of three methods of measuring emphysema. Hum. Pathol. 1: 215–226.

20. LAMB, D. 1990. Chronic obstructive airways disease. In Respiratory Medicine. R. A. L. Brewis, G. J. Gibson & D. M. Geddes, eds. Bailliere Tindall.

21. THURLBECK, W. M. 1967. The internal surface area of nonemphysematous lungs. Am. Rev. Respir. Dis. 95: 765–773.

22. WEST, W. W., A. NAGAI, J. E. HODGKIN & W. M. THURLBECK. 1987. The National Institutes of Health Intermittent Positive Pressure Breathing Trial—Pathology Studies III. The diagnosis of emphysema. Am. Rev. Respir. Dis. 138: 123–129.

23. GOULD, G. A., W. MACNEE, A. MCLEAN, P. M. WARREN, A. REDPATH, J. J. K. BEST, D. LAMB & D. C. FLENLEY. 1988. CT measurements of lung density in life can quantitate distal airspace enlargement—an essential defining feature of human emphysema. Am. Rev. Respir. Dis. 137: 380–392.

24. WILLIAMS, M. A. 1975. Quantitative methods in biology. In Practical Methods in Electron Microscopy, Vol 6, part 2. A. M. Glavert, ed. North Holland Publishing Co. Amsterdam.

25. AHERNE, W. A. & M. S. DUNNILL. 1982. Morphometry. Edward Arnold. Bath.

26. HOUNSFIELD, G. N. 1973. Computerized transverse axial scanning (tomography). Br. J. Radiol. **46:** 1016–1019.
27. PHELPS, M. E., M. H. GADO & E. J. HOFFMAN. 1975. Correlation of effective atomic number and electron density with attenuation coefficients measured with polychromatic x-rays. Radiology **117:** 585–588.
28. HAYHURST, M. D., W. MACNEE, D. C. FLENLEY, D. WRIGHT, A. MACLEAN, D. LAMB, A. J. A. WIGHTMAN & J. J. K. BEST. 1984. Diagnosis of pulmonary emphysema by computerized tomography. Lancet **ii:** 320–322.
29. ROBINSON, P. J. & L. KREEL. 1979. Pulmonary tissue attenuation with computed tomography: Comparison of inspiration and expiration scans. J. Comput. Assist. Tomogr. **3:** 740–748.
30. FROMSON, B. H. & D. M. DENISON. 1988. Quantitative features in the computed tomography of healthy lungs. Thorax **43:** 120–126.
31. ROSENBLUM, L. J., R. A. MAUCERI, F. D. THOMAS, D. A. BASSANO, B. N. RAASSCH, C. C. CHAMBERLAIN & E. R. HEITZMAN. 1980. Density patterns in the normal lung as determined by computed tomography. Radiology **137:** 409–416.
32. HEDLUND, L. W. & C. E. PUTMAN. 1981. Analysis of lung density by computed tomography. *In* Pulmonary Diagnosis.: C. E. Putman, ed.: 107–123. Appleton-Century-Crofts. New York.
33. LEVI, C., J. E. GRAY, E. D. McCULLOUGH & R. R. HATTERY. 1982. The unreliability of CT numbers as absolute values. Am. J. Roentgenol. **139:** 443–447.
34. CODDINGTON, R., S. L. MERA, P. R. GODDARD & J. W. B. BRADFIELD. 1982. Pathological evaluation of computed tomography images of lungs. J. Clin. Pathol. **35:** 536–540.
35. HRUBAN, R. H., M. A. MEZIANE, E. A. ZERHOUNI, N. F. KHOURI, E. K. FISHMAN, P. S. WHEELER, J. S. DUMLER & G. M. HUTCHINS. 1987. High resolution computed tomography of inflation fixed lungs. Pathologic-radiologic correlation of centrilobular emphysema. Am. Rev. Respir. Dis. **136:** 935–940.
36. FOSTER, W. L., P. C. PRATT, V. I. ROGGLI, J. D. GODWIN, R. A. HALVORSEN & C. E. PUTMAN. 1986. Centrilobular emphysema: CT-pathological correlation. Radiology **159:** 27–32.
37. GODDARD, P. R., E. M. NICHOLSON, G. LASZLO & I. WATT. 1982. Computed tomography in pulmonary emphysema. Clin. Radiol. **33:** 379–387.
38. BERGIN, C., N. MÜLLER, D. M. NICHOLS, G. LILLINGTON, J. C. HOGG, B. MULLEN, M. R. GRYMALOSKI, S. OSBORNE & P. D. PARE. 1986. The diagnosis of emphysema. A computed tomographic-pathologic correlation. Am. Rev. Respir. Dis. **133:** 541–546.
39. MILLER, R. R., N. L. MÜLLER, S. VIDAL, N. J. MORRISON & C. A. STAPLES. 1989. Limitations of computed tomography in the assessment of emphysema. Am. Rev. Respir. Dis. **139:** 980–983.
40. MORRISON, N. J., R. T. ABBOUD, N. L. MÜLLER, R. B. MILLER, N. N. GIBSON, B. NELEMS & K. G. EVANS. 1990. Pulmonary capillary blood volumes in emphysema. Am. Rev. Respir. Dis. **141:** 53–61.
41. BERGIN, C. J., N. L. MÜLLER & R. R. MILLER. 1986. CT in the qualitative assessment of emphysema. J. Thorac. Imaging **1:** 94–103.
42. MURATA, K., H. ITOH, G. TODO *et al.* 1986. Centrilobular lesions of the lung: Demonstration by high-resolution CT and pathological correlation. Radiology **161:** 641–645.
43. MÜLLER, N. L., C. A. STAPLES, R. R. MILLER & R. T. ABBOUD. 1988. "Density mask." An objective method to quantitate emphysema using computed tomography. Chest **94:** 782–787.
44. THURLBECK, W. M. 1967. Internal surface area and other measurements in emphysema. Thorax **22:** 483–496.
45. NAGAI, W., I. YAMAWAKI, W. M. THURLBECK & T. TAKIZAWA. 1989. Assessment of lung parenchymal destruction by using routine histologic tissue sections. Am. Rev. Respir. Dis. **139:** 313–319.
46. SAETTA, M., R. J. SHINER, G. E. ANGUS, W. D. KIM, N.-S. WANG, M. KING, H.

GHEZZO & M. G. COSIO. 1985. Destructive index: A measurement of lung parenchymal destruction in smokers. Am. Rev. Respir. Dis. **131:** 764–769.

47. SAITO, J., P. CAGLE, N. BEREND & W. M. THURLBECK. 1989. The "destructive index" in non-emphysematous and emphysematous lungs. Morphologic observations and correlation with function. Am. Rev. Respir. Dis. **139:** 308–312.

48. GOULD, G. A., A. T. REDPATH, M. RYAN, P. M. WARREN, J. J. K. BEST, D. C. FLENLEY & W. MACNEE. Lung CT density correlates with measurements of airflow limitation and the diffusing capacity. Eur. Respir. J., in press.

49. GUGGER, M., G. A. GOULD, W. MACNEE, P. K. WRAITH & D. C. FLENLEY. 1988. Correlation between the loss of elastic recoil and the extent of emphysema in man measured in vivo. Eur. Respir. J. **1** (Suppl 2): 216s.

50. GOULD, G. A., W. MACNEE, A. T. REDPATH, M. F. SUDLOW, J. J. K. BEST & D. C. FLENLEY. 1987. Correlation of lung CT density with respiratory function in smokers and patients with bullous emphysema. Thorax **42:** 706.

51. FRANS, A., D. C. STANESCU, C. VERITER, T. CLERBAUX & L. BRASSEUR. 1975. Smoking and the pulmonary diffusing capacity. Scand. J. Respir. Dis. **56:** 165–183.

52. BIERNACKI, W., W. MACNEE, M. RYAN, J. J. K. BEST & D. C. FLENLEY. 1989. Does emphysema occur in asthma? Thorax **44:** 877–878p.

53. BIERNACKI, W., M. RYAN, J. J. K. BEST & W. MACNEE. 1990. Computerized tomography in asthmatic patients. Eur. Respir. J. **3**(suppl 10): 626.

54. BIERNACKI, W., G. A. GOULD, K. F. WHYTE & D. C. FLENLEY. 1989. Pulmonary hemodynamics, gas exchange and the severity of emphysema as assessed by quantitative CT scan in chronic bronchitis and emphysema. Am. Rev. Respir. Dis. **139:** 1509–1515.

55. BIERNACKI, W., M. RYAN, W. MACNEE & D. C. FLENLEY. 1989. Can quantitative CT scan detect the progression of emphysema? Am. Rev. Respir. Dis. **139:** A120.

56. BIERNACKI, W., M. RYAN, W. MACNEE & D. C. FLENLEY. 1989. Does the severity of emphysema as assessed by CT scan predict survival in patients with chronic bronchitis and emphysema? Thorax **44:** 342p.

The Feasibility of an Outcome Trial in the Preventive Therapy of Emphysema

MITCHELL GLASS

ICI Americas
Wilmington, Delaware 19897

"Therefore, if the scientific base is sufficient and many practical problems can be solved, it is possible that within 12 years we may have specific agents that can stop the progression of pulmonary emphysema."[1]

At present, the single most cogent hypothesis for the pathogenesis of emphysema is that a functional excess of human neutrophil elastase (HNE) occurs in susceptible cigarette smokers, resulting in a slowly progressive destruction of terminal respiratory bronchioles and the anatomic units they serve.[2]

This paper briefly reviews the problems that surround a definitive clinical and statistical test of this hypothesis, to be accomplished by performing an outcome study of preventive therapy of emphysema. It is important to note that an outcome trial is possible and was previously defined.[3] An intervention trial is feasible and ongoing (Lung Health Study)[4] to test treatment of cigarette smokers with standard therapy versus standard therapy plus intensive counseling with or without the addition of ipratropium bromide. This paper, therefore, addresses feasibility issues surrounding the planning and execution of such a trial using an HNE inhibitor. Feasibility is defined as practicability[5] as distinct from possibility. Any uncertainty, whether expressed in alternate hypotheses, patient selection criteria, or therapeutic options, correlates inversely with feasibility.

The remainder of this paper will be based on the elements of a standard clinical trial designed to show the efficacy of a pharmaceutical compound (TABLE 1). In some areas of the trial, alternative approaches may render the trial more or less feasible. This review will not propose a specific trial, but will outline issues that affect the likelihood of a generally accepted outcome study.

OBJECTIVES

The trial to be described must have two primary objectives: testing the hypothesis and supporting the approval of a medicine for this indication.

The trial must test the hypothesis that intrapulmonary extracellular HNE contributes significantly to the rate of decline of 1-second forced expiratory volume (FEV_1) in patients who continue to smoke cigarettes. The Lung Health Study may make this somewhat more feasible if it demonstrates that patients may not need to be counseled to quit smoking throughout the course of the trial.

For any trial to be of more than academic interest, it must also contribute to the approval of a medicine with which to treat patients who continue to smoke cigarettes and who have abnormally rapidly declining lung function. Otherwise, acceptance of the hypothesis would have no greater impact than to influence health care providers to convince their patients to quit cigarette smoking, an imperative that has already been defined as being among the highest physician priorities in the US by the Surgeon General.[6]

195

TABLE 1. Protocol Outline

Objectives
Background and Rationale
Study Design
Patient Definition
Patient Selection
Inclusion Criteria
Exclusion Criteria
Prohibitions and Restrictions
Treatment Definition
Concurrent Treatment
Clinical and Laboratory Measures
Screening
Evaluations during Study
Patient Dropouts
Data Analysis
Payment of Grant

A successful study to test the elastase : antielastase hypothesis must minimize the likelihood of Type I or Type II errors. For undertaking a study of the magnitude that will be described subsequently, prospective investigators should be willing to reject the elastase : antielastase hypothesis on the basis of a statistically or clinically negative outcome, rather than seek design flaws to be corrected subsequently. The likelihood of an equivocal trial must be minimized by the investigators *a priori*. If no such study is presently able to be designed, then the outcome trial is not yet feasible.

The purpose of a study should also be to support a therapeutic intervention for millions of cigarette smokers at risk of developing emphysema, which will manifest clinically as chronic obstructive pulmonary disease (COPD). This aim has ramifications for the size and scope of a study. A change in a physiologic or pathologic outcome will be required, because regulatory acceptance tends to be based on conservative endpoints. The following constitutes a list of feasibility issues for meeting this specific objective:

selection of a physiologic endpoint (FEV_1 or mortality)
simple, conservative analysis
safety data collection
avoidance of selection bias
selection of a clinically relevant difference
demonstration of dose-response relationships for efficacy
demonstration of dose-response relationships for adverse events
assessment of quality of life
assessment of health economics impact

BACKGROUND AND HYPOTHESIS

This outcome trial will indirectly test the elastase hypothesis[7] by using the rate of decline of FEV_1 as a physiologic surrogate endpoint for the pathologic emphysema. However, the straightforwardness of inhibiting HNE and thereby affecting the progression of emphysema may be challenged from two separate directions which converge in an intervention trial of HNE inhibition.

The key evidence for indicting excess elastase has long been the demonstration of pulmonary emphysema in individuals with alpha$_1$-antitrypsin deficiency (PiZZ genotype).[8] Investigators have cited varying rates at which such patients develop emphysema.[9,10] However, recent work by Silverman et al.[11] suggests that such a percentage may be grossly overestimated, with the rate 50% in life-long PiZ nonsmokers.

Alternative hypotheses have also been proposed for the pathogenesis of emphysema (TABLE 2). Other cellular products, from polymorphonuclear[12] and mononuclear cells[13] as well as epithelial cells,[14] have been implicated in extracellular proteolysis. Products directly derived from cigarette smoke, including radicals and prooxidants, may directly or in combination cause enhanced proteolysis.[15] Nitrogen dioxide has been shown to lead to the enhancement of a polymorphonuclear leukocyte chemotaxin harvested by bronchoalveolar lavage.[16] Long-lived radicals can be derived directly from cigarette smoke.[15] The validity of each alternative hypothesis contributes inversely to the likelihood of a specific HNE inhibition trial being successful.

A separate problem lies with newer hypothesized roles for HNE. First, HNE acting on cells[17] or glands[18] was shown to be the most potent secretagogue ever measured. HNE is capable of producing greater mucus secretion from glands than can mediators such as histamine or leukotrienes.[18] Second, HNE in patients with cystic fibrosis was shown to be responsible for cleavage of the CR-1 complement receptor critical for bacterial phagocytosis and killing.[19] Recurrent infections have been hypothesized to contribute to the pathogenesis of COPD.[20] No analysis of the CR-1 receptor has been described in patients with COPD. Third, HNE may play a role in wound healing or tissue remodeling.[21]

These issues may converge if a specific HNE inhibitor misses some target relevant to COPD, while simultaneously causing an effect on some nonproteolytic physiologic function. The following scenario is offered merely as an example of the potential scientific developments that may play a role in COPD and confound the analysis of a trial planned and executed over 3–6 years. An independent effect on mucus hypersecretion may enhance the effect of inhibitors on symptoms or progression of disease in patients with bronchitis. The restoration of CR-1 receptors could decrease the susceptibility of patients to recurrent infections. Wound

TABLE 2. Factors Hypothesized to Contribute to the Development of Emphysema

Alveolar Macrophage Derived	Cigarette Derived
Metalloproteinases	Long-Lived Radicals
Interstitial Collagenase	Nitrogen Oxides
Gelatinase/Type IV Collagenase	Depression of Elastin
Gelatinase/Type V Collagenase	Synthesis
Induced Metalloproteinase (Unknown)	Recurrent Infections
Stromelysin	Pi Status
Cysteine Protease	Bronchial Epithelial
Cathepsin L	Metalloprotease
Plasminogen Activator	Fibroblast Elastase
PMN Derived Factors	
Elastase (HNE)	
Cathepsin G	
Oxygen Free Radicals	
Proteinase III	

healing could be delayed or altered in patients receiving an HNE inhibitor systemically. A requirement to monitor patients for any such effect would decrease trial
feasibility. The reader is referred to later symposium papers for a more complete
analysis of other HNE physiologic roles.

STUDY DESIGN

The double-blind, randomized, placebo-controlled parallel group study remains the standard design for a large multicenter trial to evaluate a novel therapy
for efficacy versus a disease target.[22]

The issues surrounding the study design may be divided into three categories:
parameters (endpoints); study structure; and size and scope of the trial. The key
parameter of efficacy is used to determine group size, with adequate margin to
allow for dropouts including deaths. The study structure must be maintained as
simple as possible. Therefore, the easiest trial to perform is envisaged with a
single screening visit, no run-in period, visits as infrequently as possible, and
minimum duration per patient. FEV_1 has been used previously to calculate study
size for proposed trials and wound appear to continue to be the endpoint with
maximum scientific and regulatory acceptability. This trial could have a single
screening visit, no run-in, visits every 6 months, and last 3 years per patient based
on a simple analysis of FEV_1 decline across patients. The size and scope of the
trial will entail minimizing the number of sites, the monitoring costs, and the
number of doses tested versus placebo.

PATIENT DEFINITION

It is expected that the trial will enroll and dose cigarette smokers of both sexes
who continue to smoke daily. Patients will have demonstrated COPD by historical, physical, and diagnostic laboratory criteria. Patient definition is not meant to
provide specific inclusion or exclusion criteria, which follow, but to define a
population that is appropriate to study as a scientific and future therapeutic target.
For the trial, a standardized definition of COPD, cigarette smoking (minimum
daily and yearly consumption), level of impairment, coexistence of mucus hypersecretion, age, gender, O_2 requirement (degree of hypoxemia), Pi studies, and
reversibility are issues within patient definition. None of these issues is insur-
mountable in defining a patient population to study. However, any need to stratify
patients *a priori* or *post hoc* according to any of these parameters adds to the
complexity of executing or analyzing the trial and therefore decreases the overall
feasibility of performing a clinical trial with a generally accepted outcome.

PATIENT SELECTION

Inclusion criteria affect the feasibility of any study directly by limiting participation to a subset of the future target population. Inclusion criteria have a direct
effect on the recruitment and dosing phases of any trial. The following criteria for

a trial with COPD are presented as the basis for future discussion:

Male or female patient with a diagnosis of chronic bronchitis or emphysema
Active cigarette smoker who has failed at least one intensive counseling effort
$35 \leq age \leq 65$
$40\% < FEV_1 < 65\%$ predicted

Broad inclusion criteria risk damaging the opportunity to show therapeutic differences by introducing patients whose condition is too mild to show protection and too severe to show progression. Narrow inclusion criteria, such as men only or prior successful participation in an arm of the Lung Health Study, cost money to document, adversely affect recruitment, and may introduce selection bias.

Exclusion criteria are frequently more contentious. A recommended minimum could exclude PiMZ or PiZZ genotypes and exsmokers or nonsmokers. In a drug registration trial, exclusion of anyone who has participated in another investigational drug study within 4 weeks would be required, and measurement of a standard battery of laboratory tests to exclude patients with significant laboratory abnormalities other than those due to COPD or with any concurrent illness requiring therapy would be likely.

Although inclusion and exclusion criteria impinge on the feasibility of every protocol, there is nothing that specifically renders unfeasible an HNE inhibition trial in patients at risk for emphysema.

TREATMENT DEFINITION

Treatment definition and drug administration may be divided into four sets of feasibility issues: compliance, toxicity, dosing regimen, and number of treatment groups. (See section on Data Analysis.)

The inhibition of HNE may have beneficial or adverse side effects that impinge on the compliance and side effect rates of subsets of trial patients.[23] Inhibition of the mucous secretagog effect may lead to amelioration of the troubling symptoms of chronic bronchitis and lead to increased compliance among bronchitic patients relative to those with "dry" emphysema. CR-1 receptor restoration in patients with recurrent infections may result in a winter marked by fewer or milder bacterial bronchitides which may positively reinforce compliance with drug regimens. Extracellular HNE inhibition may diminish the acute phase response or retard wound healing, either of which may lead to decreased compliance from patients receiving systemically active drugs.

A second major feasibility issue lies in drug delivery and regimen. Aerosol medications carry the theoretic advantage of site specificity and fewer systemic side effects. Aerosol therapy targeted to the distal airways and acini would best be delivered by nebulization; the MDI or dry powder inhaler is likely to miss the target within the lungs.[24] With nebulized therapy, larger doses mean more frequent dosing or longer treatments. Compliance, especially for a drug that provides no symptomatic relief and is given by nebulizer over the course of minutes, is likely to be lower than that for a tablet taken once or twice daily.

Toxicity may cause feasibility problems by requiring trials to be curtailed, by requiring extensive safety data collection, and by affecting compliance or the numbers of dropouts. The issues of toxicity, tolerance, and side effects will vary depending on the need to test a novel therapy versus the introduction of a new claim for an older medicine. Older therapies advanced as candidates have, in some cases, been used to treat more than 1,000,000 patients (colchicine[25]), but they lack specificity and potency for testing this hypothesis. These medicines have the advantages of lessening the need for dose ranging, frequent safety measurements, and side effect profiling. In large trials of well-tolerated medicines such as ACE inhibitors, the incidence of adverse experiences (cough) exceeds 15%.[26] A well-described, drug-specific effect such as taste, diarrhea, photosensitivity, or change in urine color may break the blind in trials.

Novel therapies, including all specific HNE inhibitors, have yet to be exposed to 1,000 patients. These therapies will have unpredictable toxicities and safety and side effect profiles that may alter the need to collect data during the trial. The effect of an exogenous natural or synthetic HNE inhibitor on the regulation of $alpha_1$-antitrypsin, $alpha_2$-macroglobulin, tissue metalloprotease inhibitor, or secretory leukocyte protease inhibitor is unknown, but it may become known for animals or man during the course of this trial. The administration of drug by mouth or aerosol can be predicted to have an effect on the safety profile. Aerosols for pulmonary indications are generally used to avoid side effects.[27] Medications[28] that are given at $1/10$ dose of systemic medication include albuterol, corticosteroids, anticholinergics (ipratropium), and antiinfectives (pentamidine and tobramycin).

Dose ranging is generally accomplished in shorter studies of fewer patients using endpoints that may be surrogates for the physiologic endpoint. These Phase II trials require a measurement that allows fewer numbers of patients or shorter trials. The key feasibility issue with respect to dosing pertains to novel medicines for which dose ranging will be required.

PiZ patients with progressive deterioration of lung function had been targeted as the Phase II population for the dose-ranging aspects of a preventive therapy. PiZ emphysema offered the theoretic advantage of a better defined population, less variability in rate of FEV_1 decline, and an orphan indication.[1,29,30] Even with these advantages, a trial of 2 years per patient was envisaged.[30] Presently, the targeting of this population for Phase II work has run into three major feasibility issues. First, Silverman, Campbell et al.[11] challenged whether this population is as homogeneous as was believed when this approach was suggested. Second, to the extent that agents other than HNE play a role in the pathogenesis of emphysema, this population may not be predictive. Third, the approval of Prolastin® on biochemical evidence alone[28,31] has rendered the use of this population less feasible. With an approved alternative medication, participation in a 2-year double-blind, randomized trial with a 25–50% likelihood of receiving placebo was diminished. Selection bias would be a problem in restricting trials to patients who would not or could not be treated with Prolastin®.

Two other approaches continue to be followed. The first is to seek other HNE-induced diseases that run their course more acutely but are relevant to cigarette smoke-induced COPD. These diseases could be used as dose-ranging alternatives. Alternatively, biochemical markers that reflect excess elastase and its inhibition could be used to choose doses for subsequent studies whether or not they are accepted as true Phase II efficacy endpoints. Putative markers include elastase-generated products of fibrinogenolysis,[32] elastin degradation,[33] fibronectin,[34] laminin,[35] and collagen degradation.[36] However, none has been reproducibly developed and validated in blinded samples across patient populations or been shown to reflect therapy. This creates a major feasibility issue in terms of dose ranging for shorter periods or in smaller populations.

The number of doses also decreases trial feasibility. The attractions of testing only the largest dose versus placebo includes lower cost per trial, fewer patients, and less likelihood of a center-specific effect. The disadvantages include increased likelihood of side effects or toxicity problems and compliance for nebulized compounds (more frequent dosing or longer duration) and failure to show a dose-response effect. The latter is among the most commonly faulted aspect of an NDA (new drug application) submitted to the FDA for approval of a new therapy or indication.[37] An alternative approach is to start all patients on one dose with subsequent decreasing of the dose in those with side effects and increasing of the dose in those without obvious (biochemical) efficacy. This approach has the advantage of keeping a greater percentage of patients in the trial for safety evaluation and potentially showing a dose response relationship. However, sufficient numbers of patients must complete the trial in each regimen to allow statistical comparison for this approach to increase feasibility. From the standpoint of new chronic therapy, at least two doses of compound must be compared to placebo.

CONCOMITANT THERAPY

Therapy must be standardized to the greatest possible extent *a priori* if the patient groups are to be comparable without multiple *post hoc* stratifications. Certain therapies (O_2, steroids) may be eliminated by investigator consensus or regulators. Corticosteroids have been described to affect leukocyte functions,[38] which may necessitate their exclusion as concomitant therapy. This may become a major issue if an unexpected beneficial effect of steroid treatment of COPD is shown during the dosing phase of the trial.

It is also expected that the HNE inhibitor will be added to baseline therapy. Baseline therapy with bronchodilators, mucolytics, antibiotics, or theophylline therapy must be standardized (and minimized) to increase the feasibility of an outcome trial. The selection of an anticholinergic rather than β-agonist for bronchodilation[39] may require certain patients to switch therapy without benefit of a run-in period (or before informed consent).

CLINICAL AND LABORATORY MEASURES

The feasibility of any outcome trial will rely heavily on the medical and regulatory requirement to collect safety data before, during, and after the dosing period. If the trial to test the hypothesis uses an investigational and novel drug, then regular safety measurements, including blood tests, electrocardiograms, pulmonary function tests, and symptoms, will need to be collected (TABLE 3). This may be somewhat lessened by using a previously approved medicine. Cost and standardization of bronchodilators for spirometry are the two feasibility issues. The Lung Health Study has rendered this element of the study more feasible by showing that selected sites, with dedicated personnel and equipment, can enroll large numbers of appropriate patients.

PATIENT DROPOUTS

Sufficient numbers of patients should be enrolled and dosed across groups to allow for statistical and clinical evaluations of treatment effect on slope of decline of FEV_1. Enrollment targets need to be adjusted for likely numbers of dropouts. Inadequate study size to support claims of efficacy was the single most commonly cited failing in rejected new drug applications in a recent 5-year review of the FDA.[37] Patient dropouts will be affected by four factors which can be predicted to some extent. These factors must be assessed *a priori* for effect on patient selection, recruitment, safety evaluation, and analysis. These feasibility issues are deaths, compliance, protocol deviations, and adverse effects.

Patients who die or are dropped from the trial may be analyzable for efficacy if they have been dosed for a certain minimum period. An analysis of survival rates may be feasible if patient groups are comparable *post hoc* with respect to age, FEV_1, concurrent illness, and causes of death. In patients under 65 years old, with FEV_1 greater than 40% predicted, mortality was 15% in 3 years.[40] This approached 40% for older, sicker patients.[40,41]

Lack of compliance was measured in the IPPB trial; 113 of 985 (11.4%) pa-

TABLE 3. Clinical and Laboratory Measures

Screening
History, physical examinations, vital signs
Laboratory tests (safety): will include hematology, chemistry, liver function tests, uric acid, urinalysis
Electrocardiogram
Pulmonary function tests after bronchodilator
Subjective symptomatology
Chest X-ray
Pi genotype
Reversibility of airflow obstruction
Quantified sputum production
Evaluations during study at 6-month intervals
Laboratory tests (safety)
Physical examinations, vital signs
Pulmonary function tests after bronchodilator
Subjective symptomatology
Electrocardiograms (yearly)
Urinary cotinine

tients were noncompliant.[41] As has been described, the ease of drug administration and any symptomatic benefit to be derived from therapy will render patients more compliant. The protocol deviations most likely to occur are patients quitting smoking or failure to meet criteria (on evaluation). 0.75% of smokers will quit each year.[42] Deviations will be minimized by highly standardized (prescriptive) patient selection. (See section on Patient Selection.) These numbers may well be additive to noncompliance and dropout figures, suggesting that an overall rate of greater than 25% of patients will fail to complete the trial for any of these three reasons.

The fourth feasibility issue, adverse effects, will be drug class and dose specific. Medicines such as colchicine, the choice for an earlier trial, may produce gastrointestinal side effects (diarrhea) in an unacceptable percentage of patients receiving a drug for an asymptomatic chronic disorder.

DATA ANALYSIS

Perhaps no area of the trial presents more critical feasibility issues than does that of data analysis. The trial needs to reflect medical and regulatory acceptance of the selected analytic methods and key parameters if it is to result in acceptance or rejection of the hypothesis and claim of efficacy.

In the 1978 NHLBI study group, Dr. Margaret Wu presented an ordinary least-squares analysis based on a 50% regression toward the mean of the slope of decline of FEV_1, which showed that group sizes of approximately 1,000 analyzable patients per group per intervention would be required for a trial in preventing COPD.[30] These numbers were subsequently confirmed by an independent analysis at ICI. There are four advantages of an ordinary least-squares analysis. First, the analysis has been well described. Second, the basis for the calculation is well regarded as being statistically robust and medically significant. Third, this analysis is very conservative. Fourth, because a cross-sectional analysis is so conservative, it is widely accepted and therefore resistant to changes in personnel. This analysis, if being used in conjunction with a drug whose dose : response and therapeutic : toxic ratios are well known, could allow for the fewest (≤ 2) interim visits per patient.

Its two primary disadvantages have also been described. Given 25% of patients who would not complete a 3-year dosing regimen, this trial would require a minimum of 3,750 patients for two doses of drug versus placebo. Second, it may disallow all patients who drop out, because it is based on endpoints after a fixed interval.

Similarly, if an analysis could be based on a surrogate endpoint with less variability than FEV_1, the use of the most conservative analytic technique would be desirable, but it would allow for fewer completed patients. A cross-sectional endpoint analysis has been used to generate patient numbers for the Lung Health Study; this has resulted in 2,000 patients being enrolled per treatment arm.

The longitudinal analysis has been compared to an ordinary least-squares analysis for FEV_1 data.[43–45] Its primary advantages appear to be the ability to utilize data from patients with missing data points or endpoints and its capacity to uncover duration-specific effects. Duration-specific effects, which encompass phenomena with delay in onset relative to treatment, are likely in therapeutic trials with elastase inhibitors in treating the progression of FEV_1 decline. Three potential disadvantages and, therefore, feasibility issues accrue to longitudinal analysis. First, it is still in development.[46] Second, it will require regulators, clinicians, statisticians, and sponsors, with varying degrees of statistical sophistication to accept it. Third, because it is less conservative, longitudinal analysis may be more susceptible to changes in personnel within the scientific or regulatory community responsible for the outcome trial. Given that a negative outcome should result among participants in the rejection of the protease : antiprotease hypothesis, a conservative statistical approach may be necessary.

Post hoc or interim analyses of the Lung Health Study may generate data sufficient to suggest that a longitudinal analysis may be equally useful in an FEV_1 outcome trial. If so, it remains to be seen whether a longitudinal analysis, based on trend per patient, may increase the feasibility of an outcome trial by requiring fewer patients and shorter therapeutic courses or including more analyzable drop-outs than does the ordinary least-squares approach.

The other alternative for data analysis is the acceptance of an endpoint other than FEV_1. To this end, nearly 15 years of research has generated a series of putative biochemical markers of *in vivo* elastase excess. These markers include previously cited proteolytic products of elastin, collagen, laminin, fibronectin, and fibrinogen. These products are more or less specific degradation products caused by human neutrophil elastase. The theoretic advantage of these assays lies in three characteristics. First, elastase inhibitory capacity (EIC) was sufficient to register Prolastin® by showing restoration of a chemical imbalance. Second, such assays, if developed and validated by analytic chemical techniques, may have less variability than does the FEV_1 rate of decline. Third, these assays may reflect drug activity more directly. As such, they may be useful in dose ranging or as endpoints to demonstrate chemical effect with or without physiologic effect. Complete disappearance of chemical evidence of elastase excess without an effect on FEV_1 slope could be regarded as potent evidence against the role of HNE in the pathogenesis of COPD. Finally, if accepted as a surrogate endpoint, a biochemical marker could allow a trial in fewer patients or for a shorter duration of therapy.

Two feasibility issues remain at the core of the biochemical marker dilemma. These stem from the central disappointment that no biochemical marker has been shown to be reproducibly stable, robust, and reflective of the state of COPD across a group of patients. First, a biochemical marker remains a long way from general scientific (or regulatory) acceptance as a true surrogate to FEV_1 for the purpose of determining group size and power calculations. Second, if the biochemical marker behaves radically different from FEV_1, it is unlikely that its accuracy would remain unchallenged.

Inclusion and exclusion criteria will also impact on the analysis. Accepting FEV_1, less than 35% predicted, or patients with chronic COPD and over 65 years of age may effectively result in a survivor analysis. This effect has previously been described for FEV_1 trials.[41] Similarly, patients with COPD who do not smoke may have a different pathogenesis from those who do. Mucous hypersecretion may respond separately and more or less dramatically than may FEV_1 and, therefore, represent a stratification for analysis. Patients with a major (>15%) reversible component may meet a definition for asthma and have an overlapping pathophysiology with chronic bronchitis or emphysema. The effect of an HNE inhibitor on asthma is unknown, but there are hypothesized roles of proteases in asthma.[47]

At present, a trial with 1,000–1,500 patients enrolled per treatment arm for 3 years, with the FEV_1 rate of decline measured cross-sectionally, would appear to be the basis for power and significance calculations.

PAYMENT

Successful, safe, and specific amelioration of COPD will likely be a very successful commercial venture. More than 40 million Americans[42] and more than 5 million Canadians smoke regularly.[48] The incidence in Japan is increasing; the incidence in China, Eastern Europe, and the Soviet Union is high, but more difficult to assess exactly. It could be argued cogently that COPD is the largest untreated disorder in the Western world. Therefore, economic risk *per se* should not be a barrier to the feasibility of a sponsored trial.

However, the costs are worth examining for two reasons. First, the size and scope of the trial, as reflected in its costs, should engender a risk : benefit analysis of the undertaking. If the trial were to be financed by the NHLBI, then it should be placed in the context of the number of R0-1 grants that could alternatively be funded. Second, for a drug presently marketed, during a trial of 3–5 years' duration, the medicine would become generic for its earliest indication. Therefore, the costs borne by the sponsor would have to be considered in light of a modest revenue gain to the sponsor (and potential windfall to the generic manufacturers). This suggests that the most feasible scenario is one in which a patent-protected compound, in development as a specific and selective HNE inhibitor, will be used to test the protease : antiprotease hypothesis. A conservative estimate of this trial follows, with certain assumptions (TABLE 4):

15% dropout rate (compliance, toxicity, protocol violations, mortality; 3,450 patients)
no professional fees (investigators, interpretations)
assays done in house (drug levels, biochemical markers)
no compensation to patients
no overhead charges
no extra tests
highly successful (>50%) screening program (6,000 screens).

TABLE 4. Payment of Grant: Trial Costs[a]

	Visit Frequency		
Tests	Screen	6 Months	1 Year
Safety labs	—	75	—
ECG	—	—	100
Routine physical	—	60	—
Spirometry (post-bronchodilator)	—	190	—
Chest X-ray	100	—	—
Reversibility	250	—	—
Pi genotype	300	—	—
Total	650	325	100
Cost of Trial:	$11,700,000		

	Possible Extra Charges	
	Screening	Yearly Visits
O_2 saturation	55	—
D_LCO	—	135
Lung volumes	250	—
Chest CT	—	475
Exercise testing	600	—

[a] The costs of tests represent the median cost of five Philadelphia area hospitals.

With these assumptions, and safety and spirometric evaluation every 6 months, this trial will cost $11,700,000 in outside grant service over 5 years in 1990 dollars. The Lung Health Study suggests that some significant savings may be realized by concentrating the program at fewer hospitals, and paying for technicians' salaries plus benefits and dedicated equipment. On the basis of this approach for 10 centers (no overhead or MD fees), the costs would equal $8,400,000.

The feasibility of the trial could be affected by having a nonprofit or minimal profit endpoint.

Two final feasibility issues should be mentioned. The approval of a specific HNE inhibitor for an indication other than progression of COPD may render unfeasible a double-blind trial with FEV_1 endpoint for outcome. Therefore, this approach for any HNE inhibitors must be viewed with caution if the long-term goal is amelioration of COPD with individuals who cannot quit smoking. Physician-prescribing habits routinely include patients beyond those specifically evaluated in a given trial.

In the case of a specific HNE inhibitor, the patent life of the compound will also be a feasibility issue. Patent life is limited to 17 years with partial restoration possible in some countries (5 years in US).[49] The average US registration program consumes most of a compound patent life. A phase II-III study for dose ranging and efficacy that would be built on early safety and efficacy studies could take 5–6 years in dosing and analysis. This trial could easily absorb the patent life of a compound, rendering its profitability marginal without some extension of present (US) law. New medicines that do not provide symptomatic relief for targeted chronic disorders are expected to penetrate slowly into the target population; 1 or 2 years of lost exclusivity may adversely affect the risk : benefit ratio of a sponsor undertaking such a trial, depending on resource requirements and scientific feasibility.

CONCLUSION

At present, the number of identifiable issues and alternatives in each element of the trial, as well as present and future surprises, relegate as unfeasible a preventive outcome trial in the pathogenesis of emphysema. However, there is room for optimism that such a trial may soon be feasible. First, after 25 years, the hypothesis that excess HNE contributes to the pathogenesis of emphysema remains the single most cogent hypothesis. Second, there are specific and selective HNE inhibitors in development.[50] Third, analytic chemical techniques are being applied to biochemical markers. These may provide a means for dose ranging, a surrogate endpoint, or a "positive control" for a chemically effective dose to be judged clinically. Fourth, the Lung Health Study may allow an alternate analysis or parameter that abbreviates the trial or requires fewer patients.

Dr. Cohen's comments of 1979 may be premature for a therapy, but timely for the initiation of a definitive test of the elastase hypothesis for cigarette smoke-induced emphysema.

REFERENCES

1. COHEN, A. B. 1979. Am. Rev. Res. Dis. **120:** 723–727.
2. CHRZANOWSKI, P., S. KELLER, J. CERRETA, I. MANDL & G. M. TURINO. 1980. Am. J. Med. **69:** 351–359.
3. SNIDER, G. L. 1983. Am. Rev. Respir. Dis. **127**(2): S39–40.
4. ANTHONISEN, N. R. 1989. Am. Rev. Respir. Dis. **140:** 871–872.
5. Oxford University Press. 1987. Compact Edition of the Oxford English Dictionary. Complete Text Reproduced Micrographically. Oxford University Press. Cambridge.
6. The Health Consequences of Smoking: Chronic Obstructive Lung Disease. Rockville, MD: Centers for Disease Control, Office on Smoking and Health. 1984. DHHS publication (PHS) 84-50205.

7. JANOFF A., B. SLOAN, G. WEINBAUM *et al.* 1977. Am. Rev. Respir. Dis. **115:** 461–478.
8. KUEPPERS, F. & L. F. BLACK. 1974. Am. Rev. Respir. Dis. **110:** 176–194.
9. BLACK, L. F. & F. KUEPPERS. 1978. Am. Rev. Respir. Dis. **117:** 421–428.
10. LIEBERMAN, J., B. WINTER & A. SASTRE. 1986. Chest **89**(3): 370–373.
11. SILVERMAN, E. K., J. A. PIERCE, M. A. PROVINCE, D. C. RAO & E. J. CAMPBELL. 1989. Ann. Int. Med. **111:** 982–991.
12. KAO, R. C., N. G. WEHNER, K. M. SKUBITZ, B. H. GRAY & J. R. HOIDAL. 1988. J. Clin. Invest. **82:** 1963–1973.
13. SENIOR, R. M., N. L. CONNOLLY, J. D. CURY, H. G. WELGUS & E. J. CAMPBELL. 1989. Am. Rev. Respir. Dis. **139:** 1251–1256.
14. COLLIER, I. E., S. M. WILHELM, A. Z. EISEN, B. L. MARMER, G. A. GRANT, J. L. SELTZER, A. KRONBERGER, C. HE., E. A. BAUER & G. I. GOLDBERG. 1988. **263**(14): 6579–6587.
15. PRYOR, W. A., D. G. PRIER & D. F. CHURCH. 1983. Env. Health Perspect. **47:** 345–355.
16. SILVERMAN, S. P., D. M. OPPENHEIM, M. GLASS & G. A. WEINBAUM. 1987. Am. Rev. Respir. Dis. **135**(4): A494.
17. BOAT, T. F., P. W. CHENG, J. D. KLINGER, C. M. LIEDTKE & B. TANDLER. 1984. Ciba Found. Symp. **109:** 72–87.
18. SOMMERHOFF, C. P., J. A. NADEL, C. B. BASBAUM & G. H. CAUGHEY. 1990. J. Clin. Invest. **85:** 682–689.
19. BERGER, M., R. U. SORENSEN, M. F. TOSI, D. G. DEARBORN & G. DORING. 1989. J. Clin. Invest. **84:** 1302–1313.
20. REID, L. 1980. Chronic obstructive lung diseases. *In* Pulmonary Disease and Disorders. A. P. Fishman, ed. Vol. 1: 503–525. McGraw-Hill Book Co. New York, NY.
21. CAMPBELL, E. J. 1986. Am. Rev. Respir. Dis. **134:** 435–437.
22. FINKEL, M. J. 1977. General Considerations for the Evaluations of Drugs. Rockville, MD. HEW publication (FDA) 77–3040.
23. HAYNES, R. B. & R. DANTES. 1987. Controlled Clinical Trials **8:** 12–18.
24. NEWMAN, S. P., D. PAVIA, N. GARLAND & S. W. CLARKE. 1982. Eur. L. Respir. Dis. **63**(119): 57–65.
25. COHEN, A. B. 1990. Personal communication.
26. JUST, P. M. 1989. Pharmacotherapy **9**(2): 82–87.
27. DOHERTY, M. J., C. PAGE, C. BRADBEER, S. THOMAS, D. BARLOW, T. O. NUNAN & N. T. BATEMAN. 1988. The Lancet Dec. 3: 1283–1286.
28. BARNHART, E. R. 1990. Physicians' Desk Reference. 44th Ed: 604, 683, 1234, 1249, 1861. Medical Economics Company. NJ.
29. BUIST, A. S., B. BURROWS, S. ERIKSSON, C. MITTMAN & M. WU. 1983. Am. Rev. Respir. Dis. **127**(2): S43–S45.
30. BURROWS, B. 1983. Am. Rev. Respir. Dis. **127**(2): S42–S43.
31. GADEK, J. E. & R. G. CRYSTAL. 1983. Am. Rev. Respir. Dis. **127**(2): S45–S46.
32. WEITZ, J. I., K. A. CROWLEY, S. L. LANDMAN, B. I. LIPMAN & J. YU. 1987. Ann. Int. Med. **107:** 680–682.
33. KUCICH, U., P. CHRISTNER, M. LIPPMANN, P. KIMBEL, G. WILLIAMS, J. ROSENBLOOM & G. WEINBAUM. 1985. Am. Rev. Res. Dis. **131:** 709–713.
34. SUTER, S., U. B. SCHAAD, J. J. MORGENTHALER, I. CHEVALLIER & H. P. SCHNEBLI. 1988. J. Infect. Dis. **158**(1): 89–100.
35. GRESSNER, A. M. & W. TITTOR. 1986. Klin Wochenschr. **64:** 1240–1248.
36. HERBST, T. J., J. B. MCCARTHY, E. C. TSILIBARY & L. T. FURCHT. 1988. J. Cell Biol. **106:** 1365–1373.
37. UXXELL, J. K. & G. F. MEYER. 1990. Pharmaceutical Executive **10**(6): 82–86.
38. BARNES, P. J. 1989. N. Engl. J. Med. **321**(22): 1517–1527.
39. GROSS, N. J. & M. S. SKORODIN. 1984. N. Engl. J. Med. **311**(7): 421–425.
40. ANTHONISEN, N. R., E. C. WRIGHT, J. E. HODGKIN & The IPPB Trial Group. 1986. Am. Rev. Respir. Dis. **133:** 14–20.
41. BURROWS, B., J. W. BLOOM, G. A. TRAVER & M. G. CLINE. 1987. N. Engl. J. Med. **317**(21): 1309–1314.

42. PIERCE, J. P., M. C. FIORE, T. E. NOVOTNY, E. J. HATZIANDREU & R. M. DAVIS. 1989. J. Am. Med. Assoc. **261:** 61–65.
43. BUIST, A. S. & W. M. VOLMER. 1988. Stat. Med. **7:** 11–18.
44. VOLLMER, W. M., L. R. JOHNSON, L. E. McCAMANT & A. S. BUIST. Stat. Med. **7:** 685–696.
45. DIEM, J. E. & J. R. LIUKKONEN. 1988. Stat. Med. **7:** 19–28.
46. LOUIS, T. A. 1988. Stat. Med. **7:** 29–45.
47. SCHICK, B., K. F. AUSTEN & L. B. SCHWARTZ. 1984. J. Immunol. **132:** 2571–2577.
48. SCHWARTZ, J. L. 1987. Review and Evaluation of Smoking Cessation Methods: The United States and Canada, 1978–1985. Division of Cancer Prevention and Control. National Cancer Institute. Washington, DC.
49. MATHIEU, M. P. (Ed.) 1987. New Drug Development: A Regulatory Overview. OMEC International, Inc. Washington, DC.
50. TRAINOR, D. A. 1987. TIPS **8:** 303–307.

Lung Transplantation for Chronic Obstructive Lung Disease

JOEL D. COOPER

Department of Surgery
Washington University School of Medicine
St. Louis, Missouri 63110

In 1983, following several years of laboratory research, we initiated a lung transplant program for patients with end-stage chronic pulmonary disease. Between 1963, when the first human lung transplantation was performed, and 1983, approximately 44 lung or lobe transplantations had been attempted world wide with no long-term clinical success. At the outset, we selected patients with end-stage pulmonary fibrosis for single lung transplantation. We reasoned that with this condition, both ventilation and perfusion would preferentially be directed to the transplanted lung, avoiding the ventilation-perfusion mismatch that had previously been demonstrated after single lung transplantation for chronic obstructive lung disease.[1,2] After several successful lung transplantations had been performed for end-stage pulmonary fibrosis, we sought ways to extend lung transplantation to individuals with bilateral sepsis (cystic fibrosis) and patients with chronic obstructive lung disease. In both of these conditions, we were concerned that unilateral lung replacement would be inappropriate. For patients with bilateral sepsis, retention of an infected lung in an immunosuppressed patient would clearly be unwise. In obstructive lung disease, we were concerned that the remaining, emphysematous lung might further expand, crowd the transplanted lung, and create a situation where the majority of perfusion would go to the transplanted lung, but the majority of ventilation went to the native lung.

We decided to employ combined heart-lung transplantation for patients with end-stage pulmonary disease in whom single lung transplantation was judged to be inappropriate. It soon became apparent, however, that most patients with end-stage lung disease had adequate or recoverable right ventricular function, and that with the combined heart-lung transplant, the heart was being used not out of physiologic necessity but because of technical requirements. Also, the requirement for both a heart and lungs for transplantation greatly restricted the number of organs available, as most donor hearts are used for cardiac transplantation and, hence, are not available for combined transplantation. We returned to the laboratory to develop a procedure for bilateral lung replacement without use of the heart, the so-called "double lung transplant."[3,4] This procedure was initially utilized in patients with chronic obstructive lung disease with good results.[5] With further experience, however, the double lung procedure proved to be less than ideal. Twenty-five percent of recipients died from complications due to ischemia of the donor airway. The procedure was complicated and required several hours of cardiopulmonary bypass and a period of total ischemic cardiac arrest. This was poorly tolerated by patients with preexisting cardiac problems, especially those with right ventricular hypertrophy. Intraoperative and postoperative hemorrhage was also a major problem especially when chronically infected lungs were extracted under full anticoagulation therapy as required by the use of cardiopulmonary bypass.

Because of these problems, we sought alternatives for transplantation in patients with end-stage chronic obstructive lung disease. Following a preliminary report from Paris[6] demonstrating that unilateral lung transplantation could be performed for chronic obstructive disease, we initiated a program of single lung transplantation for patients with end-stage emphysema over the age of 50 as the double lung transplant procedure was particularly associated with complications in this age group.

To date, we have performed 13 single lung transplants in patients with emphysema. All recipients have been discharged well, and restoration of lung function and exercise capacity has been gratifying. One advantage of utilizing single lung transplantation for obstructive lung disease is the ability to use each donor lung for separate recipients, thus significantly increasing the supply of available organs. We refer to this sharing of donor lungs as "twinning," a procedure that we have employed on nine occasions either using each lung at our own institution or sharing lungs with another transplant center.

Following single lung transplantation for obstructive lung disease, the mediastinum often shifts towards the transplant side. This is a compensatory mechanism and is due to the discrepancy between the large size of the thoracic space and the normal sized lung placed within it. Apparently such a shift in no way restricts the ventilation of the transplanted lung. Initial results with six patients evaluated 6 months following transplantation demonstrate that mean perfusion to the transplanted lung is 75%, whereas mean ventilation is 60%. This "mismatch" does result in some wasted ventilation, but this appears to be of little physiologic significance.

At the same time that single lung transplants were being extended to patients with chronic obstructive lung disease, ways were being sought to improve the procedure for bilateral lung replacement. We recently revised our approach completely and now use a transverse bilateral thoracosternotomy incision which gives superb exposure of both pleural spaces from apex to diaphragm.[7] With this exposure, the right lung is removed and replaced as for single lung transplantation. The left lung is then excised and replaced in similar fashion. The procedure utilizes bilateral bronchial anastomoses rather than the tracheal anastomosis used for the double lung transplant; thus, airway complications have virtually been eliminated. Furthermore, there is no need for total cardiopulmonary bypass, and in many cases no bypass whatsoever is required. As an epidural catheter is used for postoperative pain management, morbidity with this incision has been minimal. We have now performed nine procedures using this approach in patients with emphysema, cystic fibrosis, and other conditions. All patients have had satisfactory blood gas levels on room air by the third postoperative day. All patients have been discharged well from the hospital, and lung function has been excellent.

We recently compared results of single lung transplantation for obstructive lung disease (11 patients) with those of bilateral lung transplantation (6 patients). Preoperatively the single lung group had a mean FEV_1 of 0.47 liters versus 0.49 for the bilateral group. At 12 weeks, the FEV_1 in the single lung group had increased to 1.63 liters versus 3.44 for the bilateral lung group.

Using the 6-minute walk test to evaluate the two groups, the single lung group showed a preoperative 6-minute walk distance of 780 feet which increased to 1,729 feet at 12 weeks after the procedure. The corresponding figures for the bilateral group were 829 feet preoperatively and 2,195 feet postoperatively.

Using a 30-minute treadmill walking test to evaluate performance, the single lung transplant group walked a mean of 1.1 miles in 30 minutes versus 1.8 miles for the bilateral transplant group. Thus, although lung function and exercise performance have been very satisfactory after single lung transplantation for chronic obstructive pulmonary disease, results with bilateral replacement are even better. Part of the difference might be explained by the average age of the two groups; it was 15 years younger in the bilateral transplant group.

It now appears that both unilateral and bilateral lung replacement can be performed satisfactorily in patients with chronic obstructive lung disease with very acceptable morbidity and mortality. In our last consecutive 32 single and bilateral transplantations for a variety of conditions, there was one hospital death and one late death. We continue to employ the unilateral procedure for patients with obstructive lung disease over the age of 50 and the bilateral procedure for those under the age of 50 as well as for patients with cystic fibrosis. Additional experience and followup with these two groups of patients will be required before the most appropriate role for each type of transplantation can be clearly delineated.

REFERENCES

1. STEVENS, P. M., P. C. JOHNSON, R. L. BELL, A. C. BEALL & D. E. JENKINS. 1970. Regional ventilation and perfusion after lung transplantation in patients with emphysema. N. Engl. J. Med. **282:** 245–249.
2. COOPER, J. D. 1989. The other lung—Revisited. Chest **96:** 707–708.
3. PATTERSON, G. A., J. D. COOPER, J. H. DARK et al. 1988. Experimental and clinical double lung transplantation. J. Thorac. Cardiovasc. Surg. **95:** 70–74.
4. PATTERSON, G. A., J. D. COOPER, B. GOLDMAN et al. 1988. Technique of successful clinical double-lung transplantation. Ann. Thorac. Surg. **4:** 626–633.
5. COOPER, J. D., G. A. PATTERSON, R. GROSSMAN et al. 1989. Double-lung transplant for advanced chronic obstructive lung disease. Am. Rev. Respir. Dis. **139:** 303–307.
6. MAL, H., B. ANDREASSIAN, F. PAMELA et al. 1989. Unilateral lung transplantation in end-stage pulmonary emphysema. Am. Rev. Respir. Dis. **140:** 797–802.
7. PASQUE, M. K., J. D. COOPER, L. R. KAISER et al. 1990. Improved technique for bilateral lung transplantation: Rationale and initial clinical experience. Ann. Thorac. Surg. **49:** 785–791.

Recombinant Elastase Inhibitors for Therapy

HANS PETER SCHNEBLI

Pharmaceuticals Division
Ciba-Geigy Ltd.
CH-4002 Basel, Switzerland

For the last few years we have worked with three polypeptide elastase inhibitors, all of which were produced by recombinant technology. Two of these (SLPI and rAAT) are well advanced in the development towards therapeutic use, whereas one (Eglin c) had to be abandoned because of allergenicity.

I will retrace some of the history of these inhibitors, review their properties, and discuss their place in therapy and prevention of lung diseases.

HISTORY AND BACKGROUND

As is demonstrated in many of the chapters in this book, leukocyte elastase, probably in concert with (leukocyte-derived) reactive oxygen species, is implicated in connective tissue damage, particularly in chronic bronchitis, emphysema, and cystic fibrosis. This is the reason that we, and many others, have set out to develop elastase inhibitors as drugs aimed at retarding the destruction of lung tissue in these diseases. Inasmuch as the lung appears to be the most sensitive organ for elastase attack, aerosol application of elastase inhibitors to the site where their action presumably is required, naturally suggested itself. It was this possibility of delivering the inhibitors to their target site that allowed us to focus on protein proteinase inhibitors produced by recombinant technology; this class of compounds, at least with current technology, cannot be delivered orally. In addition, this "site-specific" delivery of large, poorly diffusible molecules to the lung may offer a higher degree of safety, that is, fewer systemic side effects, than may be expected with orally applied drugs or with aerosolized, low molecular weight drugs that may more freely diffuse throughout the body.

There is, however, another (historical) reason that Ciba-Geigy has focused on recombinant inhibitors rather than synthetic inhibitors. In 1980, when we decided to get serious about elastase inhibitors, many low molecular weight "leads" were described already, but all of the potent agents were overtly toxic and/or biologically unstable because of their peptidic nature. So, when we became aware of Eglin c, the potent and exceedingly stable inhibitor from the leech, we embarked on a long, interesting, but admittedly also frustrating development project on recombinant elastase inhibitors.

At first, everything went well: the gene for Eglin c was chemically synthesized, cloned into *Escherichia coli*, and expressed[1] at very acceptable levels. The molecule, because of its unusual acid and heat stability, was easy to purify and it was potent and selective. Also, our choice appeared to be "the right one," because for many years none of the synthetic inhibitors continually made by our colleagues from competitors and at universities approached the potency, selectivity, and biologic stability of Eglin c. Naturally, when after this successful development it became apparent that Eglin c caused some occupational allergies, frustration was rampant.

To avoid similar problems again, we next chose to develop a human inhibitor. Both of the two known human elastase inhibitors, *secretory leukocyte proteinase inhibitor* (SLPI = bronchial mucus inhibitor = BMI = antileukoproteinase = ALP = HUSI I = MPI etc.) and α_1-proteinase inhibitor (α_1-PI = α_1-antitrypsin = AAT) had already been cloned by others at that time. At first, we collaborated with Synergen, Boulder, Colorado (Robert Thompson and Mark Young) on the development of recombinant SLPI (rSLPI), but more recently we decided to "cover all bases" and in addition acquired the rAAT, this time from Cooper, Mt. View, California.

EGLIN C

Native Eglin c was discovered and isolated from the medicinal leech *Hirudo medicinalis* in 1977.[2] It is a polypeptide of 70 amino acids. Although it does not contain any disulfide bridge, it is an unusually stable polypeptide. Eglin c is a member of the potato inhibitor I family of serine proteinase inhibitors that lack disulfide bonds; the great similarity of the peptide sequences and the three-dimensional structures of these inhibitors is perhaps surprising, given the extreme evolutionary divergence of the organisms in which they occur.[3] The "active site" of Eglin c is a Leu_{45}-Asp_{46} bond.[4] Complex formation with proteinases (elastase, subtilisin, etc.) is reversible and does not involve cleavage of the active site bond.[5,6]

Recombinant Eglin c had the same amino acid sequence as did the leech-derived material, except that the N-terminus in the recombinant material is blocked by an acetyl group.[7] The recombinant material is indistinguishable from the naturally occurring Eglin c with respect to enzyme inhibitory properties, stability, and recognition by a number of monoclonal antibodies. Interaction of elastase with Eglin c is very rapid (association rate constant approximately 10^7 $M^{-1} s^{-1}$), nearly as rapid as the interaction of elastase with α_1-PI; with an equilibrium dissociation constant of less than 10^{-10} M, the complexes formed are very stable.[8] Eglin is a very selective inhibitor; except for leukocyte elastase, cathepsin G, mast cell chymase, chymotrypsin, and pancreatic elastase, no other human enzyme is affected by Eglin c. Eglin c was also tested for its ability to inhibit the lysosomal cysteine proteinases cathepsins B and H and the acid proteinase cathepsin D. None of these enzymes was inhibited (at 10^{-5} M) and, even more importantly, none of these enzymes destroyed the elastase inhibitory activity of Eglin c. The fact that undenatured Eglin c is virtually not attacked by any of the numerous proteinases tested to date may also explain its unusual biologic stability. In terms of safety, it was most important to show that Eglin c does not interact with the blood-borne proteinases of the clotting, the fibrinolysis, and the complement cascades; none of these systems is affected by Eglin c at concentrations of up to 10^{-5} M; very high doses of Eglin c (100 mg/kg intravenously) did not affect clotting or fibrinolysis *in vivo* (R. Wallis, unpublished observations).

Eglin c was subjected to a number of pharmacologic test systems, mostly models of lung diseases. When given intratracheally, it was effective in preventing the development of emphysema and secretory cell metaplasia induced by human leukocyte elastase in hamsters,[9] even if the inhibitor was given as many as 8 hours before the insult.[10] Leukocyte elastase has long been implicated in the pathogenesis of shock and more recently adult respiratory distress syndrome. For these reasons, Eglin c was also tested in animal models in which shock or shock-related syndromes were induced experimentally. Although earlier studies in septic and endotoxin shock in pigs[11] and in traumatic shock in rats[12] were encouraging, in a very well-controlled study in baboons under "realistic clinical" settings (polytraumatic hypovolemic shock), Eglin c was without effect.[25]

Pharmacokinetics and disposition of Eglin c[13] have been investigated after intravenous bolus injection using both a radioassay (after administration of ^{3}H-labeled compound) and an ELISA procedure. In all species investigated, plasma Eglin c levels decreased rapidly, the approximate elimination half-times being 30 minutes in rats, 40 minutes in rabbits, and 2.5 hours in baboons. In all species Eglin c was excreted mainly with the urine. In the urine of rats, Eglin c is recovered practically quantitatively as unchanged substance.

Preceding the toxicology studies in rats and baboons, a number of pharmacologic investigations were undertaken to elucidate the potential of Eglin c to cause unwanted effects. Eglin c proved to be extremely well tolerated; no significant effects were observed, even at large doses (usually 100 mg/kg iv), on the cardiovascular and central nervous systems or on the clotting, the fibrinolysis, or the complement system. Eglin c did not cause any adverse or toxic effects in rabbits and baboons after a single intravenous bolus injection of 400 mg/kg (maximal injectable dose). Daily intravenous administration of Eglin c to baboons for 14 consecutive days at doses of up to 100 mg/kg only induced occasional vomiting in all dose groups. No changes were noted in body and organ weights, blood, and urine chemistry and cytology, and no gross or microscopic changes of the tissues were attributable to the compound. Daily intravenous administration of Eglin c to rabbits for 1 month at doses up to 100 mg/kg did not induce any clinical signs or any changes in body and organ weights and blood parameters, or any effects in ophthalmoscopic, auditory, and microscopic examinations (W. Classen, unpublished observations).

The main safety problem with Eglin c was its potential to cause allergic reactions. Although none of the animals in the subchronic toxicity studies just mentioned developed signs of hypersensitivity (no positive skin reactions and no anti-Eglin c antibodies in their sera), in another study in dogs, serious allergic reactions were noted. More importantly, three persons involved in the development of Eglin c (all of whom suffered from multiple allergies before) developed hypersensitivities to the compound. After this observation, development of Eglin c, at least for chronic application, was no longer feasible, and the project was abandoned in 1986.

rSLPI (RECOMBINANT SECRETORY LEUKOCYTE PROTEINASE INHIBITOR)

SLPI has been shown to occur in all human mucus secretions and in very low concentrations in the plasma. In the lung, it is produced in the secretory glands of the upper airways as well as in the nonciliated cells of the bronchiolar epithelium, and it is believed to be the major antielastase of the upper airways, whereas in the peripheral lung α_1-PI is the major elastase inhibitor.[14]

SLPI is a large peptide (107 amino acids) with an unusually large number (eight) of intramolecular disulfide linkages and an unusually high isoelectric point (~10.5). It is an acid-stable peptide and consists of two homologous domains[15] in which the four disulfides are found in identical positions. The recombinant molecule (rSLPI) is identical in every respect to the native (SLPI) molecule.

SLPI is a potent inhibitor of human leukocyte elastase[16] and chymotrypsin-like enzymes, but it also inhibits trypsin and some trypsin-like enzymes. Because of the sequences (Arg in domain 1 and Leu in domain 2) at the putative active sites in this molecule, it was first believed that the first domain was responsible for the inhibition of the trypsin family, whereas elastase(s) and the chymotrypsin family are inhibited by the second domain. More recently, and with the use of site-directed mutagenesis, it could be shown that all enzymes bind to the second domain and that the first domain is "mute."[26] Like with Eglin c, inhibition occurs by formation of a reversible complex in which the inhibitor exists uncleaved.[17] The association rate constant for the elastase SLPI interaction ($6 \times 10^6 \, M^{-1} \, s^{-1}$) is only slightly lower than that of the elastase-Eglin c interaction and the equilibrium dissociation constant (about 10^{-9} M) is very low (G. D. Virca et al., unpublished observations).

Colleagues at Synergen derived the sequence of SLPI by standard protein analytics[18] and confirmed it by sequencing a cDNA;[19] they synthesized the gene de novo, cloned and expressed it in yeast, and can now produce SLPI at the 100-g scale.

Oral application of rSLPI is not appropriate because of its peptidic nature. Intratracheal application of SLPI was found to significantly protect against elastase-induced experimental emphysema in hamsters for up to 8 hours;[20] rSLPI by itself did not affect lung function.

In general pharmacologic evaluation after intravenous application, rSLPI proved to be quite safe. All animals tested survived and did not experience but marginal side effects attributable to the inhibitor. rSLPI did not affect the CNS or blood glucose levels. It did not, by itself, cause histamine release from mast cells and did not interfere with the proteinases of the complement system; it did not affect thrombin or kallikrein. The only unwanted effects observed were hypotension at doses above 20 mg/kg intravenously (in the cat) and a slight inhibitory effect on coagulation (APTT was prolonged twofold at 10 μM rSLPI). Furthermore, at very high concentrations, rSLPI inhibited t-PA (IC_{50} = 10 μM). Because of its human origin, SLPI is not expected to be allergenic in man.

Pharmacokinetic studies have been carried out in rats and sheep. Intravenously applied rSLPI is rapidly eliminated ($t\frac{1}{2}$ = 40 minutes in the rat).[21] After intrapulmonary application, the elimination half-times from the lung are 4 hours (rats) and 12 hours (sheep) (Crystal et al., unpublished observations). In the sheep, 1 hour after intrapulmonary application, small amounts of rSLPI appear in the interstitium (lymph) (Crystal et al., unpublished observations). In the rat, approximately 45% of the intrapulmonary rSLPI dose appeared in the plasma, peaking at 1.5 hours after application. There is good evidence for metabolism of rSLPI once it gets into the lymph or plasma.[21]

A 1-month aerosol toxicity study in cynomolgus monkeys has been completed; no clinically relevant toxicology was observed.

The rSLPI project was reacquired by Synergen in January 1990; the compound is being developed with the aim to register SLPI in cystic fibrosis and/or α_1-PI deficiency based on clinical safety and "biochemical efficacy" data.

RECOMBINANT α_1-PROTEINASE INHIBITOR

Many laboratories have attempted and largely succeeded in cloning and expressing human α_1-proteinase inhibitor, the major human elastase inhibitor. The material we work with (and will refer to as rAAT, CGP 46 479) was cloned originally at Washington University[22] and Zymogenetics; process development was at Cooper Laboratories, who licensed the project to Ciba-Geigy in 1988.

Due to its biosynthesis in yeast, rAAT lacks the carbohydrates that account for nearly 20% of the molecular mass in the native AAT; the variant we work with contains, like the human "wild type," methionine as the active site residue in position 358; in addition (and different from the wild type), it contains N-acetyl-methionine at its N-terminus.

The inhibitory properties of rAAT are similar to those of native AAT; it rapidly forms quasi-irreversible complexes with elastase, many trypsin-like enzymes, and chymotrypsin-like enzymes. Because of the slowness of dissociation and, more importantly because of a tendency to be clipped in the active site loop (close to but not necessarily at position 358), AAT was for a long time believed to be an irreversible inhibitor; new information clearly shows that the inhibition of several serine proteinases by AAT is reversible (R. W. Carrell and J. Travis, personal communication).

Cooper managed to obtain an IND with only biochemical data (demonstrating that the elastase-inhibitory activity of rAAT is the same as that of native AAT). As native AAT is pharmacologically effective and clinically accepted (Prolastin®), it was concluded that rAAT will be too. In the meantime, Snider et al. (in press) have shown that rAAT is highly effective in the elastase-induced hamster emphysema model: it fully protected the hamster for 16 hours.

None of the toxicologic studies (2-week intravenous studies in rats and cynomolgus monkeys and 2-week inhalation studies in cynomolgus monkeys) produced findings indicative of significant compound-related toxicity. In addition, the "antigenicity" of rAAT was assessed (and compared to that of native AAT) in a baboon immunization study. Not unexpectedly, both compounds elicited immune responses; the small number of animals studied did not allow a statistical analysis.

Pharmacokinetic studies in several species including rats, mice, and rhesus monkeys have shown that the nonglycosylated rAAT has a much shorter plasma half-life (1–2 hours) than does the native AAT (2–4 days). However, as the compound will be administered by inhalation, a long plasma half-life is no prerequisite. It was shown in sheep that most of the aerosolized rAAT persists for a long time in its lung (elimination half-life = 18 hours) (R. Crystal et al., unpublished observations). Hours after the aerosol application to the sheep, rAAT could be detected in the thoracic duct lymph in therapeutically relevant concentrations, demonstrating its diffusion into and through the interstitium (R. Crystal et al., unpublished observations). In the rat, the elimination half-life of intrapulmonally applied rAAT was found to be 12 hours (ELISA) and 9.5 hours (elastase inhibitory activity) (H. P. Nick et al., unpublished observations).

In clinical research, the possible allergenic potential of rAAT (due to lack of glycosylation) was already addressed in the first clinical study. Six AAT-deficient patients were repeatedly injected subcutaneously with rAAT and assessed with regard to hypersensitivity reactions and humoral and cellular responses. No reactions occurred in any of the patients except one in whom the antibody (to rAAT) level determined by ELISA was raised slightly above background.

In a second study, 10 to 200 mg of rAAT were inhaled once by a total of 18 patients.[23,24] No unwanted effects were noted, and no hypersensitivity or humoral and cellular reactions were observed.

OUTLOOK

The two recombinant human elastase inhibitors, rSLPI and rAAT, are well advanced in their development towards clinical use. Hopefully, together with a number of interesting synthetic elastase inhibitors (fluoromethyl ketones and β-lactam type elastase inhibitors) will allow clinicians to establish the usefulness of this class of compounds in the therapy and prevention of degenerative lung diseases.

Because of its "true blue" human structure, rSLPI appears to offer the highest degree of safety. The advantage of rAAT is that it is the "appropriate" molecule for replacement therapy in α_1-PI deficiency. The low molecular weight, synthetic inhibitors will be cheaper, are technically easier to apply by aerosol than are proteins, and may even be orally active. Finally, not to be neglected is the native AAT (Prolastin®), which has the tremendous advantage of being registered already.

In short, the most exciting phase of elastase inhibitor research, the validation of the concept in the clinic, is just beginning.

REFERENCES

1. Rink, H., M. Liersch, P. Sieber & F. Meyer. 1984. Nucleic Acids Res. **12**: 6369–6387.
2. Seemüller, U., M. Meier, K. Ohlsson, H. P. Müller & H. Fritz. 1977. Hoppe-Seyler's Z. Physiol. Chem. **358**: 1105–1117.
3. Seemüller, U., H. Fritz & M. Eulitz. 1981. Methods Enzymol. **80**: 804–816.
4. Seemüller, U. & M. Eulitz. 1982. Chem. Peptides Proteins **1**: 39–45.
5. Grütter, M. G., W. Marki & H.-P. Walliser. 1985. J. Biol. Chem. **260**: 11436–11437.
6. McPhalen, C. A., H. P. Schnebli & M. N. G. James. 1985. FEBS Lett. **188**: 55–58.
7. Marki, W., F. Raschdorf, W. Richter, H. Rink, P. Sieber, H. P. Schnebli & M. Liersch. 1985. In Peptides: Structure and Function. C. M. Deber, V. J. Hruby & K. D. Kopple, eds.: 385–388. Pierce Chemical Company. Rockford, USA.
8. Braun, N. J., J. L. Bodmer, G. D. Virca, G. Metz-Virca, R. Maschler, J. G. Bieth & H. P. Schnebli. 1987. Biol. Chem. Hoppe Seyler **368**: 299–308.
9. Snider, G. L., P. J. Stone, E. C. Lucey, R. Breuer, J. D. Calore, T. Seshadri, A. Catanese, R. Maschler & H. P. Schnebli. 1985. Am. Rev. Respir. Dis. **132**: 1155–1161.
10. Lucey, E. C., P. J. Stone, T. G. Christensen, R. Breuer, J. D. Calore & G. L. Snider. 1986. Am. Rev. Respir. Dis. **134**: 471–475.
11. Jochum, M., H. F. Welter, M. Siebeck & H. Fritz. 1986. In Proteinases in Inflammation and Tumor Invasion. H. Tschesche, ed.: 53–59. William de Gruyter. Berlin.
12. Hock, C. E. & A. M. Lefer. 1985. Pharmacol. Res. Commun. **17**: 217–220.
13. Nick, H. P., A. Probst & H. P. Schnebli. 1988. In Proteinases in Health and Disease. W. H. Hörl and A. Heidland, eds.: 83–88. Plenum Press. New York.
14. Kramps, J. A., C. Franken & J. H. Dijkman. 1988. Clin. Sci. **75**: 351–353.
15. Thompson, R. C. & K. Ohlsson. 1986. Proc. Natl. Acad. Sci. USA **83**: 6692–6696.
16. Gauthier, F., U. Fryksmark, K. Ohlsson & J. G. Bieth. 1982. Biochim. Biophys. Acta **700**: 178–183.
17. Grütter, M. G., G. Fendrich, R. Huber & W. Bode. 1988. EMBO J. **7**: 345–351.

18. THOMPSON, R. C. & K. OHLSSON. 1986. Proc. Nat. Acad. Sci. USA **83:** 6692–6696.
19. OHLSSON, K. & M. ROSENGREN. 1987. *In* Pulmonary Emphysema and Proteolysis. J. C. Taylor & C. Mittman, eds.: 307–322. Academic Press, Inc. Orlando, FL.
20. SNIDER, G. L., P. J. STONE, E. C. LUCEY, R. BREUER, J. D. CALORE, T. SESHARDI, A. CATANESE, R. MASCHLER & H. P. SCHNEBLI. 1985. Am. Rev. Respir. Dis. **132:** 1155–1161.
21. GAST, A., W. ANDERSON, A. PROBST, H. P. NICK, R. C. THOMPSON, S. P. EISENBERG & H. P. SCHNEBLI. 1990. Am. Rev. Resp. Dis. **141:** 889–894.
22. LONG, G. L., T. CHANDRA, S. L. C. WOO, E. W. DAVIE & K. KURACHI. 1984. Biochemistry **23:** 4828–4837.
23. HUBBARD, R. C., M. A. CASOLARO, M. MITCHELL, S. E. SELLERS, F. ARABIA, M. A. MATTHAY & R. G. CRYSTAL. 1989. Proc. Natl. Acad. Sci. USA **86:** 680–684.
24. HUBBARD, R. C., N. G. MCELVANEY, S. E. SELLERS, J. T. HEALY, D. B. CZERSKI & R. G. CRYSTAL. 1989. J. Clin. Invest. **84:** 1349–1354.
25. JUNGER, W. G., C. LIENERS, H. REDL & G. SCHLAG. 1989. Reasons for the ineffectiveness of Eglin c to ameliorate endotoxin shock in sheep. *In* Progress in Clinical and Biological Research. G. Schlag and H. Redl, eds. Vol. 308: 1142p. Alan R. Liss, Inc. New York.
26. EISENBERG, S. P., K. K. HALE, P. HEIMDAL & R. C. THOMPSON. 1990. J. Biol. Chem. **265:** 7976–7981.

The Discovery and Biologic Properties of Cephalosporin-Based Inhibitors of PMN Elastase

PHILIP DAVIES,[a,b] B. M. ASHE,[a] R. J. BONNEY,[b]
C. DORN,[a] P. FINKE,[a] D. FLETCHER,[a] W. A. HANLON,[a]
J. L. HUMES,[a] A. MAYCOCK,[c] R. A. MUMFORD,[a]
M. A. NAVIA,[d] E. E. OPAS,[a] S. PACHOLOK,[a] S. SHAH,[a]
M. ZIMMERMAN,[a] AND J. B. DOHERTY[a]

[a]Department of Immunology & Inflammation
Merck Sharp & Dohme Research Laboratories
Rahway, New Jersey 07065

[b]Bristol-Myers Co.
Buffalo, New York 14213

[c]Sterling Research Group
Malvern, Pennsylvania 19355

[d]Vertex Pharmaceutical Inc.
Cambridge, Massachusetts 02139

Elastase (E.C. 3.4.21.37) is one of several serine proteinases present at high concentrations in the azurophilic granules of the polymorphonuclear leukocyte (PMN). Other members of the family include cathepsin G and a recently described enzyme designated variously as PR3,[1] AGP7,[2,3] or PR4.[4] These three enzymes show considerable sequence homology with N-terminal sequences characteristic of serine proteinases.[3] In addition, antibacterial proteins with sequence characteristics of the family, but lacking detectable enzyme activity, have been described and named azurocidin[2,5] and CAP37,[6,7] respectively. Sequence analysis of these two entities show very close homology.[3,6,7] However, Wilde et al.[3] have indicated that the active site serine of azurocidin is replaced by a glycine, whereas Pereira et al.[7] have found the additional substitution of a serine for the catalytic histidine[41] in CAP37.

PMN elastase and cathepsin G are synthesized in promyelocytes in the bone marrow[8,9] and are packaged and released from the concave face of the Golgi complex in these cells.[10,11] The genes for PMN elastase and cathepsin G have been characterized.[9,12–17] The mature protein of PMN elastase contains 220 amino acid residues[8,18] with a prepro domain of 29 residues and a C-terminal extension containing 20 residues. The enzyme is packaged in the azurophilic granules of PMN at extremely high concentrations.[19] Current opinions regard the primary physiologic functions of PMN elastase as being intracellular ones, primarily in host defense against infectious agents. It is thought to act in concert with other mechanisms, including oxidative processes and antibacterial proteins, in the killing and digestion of ingested microorganisms. Although no specific defects in the expression of the PMN elastase gene have been described, a failure to maintain normal levels of elastase and cathepsin G in mature PMNs in the beige mouse[20,21] has been described which may be due to inactivation of enzyme by inhibitors present in PMNs of these mice.[22] Bacterial killing in beige mice and their human counterparts, individuals with Chediak-Higashi syndrome, is deficient. It has been suggested that this may be a consequence of the lack of elastase and cathepsin G. However, other defects in this syndrome, particularly defective secondary lysosome formation, may account for the increased susceptibility to infection.

POSSIBLE ROLES OF PMN ELASTASE IN TISSUE DESTRUCTION ASSOCIATED WITH INFLAMMATORY DISEASE

Several lines of circumstantial evidence indicate that PMN elastase contributes to tissue damage in inflammatory diseases. First, elevated levels of alpha$_1$-proteinase inhibitor (α_1-PI)-PMN elastase complexes have been observed in a wide range of acute inflammatory diseases (TABLE 1) in which local accumulation and activation of PMNs occur. Plasma levels of α_1-PI-PMN elastase of approximately 80 ng/ml are found in healthy individuals. No clear-cut function has been established for the enzyme in the extracellular environment, and the site(s) of formation of these complexes has not been established. As is seen in TABLE 1, the levels of complexes have been correlated with the severity of disease in a number of instances, raising the possibility that active enzyme may contribute to pathology before its inactivation by α_1-PI. Endogenous free PMN elastase is detectable in the sputum of patients with cystic fibrosis[31] and may contribute to tissue damage and impairment of host defense functions mediated by phagocytes in the large airways of these individuals.[32] Second, it is widely documented that susceptibility to the early development of emphysema is increased in individuals with the PiZZ and null phenotypes for α_1-PI.[33] In PiZZ individuals the plasma level of α_1-PI is approximately 10% of normal, whereas the inhibitor is completely absent in null individuals. It has been widely argued that the predisposition to the early development of emphysema in such individuals results from a lack of sufficient inhibitor in the microenvironment of the terminal airways, resulting in the unhindered attack of elastase on interstitial connective tissue, including elastin. In this context, studies on the activity of α_1-PI against a variety of serine proteinases show that it is most potent against PMN elastase and, more importantly, its association constant for the enzyme is sufficient for it to be effective in a biologic milieu in the presence of natural substrates.[34]

Third, *in vitro* studies indicate that PMNs can be stimulated by inflammatory stimuli to release the content of their azurophilic granules, including elastase. Substrate degradation can be demonstrated when PMNs are attached to a suitable substrate matrix in the presence of α_1-PI, suggesting that degranulation of PMNs at sites of inflammation can result in transient extracellular elastase activity.[35-37] This possibility is further supported by the observations that beige mice fail to develop albuminuria associated with experimental antiglomerular basement membrane-induced glomerulonephritis despite an apparently normal recruitment of PMNs to the site of immune-complex formation.[39] Starcher and Williams[40] have shown that intratracheally administered endotoxin fails to induce lung injury seen in normal C57 black mice when administered to beige mice.

Fourth, exogenous PMN elastase can induce acute *in vivo* tissue destruction in the lung,[41,42] kidney,[43] skin, and other tissue sites when administered locally. This phenomenon has been most extensively studied in animal models of emphysema and is reviewed in this volume and elsewhere by Snider and his colleagues.[44]

Thus, it appears that a number of situations exist in which elastase is not fully inhibited by natural inhibitors in extracellular environments. This provides a rationale for developing low molecular weight synthetic inhibitors for evaluation of their efficacy to supplement natural inhibitors in disease involving PMN activation.

TABLE 1. Elevations in Plasma α_1-PI-PMN Elastase Complex Levels in Acute Inflammatory Disease

Disease	Plasma Levels α_1-PI-PMN Elastase (ng/ml)	Reference
Septicemia associated with surgery:		
Survivors	647 ± 117	
Nonsurvivors	985 ± 155	
Noninfected	209 ± 26	
Controls	87 ± 26	23
Rheumatoid arthritis	425	
Controls	75	24
Rheumatoid arthritis	423 ± 323	
Controls	126 ± 57	25
Pancreatitis:		
Mild	348 ± 39	
Severe	897 ± 183	
Lethal	799 ± 244	
Control	67 ± 31	26
Septicemia:		
Nonbacteremic	341	
Bacteremic	773	27
Crohn's disease	169	
Ulcerative colitis	119	
Controls	55 ± 14	28
Severe multiple trauma	>2,500	
Controls	60–110	29
Acute promyelocytic leukemia	Up to 1,600	30

CRITERIA FOR THE DEVELOPMENT AND EVALUATION OF THE EFFICACY AND SAFETY OF SYNTHETIC PMN ELASTASE INHIBITORS

The initial criteria to be satisfied are ones of specificity and potency. These are necessary for unambiguous demonstration and interpretation of the significance of the biologic activities of such inhibitors. As synthetic elastase inhibitors should be efficacious under circumstances in which natural inhibitors have failed to fulfill their primary function of inhibiting this enzyme, it is important that their activity be demonstrable against elastase in the presence of their natural counterparts.

Upon demonstration of activity in a biologic milieu it is then possible to proceed with functional studies of inhibitors *in vivo*. *In vivo* models should allow a pharmacokinetic profile of compounds to be established as well as the demonstration of their activity against tissue damage mediated by endogenous elastase.

DISCOVERY OF THE POTENTIAL OF CEPHALOSPORIN-BASED COMPOUNDS TO INHIBIT MAMMALIAN SERINE PROTEINASES

The discovery that neutral cephalosporins are inhibitors of mammalian serine proteinases was based on the knowledge that antibiotics such as cephalosporins are acylating inhibitors of bacterial serine proteinases. This led Zimmerman (see ref. 45) to evaluate the activity of clavulanic acid and its benzyl ester as inhibitors of PMN elastase. Evaluation of the two compounds revealed an important dichot-

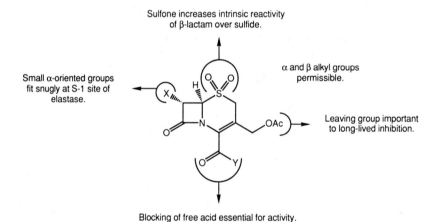

FIGURE 1. Structural determinants of the PMN elastase inhibitory capacity of cephalosporins.

omy. The free acid, known to inhibit the bacterial enzyme, was inactive against PMN elastase, but its benzyl ester had good activity against the mammalian enzyme. This observation prompted Doherty and his colleagues[45] to select a large number of β-lactams, many of them ester intermediates for the synthesis of antibiotics, for evaluation as inhibitors of PMN elastase and other mammalian serine proteinases. A comprehensive structure activity relationship, some salient features of which are shown in FIGURE 1, emerged for these compounds as described initially by Doherty et al.[45] and more recently by Doherty et al.,[46] Finke et al.,[47] and Shah et al.[48] Compounds showed greater activity with 7α substitution than with 7β substituents, the latter being less potent or inactive. Doherty et al.[45] suggested that this was due to the ability of PMN elastase to cleave L-L amino acid peptide linkages in contrast to the D-D amino acid cleaved by bacterial serine proteinases. Consonant with this was the observation that smaller substituents in the 7α position provided more potent inhibitors, as would be expected from the preference of PMN elastase for amino acids with small alkyl substituents in the P1 sites of its substrates. Another important feature of the structure-activity relationship is that free acid substituents in C4 yield poor inhibitors. This is consistent with the preference of elastase for neutral molecules at its sites of cleavage. Also, the presence of a leaving group at the C3 position was desirable. The reason for this became apparent from mechanism-of-action studies based on the observation that the class includes time-dependent inhibitors of elastase. Kinetic data indicate the formation of an initial, reversible complex followed by acylation of the active serine of the enzyme. Further studies by Navia *et al.*[49] provided a solution to the structure of the complex between porcine pancreatic elastase and a potent cephalosporin inhibitor. This study confirmed the formation of an ester bond between serine[195] and the inhibitor after hydrolysis of the β-lactam ring. A further covalent linkage was generated by the nucleophilic attack of His[57] on the leaving group at C3. This complex has a half-life of many hours which results in essentially irreversible inhibition of the enzyme.

PHARMACOLOGIC EVALUATION OF THE BIOLOGIC ACTIVITY OF CEPHALOSPORIN INHIBITORS OF ELASTASE

The most direct method for evaluating functional activity of PMN elastase inhibitors is to measure the potency and duration of action of a compound in models of human PMN elastase-mediated tissue damage. As is detailed elsewhere by Bonney *et al.*[41] and Fletcher *et al.*,[42] assay for inhibition of human PMN elastase-mediated lung hemorrhage has allowed the identification of several compounds with *in vivo* activity. The potency and duration of action of two of these cephalosporin-based compounds are shown in TABLE 2. In addition, Opas *et al.*[50] have shown in this volume that under conditions in which the cephalosporin inhibitor L-658,758 inhibits hemorrhage in the hamster lung, complexes of the inhibitor with the human elastase can be recovered, thereby conclusively demonstrating the successful inhibition of the enzyme.

ACTIVITY OF CEPHALOSPORIN INHIBITORS AGAINST ELASTASE BEING RELEASED FROM PMN

The inhibition of tissue damage in the lung by compounds such as L-658,758 and L-659,286 gives an indication that this class of inhibitors is active in an *in vivo* environment and allows the demonstration of their pharmacokinetics and potency in the lung. However, this does not address the potency of these compounds to inhibit endogenous elastase, which will be their target at sites of inflammation. The cartoon shown in FIGURE 2 outlines an approach to studying inhibitors under such conditions. When elastase is released by exocytosis from stimulated PMN, its local concentration far exceeds that of α_1-PI, that is, in the region of 30–50 μM in plasma, for a brief period of time. Under these transient conditions, elastase should hydrolyze its many substrate proteins present in plasma. As described elsewhere in this volume by Mumford *et al.*,[51] several specific cleavage products are generated from fibrinogen by PMN elastase. The development of a highly specific and sensitive radioimmunoassay for one of these products, the Aα(1–21) N-terminal fragment of the Aα-chain of fibrinogen, allows the assay of the catalytic activity of elastase as it is being released from PMNs. Mumford *et al.*[52]

TABLE 2. Biochemical and Biologic Potency of Cephalosporin-Based PMN Elastase Inhibitors

	L-659,286	L-658,758
Hydrolysis Suc-Ala-Ala-Pro-Ala -p-nitroanilide (k_{inact}/K_i, M^{-1} s^{-1})	12,800	4,100
Solubilization of elastin IC_{50} ($\mu g/ml$)	1–2	1–2
PMN elastase-induced lung hemorrhage (% inhibition)[a]	$ED_{50} = 15$ μg 18% at 10 μg	$ED_{50} = 5$ μg 67% at 10 μg

[a] Drug administered intratracheally 30 minutes before enzyme (for details see ref. 42).

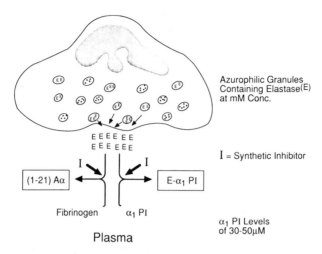

FIGURE 2. Demonstration of the inhibition of PMN elastase activity by synthetic inhibitors in whole blood. The activity of released elastase can be demonstrated by assay of the release of its specific cleavage product Aα(1–21) from fibrinogen.[51] At the same time formation of complexes with its natural inhibitor α_1-PI can be shown.[53] In the presence of a synthetic inhibitor such as L-658,758, inhibition of the formation of both Aα(1–21)[53] and α_1-PI-PMN elastase complexes[51] can be shown.

showed that whole blood incubated with stimuli such as A21387 or zymosan generate between 1 and 3 nM of Aα(1–21). At the same time released enzyme becomes complexed with α_1-PI.[53] Pacholok *et al.*[53] have shown that cephalosporin inhibitors such as L-658,758 gives a concentration-dependent inhibition of the formation of both Aα(1–21) (IC$_{50}$ = 14.5 ± 3.5 μM) and α_1-PI-PMN elastase complexes (IC$_{50}$ = 38.3 ± 4.4 μM) in this experimental protocol. Synthetic inhibitors also inhibit Aα(1–21) formation observed when isolated PMNs are stimulated to degranulate onto a matrix of fibrinogen. Under these conditions, in the absence of plasma protein, the IC$_{50}$ for L-658,758 is considerably lower.[54] The efficacy of L-658,758 in inhibiting Aα(1–21) formation can also be demonstrated after *in vivo* administration to a chimpanzee. As described in this volume by Mumford *et al.*,[51] an intravenous bolus of 10 mg/kg of L-658,758 gives up to 70% inhibition of Aα(1–21) formation *ex vivo*. This inhibition is proportional to the blood levels of L-658-758 achieved, and it declines as the compound disappears from the circulation.

These studies indicate that cephalosporin inhibitors are effective in inhibiting elastase activity in a supplementary manner to that of α_1-PI. The extension of such activity to acute inflammatory disease in which tissue damage may be caused by elastase released from PMNs at sites of inflammation and infection would be of considerable therapeutic potential. To explore this possibility with this class of inhibitors a model of acute immune-complex-mediated tissue damage in the lung was developed by Fletcher *et al.*[55] An acute reverse passive Arthus reaction was initiated by a simultaneous intravenous injection of an antigen (ovalbumin) and intratracheal instillation of a rabbit antiovalbumin antibody. Over a period of 4 hours the classical features of a reverse passive Arthus reaction, namely, the accumulation of PMNs and an increase in vascular permeability, were observed.

In addition, under the conditions developed by Fletcher et al.,[55] a hemorrhagic response was also observed with release of blood into the airspace as measured by lavage and spectrophotometric determination of hemoglobin levels (Fletcher et al., in preparation). The administration of L-658,758 approximately 1.5 hours after the formation of the immune complexes (to exclude possible effects on cell recruitment) gave a dose-dependent inhibition of the hemorrhage, with up to 90% inhibition being observed at the highest dose of drug used. The effects on vascular permeability and cell recruitment were of marginal significance, suggesting that the activity of L-658,758 is a specific one directed towards the PMN elastase-mediated destruction of basement membranes.

CONCLUSIONS

The biologic profile of the cephalosporin inhibitors of PMN elastase that we have summarized here provides a definitive rationale for the evaluation of synthetic, low molecular weight, irreversible inhibitors of PMN elastase in diseases in which tissue damage may be caused by the release of this enzyme at sites of acute inflammation and infection. By the use of specific readouts of PMN elastase activity in biologic fluids, it has been possible to demonstrate the activity of these inhibitors in environments containing physiologic levels of natural inhibitors such as α_1-PI. Apparently the compounds that we have described have the appropriate pharmacokinetic profile to express considerable functional activity in the lung, both against free human PMN elastase introduced by intratracheal administration[41,42] as well as endogenous rat PMN elastase released at a local site of acute inflammation resulting from immune complex formation in the reverse passive Arthus reaction.[55] These observations indicate that synthetic PMN elastase inhibitors may find a place in the treatment of a wide variety of acute inflammatory diseases in which tissue damage caused by the extracellular release of the enzyme may occur.

ACKNOWLEDGMENTS

The assistance of Lorena Bennett in the preparation of this manuscript is gratefully acknowledged.

REFERENCES

1. KAO, R. C., N. G. WEHNER, K. M. SKUBITZ, B. H. GRAY & J. R. HOIDAL. 1988. Proteinase 3. A distinct human polymorphonuclear leukocyte proteinase that produces emphysema in hamsters. J. Clin. Invest. **82:** 1963–1973.
2. CAMPANELLI, D., P. A. DETMERS, C. F. NATHAN & J. E. GABAY. 1990. Azurocidin and a homologous serine protease from neutrophils. Differential antimicrobial and proteolytic properties. J. Clin. Invest. **85:** 904–915.
3. WILDE, C. G., J. L. SNABLE, J. E. GRIFFITH & R. W. SCOTT. 1990. Characterization of two azurophil granule proteases with active-site homology to neutrophil elastase. J. Biol. Chem. **265:** 2038–2041.
4. OHLSSON, J., C. LINDER & M. ROSENGREN. 1990. Monoclonal antibodies specific for neutrophil proteinase 4. Production and use for isolation of the enzyme. Z. Biol. Chem. Hoppe-Seyler **371:** 549–555.

5. GABAY, J. E., R. W. SCOTT, D. CAMPANELLI, J. GRIFFITH, C. WILDE, M. N. MARRA, M. SEEGER & C. F. NATHAN. 1989. Antibiotic proteins of human polymorphonuclear leukocytes. Proc. Natl. Acad. Sci. **86:** 5610–5614.
6. SPITZNAGEL, J. K. 1990. Antibiotic proteins of human neutrophils. J. Clin. Invest. **86:** 1381–1386.
7. PEREIRA, H. A., W. M. SHAFER, J. POHL, L. E. MARTIN & J. K. SPITZNAGEL. 1990. CAP37, a human neutrophil-derived chemotactic factor with monocyte specific activity. J. Clin. Invest. **85:** 1468–1476.
8. FOURET, P., R. M. DUBOIS, J. BERNAUDIN, H. TAKAHASHI, V. J. FERRANS & R. G. CRYSTAL. 1989. Expression of the neutrophil elastase gene during human bone marrow cell differentiation. J. Exp. Med. **169:** 833–845.
9. HANSON, R. D., N. L. CONNOLLY, D. BURNETT, E. J. CAMPBELL, R. M. SENIOR & T. J. LEVY. 1990. Developmental regulation of the human cathespin G gene in myelomonocytic cells. J. Biol. Chem. **265:** 1524–1530.
10. BAINTON, D. F. & M. G. FARQUHAR. 1966. Origin of granules in polymorphonuclear leukocytes. Two types derived from opposite faces of the Golgi apparatus in developing granulocytes. J. Cell. Biol. **28:** 277–301.
11. BAINTON, D. F., J. L. ULLYOT & M. G. FARQUHAR. 1971. The development of neutrophilic polymorphonuclear leukocytes in human bone marrow. Origin and content of azurophil and specific granules. J. Exp. Med. **134:** 907–934.
12. OKANO, K., Y. AOKI, T. SAKURAI, M. KAJITANI, S. KANAI, T. SHIMAZU, H. SHIMIZU & M. NARUTO. 1987. Molecular cloning of complementary DNA for human medullasin: An inflammatory serine protease in bone marrow cells. J. Biochem. **102:** 13–16.
13. FARLEY, D., G. SALVESEN & J. TRAVIS. 1988. Molecular cloning of human neutrophil elastase. Z. Biol. Chem. Hoppe-Seyler **369:** 3–7.
14. HOHN, P. A., N. C. POPESCU, R. D. HANSON, G. SALVESEN & T. J. LEY. 1989. Genomic organization and chromosomal localization of the human cathespin G gene. J. Biol. Chem. **264:** 13412–13419.
15. SALVESEN, G., D. FARLEY, J. SHUMAN, A. PRYZYBYLA, C. REILLY & J. TRAVIS. 1987. Molecular cloning of human cathespin G: Structural similarity to mast cell and cytotoxic T lymphocyte proteinases. Biochemistry **26:** 2289–2293.
16. TAKAHASHI, H., T. NUKIWA, K. YOSHIMURA, C. D. QUICK, D. J. STATES, M. D. HOLMES, J. WHANG-PENG, T. KNUTSEN & R. G. CRYSTAL. 1988. Structure of the human neutrophil elastase gene. J. Biol. Chem. **263:** 14739–14747.
17. TAKAHASHI, H., T. NUKIWA, P. BASSET & R. G. CRYSTAL. 1988. Myelomonocytic cell lineage expression of the neutrophil elastase gene. J. Biol. Chem. **263:** 2543–2547.
18. SINHA, S., W. WATOREK, S. KARR, J. GILES, W. BODE & J. TRAVIS. 1987. Primary structure of human neutrophil elastase. Proc. Natl. Acad. Sci. USA **84:** 2228–2232.
19. CAMPBELL, E. J. 1986. Preventive therapy of emphysema: Lessons from the elastase model. Am. Rev. Respir. Dis. **134:** 435–437.
20. TAKEUCHI, K., H. WOOD & R. T. SWANK. 1986. Lysosomal elastase and cathepsin G in beige mice. Neutrophils of beige (Chediak-Higashi) mice selectively lack lysosomal elastase and cathepsin G. J. Exp. Med. **163:** 665–677.
21. TAKEUCHI, K. H., M. P. MCGARRY & R. T. SWANK. 1987. Elastase and cathepsin G activities are present in immature bone marrow neutrophils and absent in late marrow and circulating neutrophils of beige (Chediak-Higashi) mice. J. Exp. Med. **166:** 1362–1376.
22. TAKEUCHI, K. H. & R. T. SWANK. 1989. Inhibitors of elastase and cathespin G in Chediak-Higashi (beige) neutrophils. J. Biol. Chem. **264:** 7431–7436.
23. DUSWALD, K.-H., M. JOCHUM, W. SCHRAMM & H. FRITZ. 1985. Released granulocytic elastase: An indicator of pathobiochemical alterations in septicemia after abdominal surgery. Surgery **98:** 892–899.
24. ADEYEMI, E. O., L. B. CAMPOS, S. LOIZOU, M. J. WALPORT & H. J. F. HODGSON. 1990. Plasma lactoferrin and neutrophil elastase in rheumatoid arthritis and systemic Lupus erythematosus. Br. J. Rheumatol. **29:** 15–20.
25. KURAMITSU, K. & A. YOSHIDA. 1990. Plasma and synovial fluid levels of granulocyte elastase-α-1-protease inhibitor complex in patients with rheumatoid arthritis. Rheumatol. Int. **10:** 51–56.

26. GROSS, V., J. SCHOLMERICH, H.-G. LESER, R. SALM, M. LAUSEN, K. RUCKAUER, U. SCHOFFEL, L. LAY, A. HEINISCH, E. H. FARTHMANN & W. GEROK. 1990. Granulocyte elastase in assessment of severity of acute pancreatitis. Comparison with acute-phase proteins C-reactive protein, α_1-antitrypsin, and protein inhibitor α_2-macroglobulin. Dig. Dis. Sci. **35:** 97–105.

27. STRUELENS, M., J. DELVILLE, P. LUYPAERT & J. WYBRAN. 1988. Granulocyte elastase compared to C-reactive protein for early diagnosis of septicemia in critically ill patients. Eur. J. Clin. Microbiol. Infec. Dis. **7:** 193–195.

28. ADEYEMI, E. O., S. NEUMANN, V. S. CHADWICK, H. F. J. HODGSON & M. B. PEPYS. 1985. Circulating human leucocyte elastase in patients with inflammatory bowel disease. Gut **26:** 1306–1311.

29. FRITZ, H., M. JOCHUM, R. GEIGER, K. H. DUSWALD, H. DITTMER, H. KORTMANN, S. NEUMANN & H. LANG. 1986. Granulocyte proteinases as mediators of unspecific proteolysis in inflammation: A review. Folia Histochem. et Cytobiol. **24:** 99–115.

30. JOCHUM M. & H. FRITZ. 1983. Plasma levels of human granulocytic elastase-α_1-proteinase inhibitor complex (E-α_1PI) in patients with septicemia and acute leukemia. *In* Selected Topics in Clinical Enzymology. D. M. Godberg & R. M. Werner, eds.: 85–100. Walter de Gruyter & Co. Berlin, NY.

31. S. SUTER, U. B. SCHAAD, P. ROUX-LOMBARD, E. GIRARDIN, G. GRAU & J. M. DAYER. 1989. Relation between tumor necrosis factor-alpha and granulocyte elastase-α_1-proteinase inhibitor complexes in the plasma of patients with cystic fibrosis. Am. Rev. Respir. Dis. **140:** 1640–1644.

32. TOSI, M. F., H. ZAKEM & M. BERGER. 1990. Neutrophil elastase cleaves C3bi on opsonized pseudomonas as well as CR1 to create a functionally important receptor mismatch. J. Clin. Invest. **86:** 300–308.

33. CRYSTAL, R. G. 1990. α1-Antitrypsin deficiency, emphysema, and liver disease. Genetic basis and strategies for therapy. J. Clin. Invest. **85:** 1343–1352.

34. TRAVIS, J. & S. G. SALVESEN. 1983. Human plasma proteinase inhibitors. Ann. Rev. Biochem. **52:** 655–709.

35. CAMPBELL, E. J., R. M. SENIOR, J. A. McDONALD & D. L. COX. 1982. Relative importance of cell-substrate contact and oxidative inactivation of proteinase inhibitors *in vitro*. J. Clin. Invest. **70:** 845–852.

36. CAMPBELL, E. J. & M. A. CAMPBELL. 1988. Pericellular proteolysis by neutrophils in the presence of proteinase inhibitors: Effects of substrate opsonization. J. Cell. Biol. **106:** 667–676.

37. WEISS, S. J., J. T. CURNUTTE & S. REGIANI. 1986. Neutrophil-mediated solubilization of subendothelial matrix: Oxidative and non-oxidative mechanisms of proteolysis used by normal and chronic granulomatous disease phagocytes. J. Immunol. **136:** 636–641.

38. RICE, W. G. & S. J. WEISS. 1990. Regulation of proteolysis at the neutrophil-substrate interface by secretory leukoprotease inhibitor. Science **249:** 178–181.

39. SCHRIJVER, G., J. SCHALKWIJK, J. C. M. ROBBEN, K. J. M. ASSMANN & R. A. P. KOENE. 1989. Antiglomerular basement membrane nephritis in beige mice. Deficiency of leukocytic neutral proteinases prevents the induction of albuminuria in the heterologous phase. J. Exp. Med. **169:** 1435–1448.

40. STARCHER, B. & I. WILLIAMS. 1989. The beige mouse: Role of neutrophil elastase in the development of pulmonary emphysema. Exp. Lung Res. **15:** 785–800.

41. BONNEY, R. J., B. ASHE, A. MAYCOCK, P. DELLEA, K. HAND, D. OSINGA, D. FLETCHER, R. MUMFORD, P. DAVIES, D. FRANKENFIELD, T. NOLAN, L. SCHAEFFER, W. HAGMANN, P. FINKE, S. SHAH, C. DORN & J. DOHERTY. 1988. Pharmacological profile of the substituted beta-lactam L-659,286: A member of a new class of human PMN elastase inhibitors. J. Cell. Biochem. **39:** 47–53.

42. FLETCHER, D., D. OSINGA, K. HAND, P. DELLEA, B. ASHE, R. MUMFORD, P. DAVIES, W. HAGMANN, P. FINKE, J. DOHERTY & R. J. BONNEY. 1990. A comparison of alpha-1-proteinase inhibitor methoxysuccinyl-Ala-Ala-Pro-Val chloro-methylketone, and specific beta-lactam inhibitors in an acute model of human PMN elastase-induced lung hemorrhage in the hamster. Am. Rev. Respir. Dis. **141:** 672–677.

43. JOHNSON, R. J., W. G. COUSER, C. E. ALPERS, M. VISSERS, M. SCHULZE & S. J.

KLEBANOFF. 1988. The human neutrophil serine proteinases, elastase and cathepsin G, can mediate glomerular injury *in vivo*. J. Exp. Med. **168:** 1169–1174.

44. SNIDER, G. L., E. C. LUCEY & P. J. STONE. 1986. Animal models of emphysema. Am. Rev. Respir. Dis. **133:** 149–169.

45. DOHERTY, J. B., B. M. ASHE, L. W. ARGENBRIGHT, P. L. BARKER, R. J. BONNEY, G. O. CHANDLER, M. E. DAHLGREN, C. P. DORN JR., P. E. FINKE, R. A. FIRESTONE, D. FLETCHER, W. K. HAGMANN, R. MUMFORD, L. O'GRADY, A. L. MAYCOCK, J. M. PISANO, S. K. SHAH, K. R. THOMPSON & M. ZIMMERMAN. 1986. Cephalosporin antibiotics can be modified to inhibit human leukocyte elastase. Nature **322:** 192–194.

46. DOHERTY, J. B., B. M. ASHE, P. L. BARKER, T. J. BLACKLOCK, J. W. BUTCHER, G. O. CHANDLER, M. E. DAHLGREN, P. DAVIES, C. P. DORN, JR., P. E. FINKE, R. A. FIRESTONE, W. K. HAGMANN, T. HALGREN, W. B. KNIGHT, A. L. MAYCOCK, M. A. NAVIA, L. O'GRADY, J. M. PISANO, S. K. SHAH, K. R. THOMPSON, H. WESTON & M. ZIMMERMAN. 1990. Inhibition of human leukocyte elastase. 1. Inhibition by C-7-substituted cephalosporin *tert*-butyl esters. J. Med. Chem. **33:** 2513–2521.

47. FINKE, P. E., B. M. ASHE, W. B. KNIGHT, A. L. MAYCOCK, M. A. NAVIA, S. K. SHAH, K. R. THOMPSON, D. J. UNDERWOOD, H. WESTON, M. ZIMMERMAN & J. B. DOHERTY. 1990. Inhibition of human leukocyte elastase. 2. Inhibition by substituted cephalosporin esters and amides. J. Med. Chem. **33:** 2522–2528.

48. SHAH, S. K., K. A. BRAUSE, G. O. CHANDLER, P. E. FINKE, B. M. ASHE, H. WESTON, W. B. KNIGHT, A. L. MAYCOCK & J. B. DOHERTY. 1990. Inhibition of human leukocyte elastase. 3. Synthesis and activity of 3'-substituted cephalosporins. J. Med. Chem. **33:** 2529–2535.

49. NAVIA, M. A., J. P. SPRINGER, T. LIN, H. R. WILLIAMS, R. A. FIRESTONE, J. M. PISANO, J. B. DOHERTY, P. E. FINKE & K. HOOGSTEEN. 1987. Crystallographic study of a β-lactam inhibitor complex with elastase at 1.84 Å resolution. Nature **327:** 79–82.

50. OPAS, E., P. DELLEA, D. FLETCHER, P. DAVIES & J. L. HUMES. The *in vitro* and *in vivo* recovery of PMN elastase: Beta lactam inhibitor complexes by cation exchange chromatography. Ann. N.Y. Acad. Sci. This volume.

51. MUMFORD, R. A., H. WILLIAMS, J. MAO, M. E. DAHLGREN, D. FRANKENFELD, T. NOLAN, L. SCHAFFER, J. B. DOHERTY, D. FLETCHER, K. HAND, R. BONNEY, J. L. HUMES, S. PACHOLOK, W. HANLON & P. DAVIES. Measurement of the polymorphonuclear leukocyte elastase generated fibrinopeptide Aα(1–21) in chimpanzee blood stimulated with the calcium ionophore A23187. Ann. N.Y. Acad. Sci. This volume.

52. MUMFORD, R. A., H. WILLIAMS, J. MAO, M. E. DAHLGREN, R. HUBBARD, R. CRYSTAL & P. DAVIES. 1990. Production of Aα(1–21), the human PMN elastase generated cleavage product of fibrinogen, in PiZZ individuals' blood stimulated by calcium ionophore (A23187) before and after reconstitution therapy with α-antitrypsin (α_1AT). Am. Rev. Respir. Dis. **141:** A111.

53. PACHOLOK, S. G., R. J. BONNEY, P. DAVIES, J. DOHERTY, R. A. MUMFORD & J. L. HUMES. 1990. Effects of L-658,758, a substituted cephalosporin PMN elastase inhibitor, on A23187-stimulated PMN elastase: $\alpha_1$proteinase inhibitor complex formation in human blood. Am. Rev. Respir. Dis. **141:** A111.

54. HANLON, W. A., J. L. HUMES, B. KNIGHT, R. A. MUMFORD & J. STOLK. 1990. A model for the study of elastase released from azurophilic granules in polymorphonuclear leukocytes. *In* Abstracts of the Fifth International Conference, Inflammation Research Association, White Haven, PA., Sept. 23–27.

55. FLETCHER, D., D. OSINGA, R. J. BONNEY, K. KEENAN, P. FINKE, J. B. DOHERTY & P. DAVIES. 1990. The role of polymorphonuclear leukocyte elastase in immune complex-mediated tissue injury in the lung. Am. Rev. Respir. Dis. **141:** A111.

Biochemistry and Pharmacology of ICI 200,880, a Synthetic Peptide Inhibitor of Human Neutrophil Elastase

JOSEPH C. WILLIAMS, ROSS L. STEIN,[a]
RALPH E. GILES, AND ROBERT D. KRELL

Pulmonary Pharmacology Section
Department of Pharmacology
ICI Pharmaceuticals Group
ICI Americas Inc.
Wilmington, Delaware 19897

A large body of circumstantial evidence indicating that a protease-antiprotease imbalance contributes to the pathogenesis of emphysema has been accumulated. Several lines of evidence suggest that the elastase contained within polymorpho-nuclear leukocytes (PMN) may represent the most important protease limb of this hypothesis.[1] Even though the evidence for human neutrophil elastase (HNE) involvement in human emphysema is circumstantial, this hypothesis does suggest a rational approach for the design of chemicals with therapeutic potential.

Many protease inhibitors have been synthesized and biologically evaluated.[2] These inhibitors include both irreversible inhibitors such as the peptide chloro-methylketones,[3] reversible inhibitors such as the peptide boronic acids,[4] cepha-losporin inhibitors,[5] and peptide aldehydes.[6] Moreover, several biosynthetically derived natural inhibitors including alpha$_1$-proteinase inhibitor,[7] Eglin C,[8] and secretory leukoproteinase inhibitor (SLPI)[9,10] have been described both biochem-ically and pharmacologically. Variously, these compounds have inhibited HNE both *in vitro* and *in vivo* with varying degrees of success.

This communication describes the biologic profile of ICI 200,880 (FIG. 1), a selective, potent, peptide trifluoromethylketone inhibitor of HNE that prevents the acute effects and halts the progression of elastase-induced destructive lung lesions in experimental animals.

METHODS

Determination of Kinetics of Inhibition of Human Neutrophil Elastase

Substrate, Methoxysuccinyl-Ala-Ala-Pro-Val-pNA, was hydrolyzed by HNE releasing *p*-nitroanalide which was continuously measured spectrophotometri-cally by monitoring absorbance changes at 410 nm. Both substrate and inhibitor were dissolved in DMSO. Fifty microliters of both substrate and inhibitor or DMSO were added to a cuvette containing 2.895 ml of buffer I (10 mM Na phosphate, 500 mM NaCl, pH 7.6). The cuvette was placed in a thermostatically

---

[a] PRESENT ADDRESS: Merck and Co., Rahway, New Jersey 07065.

controlled, water-jacketed holder in the cell compartment of a Cary 210 spectro-photometer (Varian Techtron, PTX, Ltd, Victoria, Australia) and allowed to reach thermal equilibrium. The temperature was maintained at 25 ± 0.1°C. The reaction was initiated by the addition of 5 μl of enzyme solution (0.14 mg/ml). Absorbance was continuously monitored and stored in a DEC PDP 11/32 mini-computer. Initial and steady-state velocities were calculated by a fit of the experimental data to a linear dependence on time by linear least-squares analysis. Triplicate determinations were conducted for each inhibitor concentration.

Determination of Protease Selectivity

The protease selectivity of ICI 200,880 was determined by using minor modifications of the foregoing protocol. All substrates were dissolved in DMSO and combined with inhibitor (dissolved in the same solvent). Equal volumes (50 μl) of inhibitor and substrate were added to 2.895 ml of buffer I in a cuvette which was placed in a spectrophotometer for continuous monitoring of absorbance changes. When the reaction mixtures had reached thermal equilibrium (25°C), the reactions

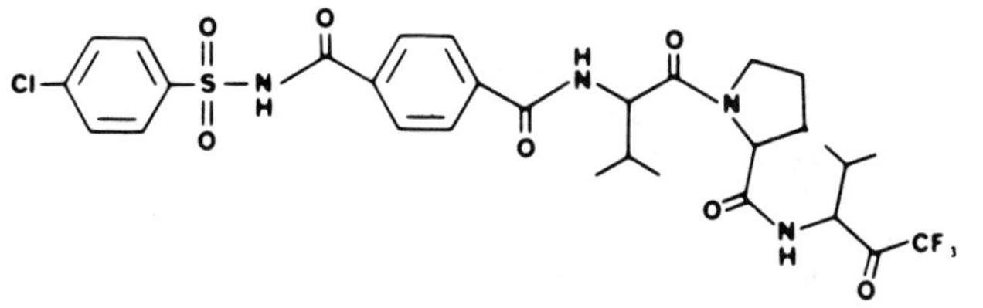

FIGURE 1. Structure of ICI 200,880. The chemical name is [4-(4-*chloro*phenyl sulfonycar-bamoyl) benzoyl-L-valyl-L-prolyl-1-(RS)-(1-trifluoroacetyl-2-methylprolyl)amide].

were initiated by the addition of 5 μl of enzyme and absorbance was monitored at either 410 or 348 nm, depending on the substrate leaving group. Assay conditions for the various enzymes were as follows:

Porcine pancreatic elastase:
  Substrate; succinyl-Ala-Ala-Ala-pNA: buffer I:
  enzyme concentration; 24 nM: wavelength; 410 nm.
Bovine pancreatic chymotrypsin:
  Substrate; succinyl-Ala-Ala-Pro-Phe-pNA: buffer I:
  enzyme concentration; 3.3 nM: wavelength; 410 nm.
Human plasma thrombin:
  Substrate; BZ-Phe-Val-Ala-pNA: buffer I:
  enzyme concentration; 5 units/ml: wavelength; 410 nm.
Acetylcholinesterase:
  Substrate; acetylcholine: buffer I:
  enzyme concentration; 14 nM: wavelength; 420 nm.

Human leukocyte cathepsin G:
  Substrate; succinyl-Ala-Ala-Pro-Phe-pNA: 0.1 M Tris,
  0.7 mM NaN_3, pH 8.3.
  enzyme concentration; 25 nM: wavelength; 410 nm.
Angiotensin-converting enzyme:
  Substrate; FA-Phe-Gly-Gly: 50 mM HEPES, 300 mM NaCl,
  pH 7.5: enzyme concentration; 10 nM: wavelength; 348 nm.
Trypsin:
  Substrate; Bz-Phe-Val-Arg-pNA: buffer I:
  enzyme concentration; 13 nM: wavelength; 410 nm.
Papain:
  Substrate; N-CBZ-Gly-pNPE: 0.2 M $NaPO_4$, 0.005 M EDTA,
  1.65% DMSO, 0.8% CH_3CN, pH 6.0: enzyme concentration; 87 nM:
  wavelength; 410 nm.
Pepsin:
  Substrate; Leu-Ser-(NO_2)-Phe-Nleu-Ala-Leu-OCH_3: 0.1 M,
  Na Citrate, 3.3% DMSO, pH 3.0.
  enzyme concentration; 9.5 nM: wavelength; 410 nm.

Pharmacokinetic Evaluation of ICI 200,880

This assay utilizes the HNE inhibitory activity of the inhibitors to assess their presence in biologic fluids. In the species evaluated, the endogenous elastase inhibitors found in biologic fluids are also trypsin inhibitors. By complexing the endogenous inhibitors with exogenous trypsin, the levels of ICI 200,880 in various biologic fluids could be quantified. The pharmacokinetic characterization of the compound was determined as follows: before and at varying times after intravenous administration, blood samples (0.2 ml) were obtained via cardiac puncture under light anesthesia. The blood was expressed into 2-ml centrifuge tubes, allowed to clot for 30 minutes, and centrifuged in a Beckman microfuge. Fifty microliters of serum were then combined with 50 μl of trypsin (5 mg/ml) and allowed to interact for 5 minutes. The trypsin-treated serum (10 μl) was then added to a 0.52-ml cuvette containing buffer with 20 nM HNE. After an additional 30 minutes of incubation, the reaction was started with the addition of substrate (350 μl of 1.6 mM Methoxysuccinyl-Ala-Ala-Pro-Val-pNA) and the reaction monitored spectrophotometrically at 410 nm in a Baxter-Rotochem centrifugal analyzer. Percentage of inhibition of HNE by trypsinized serum was calculated. The actual serum concentrations of the inhibitors were determined by comparing the percentage of inhibition of the test sample with the percentage of inhibition determined for a standard curve for ICI 200,880 prepared in control blood. Log serum drug concentration was plotted versus time after drug administration, and an approximate t½ elimination time was determined from the resultant curve.

The retention of ICI 200,880 in the lungs of hamsters after inhalation exposure was determined by exposing animals, in a muzzle only exposure system, to an aerosol of ICI 200,880. Compound was dissolved in phosphate-buffered saline solution (PBS) at a concentration of 10 mg/ml and aerosols were generated using a Retec® nebulizer. The animals were exposed to the resultant atmosphere for 60 minutes. At various times after the end of the exposure, the animals were killed, and the lungs were removed and lavaged with PBS. Lavage samples were processed and analyzed for ICI 200,880 as just described.

Induction of Acute Lung Injury with HNE

Male, Syrian hamsters (90 to 110 g) were lightly anesthetized with Brevital sodium (30 mg/kg ip) and the trachea was surgically exposed. A dose of 400 μg of HNE in 0.3 ml 0.01 M PBS was injected into the exposed trachea via a $\frac{1}{2}$-inch, 23-gauge needle. The incision was closed with stainless steel surgical staples and the animals were allowed to recover. Twenty-four hours after the injection of HNE, the animals were killed with an overdose of pentobarbital sodium. The lungs and heart were resected and the lungs and trachea carefully cleaned of extraneous material. After measurement of wet lung weight, the tracheas were cannulated and lavaged three times with 2 ml of PBS. The recovered lavages were pooled for each animal and the volume was recorded. Total red and white blood cell counts were determined using a Coulter counter. The data are expressed as lung weight/100 g body weight and total cells recovered (cells/ml × volume recovered).

Therapeutic Effect of ICI 200,880 in PPE- and HNE-Induced Emphysema in Hamsters

PPE Therapeutic Trial

On day 0, male, Syrian hamsters (90–100 g) were lightly anesthetized with Brevital sodium and received intratracheal injections of vehicle or 100 or 150 μg/animal PPE in a total volume of 0.3 ml PBS. The animals were allowed to recover and placed in cages with food and water allowed *ad libitum*. Twenty-four hours after administration of enzyme, the animals were divided into equal groups and received twice daily subcutaneous injections of either 50 or 100 μmol/kg ICI 200,880 or vehicle for either 14 or 28 days after which they were killed and the lungs prepared for morphometric analysis as will be described. Two groups of animals received an intratracheal injection of PBS followed by twice daily administration of ICI 200,880 (100 μmol/kg sc) for either 14 or 28 days after which the lungs were removed and processed for morphometric analysis as will be described.

HNE Therapeutic Trial

On day 0, animals received intratracheal injections of either vehicle or 400 μg HNE/animal as described in the previous paragraph. Twenty-four hours later the animals were divided into equal groups and were treated twice daily with subcutaneous injections of either 50 or 100 μmol/kg ICI 200,880 or vehicle. After 56 days of drug or vehicle treatment, all animals were killed and lungs were prepared as will be described. A separate group of animals received an intratracheal injection of PBS followed by subcutaneous administration of 100 μmol/kg of ICI 200,880 for the duration of the study.

Morphometric Analysis

At the indicated times after enzyme administration, the animals were killed with an injection of pentobarbital sodium, the thorax was opened, and the lungs and heart were removed. The heart and other extraneous material were carefully dissected away, and the lungs were washed in saline solution, blotted dry, and weighed. The lungs were then inflated and fixed with 10% phosphate-buffered formalin (pH 7.0) at 25 cm formalin pressure. Five-micron hematoxylin and eosin sections were prepared by JDJ Laboratories (Elkton, Maryland) and mean linear intercept determined using an Optomax Image Analyzer (Hollis, New Hampshire).

Statistical Analysis

Data were evaluated using Student's two-tailed t test with $p < 0.05$ considered statistically significant.

Source of Materials

Bovine pancreatic chymotrypsin, human plasma thrombin, acetylcholinesterase, angiotensin-converting enzyme, bovine pancreatic trypsin, pepsin, and all synthetic substrates were purchased from Sigma Chemical Co. (St. Louis, Missouri). Porcine pancreatic elastase was purchased from Worthington Biochemicals (Freehold, New Jersey), and human neutrophil elastase and cathepsin G were purchased from Elastin Products (Owensville, Missouri). All other reagents were of the highest grade commercially available. Male, Syrian hamsters were purchased from Charles River Laboratories (Wilmington, Massachusetts).

RESULTS

Biochemical Characterization of ICI 200,880

Potency and Selectivity of ICI 200,880 Against Several Hydrolases

ICI 200,880 was evaluated as an inhibitor of several hydrolases including hamster and human neutrophil elastase. ICI 200,880 is a potent, slow-binding inhibitor of human neutrophil elastase with a K_i value of $(5.0 \pm 1.0) \times 10^{-10}$ M and k_{on} of $(8 \pm 1) \times 10^{-4}$ M^{-1} s^{-1}.

In contrast, the compound was a very weak inhibitor of a variety of other proteases (TABLE 1), the exception being porcine pancreatic elastase for which a 150-fold selectivity ratio was observed.

Pharmacologic Characterization of ICI 200,880

Pharmacokinetic Evaluation

When administered as an intravenous bolus to hamsters, ICI 200,880 (20 μmol/kg) was rapidly distributed and eliminated from the circulation with an estimated t½ of 62 minutes (FIG. 2). The t½ retention time after a 60-minute inhalation exposure to ICI 200,880 (10 mg/ml) was approximately 10 hours (FIG. 3). In contrast, the ability of the lung to retain the irreversible inhibitor of HNE, Methoxysuccinyl-Ala-Ala-Pro-Val-CMK, was much less. When administered by inhalation under the same conditions as those for ICI 200,880, this inhibitor was not detectable for more than 1 hour after the end of the exposure. The estimated $t\frac{1}{2}$ retention time was 0.8 hour (FIG. 3).

Inhibition of HNE-Induced Acute Lung Injury by ICI 200,880

The ability of ICI 200,880, administered by inhalation exposure, to inhibit HNE-induced acute lung injury was evaluated. The animals were exposed to an aerosol of ICI 200,880 generated from a 10 mg/ml solution by a Retec® nebulizer for 1 hour. Twenty-four or 48 hours after the end of the aerosol exposure, the animals were challenged with a dose of 100 μg/animal of HNE intratracheally and killed 24 hours later. ICI 200,880 provided significant ($p \leq 0.05$) inhibition of the HNE-induced increases of lung weight/body weight and total red blood cells and

TABLE 1. Hydrolase Selectivity of ICI 200,880

Hydrolase	K_i(nM)
Serine	
HLE	0.5
PPE	74
Cathepsin G	180,000
Chymotrypsin	NI[a]
Thrombin	65,000
Trypsin	NI
Acetylcholinesterase	NI
Cysteine	
Papain	NI
Metallo	
Thermolysin	NI
Angiotensin-converting enzyme	NI
Acid	
Pepsin	45,000

[a] NI = no inhibition at 100 μM.

white blood cells (FIG. 4). Forty-eight hours after exposure, significant ($p <0.05$) inhibition of the HNE-induced increase of total RBCs, but not lung weight/body weight or WBCs, was apparent (data not shown).

Effect of ICI 200,880 on the Progression of Pulmonary "Emphysema-Like" Lesions Induced by PPE or HNE

The ability of ICI 200,880, administered subcutaneously, to inhibit the progression of an emphysema-like lesion induced by PPE was evaluated. Increases in mean linear intercept (MLI) were produced by intratracheal administration of either 100 or 150 μg of PPE. As is apparent in FIGURE 5A, 2 weeks after PPE administration, increases ($p <0.05$) in MLI were observed. In those animals that had received ICI 200,880 (50 or 100 μmol/kg sc), the resultant PPE-induced lesion

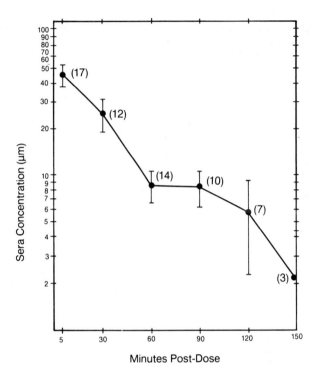

FIGURE 2. Pharmacokinetics of ICI 200,880 administered intravenously to hamsters. ICI 200,880 (0, 20.0 μmol/kg) was administered intravenously and blood samples were taken at various times and analyzed for the compound as described in *Methods*. The t½ for ICI 200,880 calculated from the 30–150-minute periods was 62 minutes ($r = 0.95$). *Symbols with vertical lines* represent the mean ± SEM for the number of animals in parentheses.

was significantly (p <0.05) reduced regardless of the dose of PPE administered. In those animals that were evaluated after 4 weeks of therapy, similar results were obtained (FIG. 5B). That is, administration of ICI 200,880 resulted in PPE-induced lesions that were significantly (p <0.05) less severe than were those observed following vehicle. Administration of ICI 200,880 at doses of 50 and 100 μmol/kg subcutaneously twice a day had no effect on MLI.

The ability of ICI 200,880 to inhibit the progression of an emphysema-like lesion induced by HNE was likewise evaluated. Increases in MLI were apparent 8 weeks after intratracheal administration of 400 μg/animal HNE (FIG. 6). However, in those animals that received ICI 200,880 therapy (either 50 or 100 μmol/kg sc twice a day) starting 24 hours after administration of the enzyme and continuing to the end of the study, no significant (p >0.05) increases were observed.

DISCUSSION

ICI 200,880 is a potent, competitive, slow-binding, and selective inhibitor of neutrophil elastase. The K_i, 0.5 nM, compares favorably with that of the substituted beta-lactam[5,11] and of the peptide aldehyde inhibitors.[6] Although ICI 200,880 is a potent inhibitor of HNE, it is a much weaker inhibitor of PPE and has little or no affinity for the other hydrolases evaluated.

The preclinical pharmacokinetic profiles of ICI 200,880 appear to be ideal for compounds that must be administered by inhalation for treatment of lung diseases. ICI 200,880 was retained in the lung for extended periods of time following inhalation exposure to hamsters. In contrast, when ICI 200,880 was administered as an intravenous bolus to hamsters, it was rapidly cleared from the circulation. This pharmacokinetic profile appears optimum, as it allows extended elastase-inhibitory coverage of the lungs, where the disease process exists, and rapid elimination from systemic circulation following absorption to minimize potential, unknown, negative consequences for inhibition of normal functions of HNE.

The ability of ICI 200,880 to prevent HNE-induced lung lesions was evaluated in a model based on the observations that intratracheal administration of HNE produces a marked hemorrhagic and inflammatory response that is apparent in the first 24 hours after enzyme administration.[12,13] ICI 200,880 provided significant inhibition of both hemorrhage and inflammation, as indicated by wet lung weight and lavageable white blood cells, when administered by inhalation 24 hours prior to intratracheal administration of 100 μg/animal HNE. Hassal *et al.*[14] reported similar inhibition of HNE-induced hemorrhage following intratracheal administra-

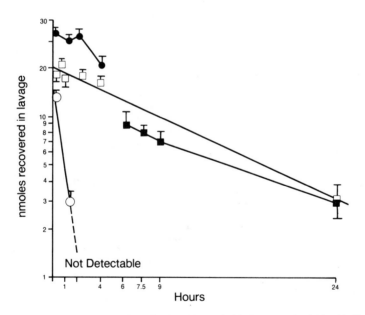

FIGURE 3. Pharmacokinetics for ICI 200,880 and Methoxysuccinyl-Ala-Ala-Pro-Val-chloromethylketone (CMK) administered by aerosol to hamsters. A 10 mg/ml solution of ICI 200,880 or CMK was aerosolized for 60 minutes with a Retec® nebulizer. At various times after administration the animals were killed, the lungs resected and lavaged, and the amount of compound in lavage fluid determined as described in *Methods*. The *symbols* (●, □, ■) represent three separate experiments with ICI 200,880. ○ represents the CMK data. *Symbols with vertical lines* represent the mean ± SEM of 6–10 animals per point.

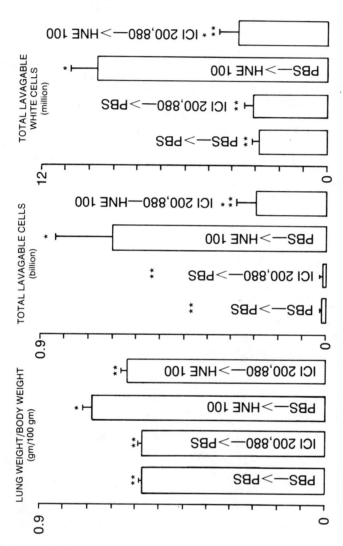

FIGURE 4. Inhibition of HNE-induced acute lung injury by aerosol administration of ICI 200,880. A 10 mg/ml solution of ICI 200,880 was aerosolized for 60 minutes prior to intratracheal administration of 100 μg/animal HNE. Twenty-four hours after HNE the animals were killed, the lungs resected, weighed, and lavaged for determination of lung weight/100 g body weight, total lavageable cells, and total lavageable white cells. Columns represent the mean ± SEM for $n = 10$–12 animals. * = $p < 0.05$ cf. PBS → PBS.

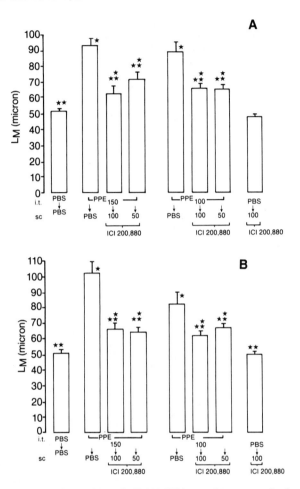

FIGURE 5. The "therapeutic" effect of ICI 200,880 in porcine pancreatic elastase-induced emphysema. At time "0", hamsters received intratracheal injections of PBS with either 100 or 150 μg/animal PPE. Twenty-four hours after elastase administration, twice daily subcutaneous administration of either 50 or 100 μmol/kg ICI 200,880 was initiated. This therapy was continued 7 days per week for the duration of the studies. Following either 2 weeks (**A**) or 4 weeks (**B**) of therapy, the animals were killed and slides prepared for determination of mean linear intercepts (Lm). $* = p < 0.05$ cf. PBS, $** = p < 0.05$ cf. relevant PPE. Columns represents mean ± SEM of $n = 9$–10 animals.

tion of another peptide-based inhibitor at doses of 10 to 1,000 μg/animal. Bonney and co-workers[11] reported that 10 and 100 μg (administered intratracheally) of L-659,286, a cephalosporin inhibitor, produce 18 and 90% inhibition of the hemorrhage response induced by 50 μg HLE, respectively.

Although the data just described indicate that ICI 200,880 is capable of inhibiting HNE both *in vitro* and *in vivo*, none of these experiments was directly related to the destruction of the architecture of the respiratory portion of the lung that is characteristic of emphysema in humans.[15] To evaluate the utility of ICI 200,880 more directly in a model of destructive lung disease, either PPE or HNE was administered directly into the lungs of hamsters and the lesions were allowed to develop prior to initiation of therapy. In contrast to the results reported by Lucey *et al.*,[16] both of these elastases produced destructive lung lesions that were temporarily progressive. In these studies, ICI 200,880 administration was initiated 24 hours after enzyme administration. This delay accomplished two things. First, it allowed time for the vast majority of the enzyme to leave the lung,[17,18] and second,

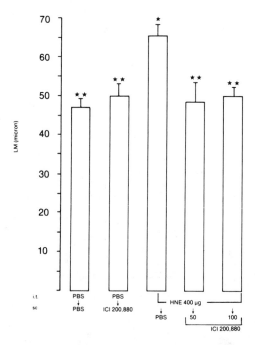

FIGURE 6. The "therapeutic" effect of ICI 200,880 in human neutrophil elastase-induced emphysema. At time "0", hamsters received either PBS or 400 μg HNE/animal 24 hours after elastase administration. Twice daily therapy with ICI 200,880 was initiated as described in FIGURE 5. After 8 weeks of therapy, the animals were killed and slides prepared for determination of Lms. * = p <0.05 cf. PBS, ** = p <0.05 cf. HLE. Columns represent mean ± SEM of n = 7–11 animals.

it provided an animal with an active lesion upon which to evaluate the effect of ICI 200,880. The latter is particularly relevant in that therapy is only applied to patients who have an active disease.

As expected, the lesion induced by PPE was both more rapid in onset and more severe than that induced by HNE.[13] The lowest doses of PPE did not progress between the first and second sampling periods and, in fact, appeared to be regressing. It has been our experience, particularly with PPE, that low doses frequently produce destructive lesions early in sequence which, based solely on MLI values, regress. The large doses of both enzymes did indeed produce temporarily progressive lesions. The parenteral administration of ICI 200,880 for 2–4 weeks in the PPE study and for 8 weeks in the HNE study appeared to prevent the development of a destructive lung lesion. The report by Giles *et al.*[19] that progesterone administered after the establishment of a papain-induced emphysema-like lesion in rats resulted in reversal of both MLI and functional residual capacity and the report by Williams *et al.*[20] that a peptide aldehyde elastase inhibitor prevented the progression of PPE- and HLE-induced increases of MLI are, to our knowledge, the only other reports of therapeutic efficacy of any treatment regimen.

HNE demonstrates a vast array of biologic activities that may contribute to a variety of diseases such as emphysema, cystic fibrosis, adult respiratory distress syndrome, and chronic bronchitis, to name but a few. Therefore, inhibitors of HNE such as ICI 200,880 may be considered to have potential for demonstration of therapeutic activity in a variety of diseases. ICI 200,880 is currently undergoing clinical evaluation in man.

SUMMARY

ICI 200,880 is representative of a new chemical class of inhibitors of human neutrophil elastase (HNE). The compound demonstrated competitive kinetics vs HNE with a K_i value of 5.0×10^{-10} M. The selectivity of ICI 200,880 for HNE versus a variety of enzymes ranged from 150- (relative to porcine pancreatic elastase [PPE]) to greater than 360,000-fold in favor of HNE. In pharmacokinetic studies ICI 200,880 displayed a long retention time when administered directly to the lung and was rapidly eliminated when administered intravenously. Aerosol pretreatment of hamsters with ICI 200,880 before intratracheal administration of HNE produced a long-lasting inhibition of enzyme-induced increases in lung weight, total lavageable red cells, and total lavageable white cells. Subcutaneous administration of either 50 or 100 μmol/kg (twice/day) of ICI 200,880 for 14 or 28 days prevented the time-dependent increase in alveolar diameter produced by a single intratracheal dose of PPE when compound dosing was initiated 24 hours following the enzyme. Treatment of hamsters with ICI 200,880 using the same protocol and doses for 8 weeks prevented the destructive lesion induced by a single intratracheal dose of HNE. It is concluded that ICI 200,880 has biochemical, pharmacokinetic, and pharmacologic profiles that make it a useful therapeutic agent for understanding the role of HNE in various diseases. ICI 200,880 is presently being evaluated in man.

ACKNOWLEDGMENTS

The authors wish to express their appreciation to S. Bailor for her assistance with the preparation of the manuscript and to Dr. J. Schwartz for the synthesis of ICI 200,880.

REFERENCES

1. JANOFF, A. 1985. Elastases and Emphysema. Current assessment of the protease-antiprotease hypothesis. Am. Rev. Respir. Dis. **132:** 417–433.
2. GROUTAS, W. C. 1987. Inhibitors of leukocyte elastase and leukocyte cathepsin G. Agents for the treatment of emphysema and related ailments. Res. Rev. **7:** 227–241.
3. POWERS, J. C., B. F. GUPTON, A. D. HARLEY, N. NIGHINO & R. J. WHITLEY. 1977. Specificity of porcine pancreatic elastase, human leukocyte elastase and cathepsin G. Inhibition with peptide chloromethyl ketones. Biochem. Biophys. Acta **484:** 156–166.
4. KETTNER, C., A. SHENVI, S. WATANABE & N. SOSKEL. 1987. A peptide boronic acid inhibitor of elastase. *In* Pulmonary Emphysema and Proteolysis: 1986. J. C. Taylor & C. Mittman, eds.: 65–72. Academic Press. New York.
5. FLETCHER, D. S., D. G. OSINGA, K. M. HAND, P. S. DELLEA, B. M. ASHE, R. A. MUMFORD, P. DAVIES, W. HAGMANN, P. E. FINKE & J. B. DOHERTY. 1990. A comparison of α_1-proteinase inhibitor methoxysuccinyl-ala-ala-pro-val-chloromethylketone and specific β-lactam inhibitors in an acute model of human polymorphonuclear leukocyte elastase induced lung hemorrhage in the hamster. Am. Rev. Respir. Dis. **141:** 672–677.
6. KENNEDY, A. J., A. CLINE, U. M. NEY, W. H. JOHNSON & N. A. ROBERTS. 1987. The effect of a peptide aldehyde reversible inhibitor of elastase on a human leukocyte elastase induced model of emphysema in the hamster. Eur. J. Respir. Dis. **71:** 472–478.
7. GADEK, J. E., H. G. KLEIN, P. V. HOLLAND & R. G. CRYSTAL. 1981. Replacement therapy of alpha-1-antitrypsin deficiency. Reversal of protease-antiprotease imbalance within the alveolar structures of PiZ subjects. J. Clin. Invest. **68:** 1158–1165.
8. NICK, H. P., A. PROBST & H. P. SCHNEBLI. 1988. Development of eglin C as a drug: Pharmacokinetics. *In* Protease II. Potential Role in Health and Disease. W. H. Horl & A. Heidland, eds.: 83–88. New York and London. Plenum Press.
9. GAUTHIER, F., U. FRYSMARK, K. OHLSSON & J. G. BIETH. 1982. Kinetics of the inhibition of leukocyte elastase by the bronchial inhibitor. Biochem. Biophys. Acta **700:** 178–183.
10. GAST, A., W. ANDERSON, A. PROBST, H. NICK, R. C. THOMPSON, S. P. EISENBERG & H. SCHNEBLI. 1990. Pharmacokinetics and distribution of recombinant secretory leukocyte proteinase inhibitor in rats. Am. Rev. Respir. Dis. **141:** 889–894.
11. BONNEY, R. J., B. ASHE, A. MAYCOCK, P. DELLEA, K. HAND, D. OGINGA, D. FLETCHER, R. MUMFORD, P. DAVIES, D. FRANKENFIELD, T. NOLAN, L. SCHAEFFER, W. HAGMANN, P. FINKE, S. SHAH, C. DORN & J. DOHERTY. 1989. Pharmacological profile of the substituted Beta-Lactam L-659,286: A member of a new class of human PMN elastase inhibitors. J. Cell Biochem. **38:** 47–53.
12. JANOFF, A., B. SLOAN, G. WEINBAUM, V. DAMIANO, R. A. SANDHAUS, J. ELIAS & P. KIMBEL. 1977. Experimental emphysema induced with human neutrophil elastase: Tissue localization of the instilled protease. Am. Rev. Respir. Dis. **115:** 461–478.
13. SENIOR, R. M., H. TEGNER, C. KUHN, K. OHLSSON, B. C. STARCHER & J. A. PIERCE. 1977. The induction of pulmonary emphysema with human leukocyte elastase. Am. Rev. Respir. Dis. **116:** 469–475.
14. HASSAL, C. H., W. H. JOHNSON, A. J. KENNEDY & N. A. ROBERTS. 1985. A new class of inhibitors of human leukocyte elastase. FEBS Lett. **183:** 201–205.
15. SNIDER, G. L., J. KLEINERMAN, W. M. THURLBECK & Z. BEGALI. 1985. The definition of emphysema: Report of a National Heart, Lung and Blood Institute, Division of Lung Diseases, workshop. Am. Rev. Respir. Dis. **132:** 182–185.
16. LUCEY, E. L., P. J. STONE, T. G. CHRISTENSEN, R. BREUER & G. L. SNIDER. 1988. An 18-month study of the hamster lungs of intratracheally administered human neutrophil elastase. Exp. Lung Res. **14:** 671–686.
17. SANDHAUS, R. A. & A. JANOFF. 1982. Elastase-induced emphysema: Retention of instilled proteinases in the rat. Am. Rev. Respir. Dis. **126:** 914–920.

18. STONE, P. J., W. PEREIRA, D. BILES, G. L. SNIDER, H. M. KAGAN & C. FRANZBLAU. 1977. Studies on the fate of pancreatic elastase in the hamster lung: ^{14}C-guanidated elastase. Am. Rev. Respir. Dis. **116:** 49–56.
19. GILES, R. E., M. P. FINKEL & J. C. WILLIAMS. 1974. The therapeutic effect of progesterone in papain-induced emphysema. Proc. Soc. Exp. Biol. Med. **147:** 489–493.
20. WILLIAMS, J. C., P. TOWNSHEND-PIALA, J. TURPIN, R. L. STEIN, A. M. STRIMPLER, B. VISCARELLO & R. D. KRELL. Biochemical and pharmacological characterization of ICI 186,756: A novel, potent and selective inhibitor of human neutrophil elastase. Exp. Lung Res. In press.

Reduction of Neutrophil Elastase Load in the Lungs of Patients with Emphysema by Reducing Neutrophil Enzyme Secretion or Chemotaxis[a]

ALLEN B. COHEN,[b,c] WILLIAM GIRARD,[b]
JERRY McLARTY,[b] BARRY STARCHER,[b] DORIS DAVIS,[b]
MICHAEL STEVENS,[b] JOEL ROSENBLOOM,[d] AND
UMBERTO KUCICH[d]

[b]Departments of Medicine, Biochemistry, and Epidemiology
University of Texas Health Center at Tyler
Tyler, Texas 75710
[d]University of Pennsylvania School of Dentistry
Philadelphia, Pennsylvania 19140

BACKGROUND

Neutrophil Elastase in the Genesis of Some Forms of Pulmonary Emphysema

In recent years, a large body of scientific data has led investigators to formulate the hypothesis that emphysema in patients with a deficiency of alpha$_1$-proteinase inhibitor or in cigarette smokers may be caused by the unrestrained action of neutrophil elastase.

In studies carried out over the last several years, we determined that neutrophils migrate through normal lungs.[2] The pool size of airspace neutrophils was 6 × 10^4 neutrophils per gram of lung,[3] and 38–42% of the pool turns over per hour.[4] These observations indicate that lungs are always exposed to a small pool of neutrophils with a rapid turnover.

Two chemotactic factors for neutrophils are secreted by human alveolar macrophages.[5] One factor has a molecular weight of 350, whereas the other factor is a peptide with an apparent molecular weight of about 8,000. Human alveolar macrophages secrete a peptide with a molecular weight of about 8,000, which is capable of releasing the azurophilic granule enzymes elastase and myeloperoxidase from human neutrophils, which were bound to membranes.[6]

There is now no question that neutrophil elastase, myeloperoxidase, and β-glucuronidase (azurophilic granule enzyme constituents) can be found in bronchoalveolar lavage fluids from normal subjects. The fate of these enzymes is

[a] These studies were supported by grant number R01-HL34745-03 from the Lung Division of the National Heart, Lung and Blood Institute and by grants from the Nobel Foundation, Ardmore, Oklahoma, The Meadows Foundation, Dallas, Texas, and the Elizabeth Gugenheim Fund.

[c] Address for correspondence: Allen B. Cohen, M.D., Ph.D., Executive Associate Director, University of Texas Health Center, POB 2003, Tyler, Texas 75710.

unknown; however, the elastase can be taken up by alveolar macrophages and excreted later.[7,8]

Use of Drugs to Decrease the Elastase Load

Mechanisms of chemotaxis, phagocytosis, and lysosomal enzyme secretion by neutrophils have numerous similarities and some differences.[9] In general, cell surface receptors initiate the processes,[10] cyclic nucleotides and other second messenger molecules transmit the signal,[11-13] cellular serine proteases and divalent cations modulate the system,[14-16] and cytoskeletal components achieve movement of the cells or subcellular particles.[17,18] These neutrophil functions can be modified by numerous drugs that are currently used to treat other diseases.

STUDIES TO DETERMINE IF CONTROL OF NEUTROPHIL FUNCTIONS MIGHT BE A FEASIBLE MEANS OF REDUCING THE NEUTROPHIL ENZYME LOAD IN THE LUNGS

Studies to Determine If Currently Available Drugs can Modify Neutrophil Chemotaxis and Enzyme Release in Vitro[19]

Although it is not possible to place drugs into totally consistent groups, we have developed a relatively consistent classification of drugs that may inhibit neutrophil functions. The classification is based on the mechanism of action on neutrophils when possible, and by drug type or action when particular actions on neutrophils are not known. The drugs in these studies were selected because they are representatives of their group, and at least one previous study claimed an effect on neutrophil function.

We analyzed 11 drugs from 8 classes of agents that modify neutrophil function to determine if they could reduce neutrophil chemotaxis or degranulation in response to stimuli which may be important in patients with emphysema.[19] The drugs that were analyzed are shown in TABLE 1.

Based on statistical considerations, a 25% depression of chemotaxis or enzyme release was selected to qualify a drug as a potentially clinically significant agent.

Results

Chemotaxis: In these studies, FMLP or human alveolar macrophage conditioned media (MCM), in separate studies, were used as stimuli for chemotaxis or degranulation. None of the 11 drugs tested reliably inhibited neutrophil chemotaxis in a dose-dependent fashion to both stimulants at concentrations achievable in human plasma.

When formyl-L-methionyl-L-leucyl-L-phenylalanine (FMLP) was used as the stimulant, sulfinpyrazone inhibited chemotaxis at the highest concentration only (FIG. 1). Auranofin had the clearest ability to inhibit chemotaxis and chemokinesis with either stimulus at concentrations higher than those achievable in human plasma (FIG. 1).

Piroxicam, ibuprofen, tetracycline, colchicine, phenylbutazone, theophylline, verapamil, ipratropium bromide, and methylprednisolone did not significantly

TABLE 1. Drugs Studied

Drug Category	Drug	References
Surface receptor competition	Sulfinpyrazone	10
	Phenylbutazone	10
Microtubule disruption	Colchicine	11–13
Nonsteroidal antiinflammatory	Auranofin	32
	Piroxicam	33
	Ibuprofen	34
Antimicrobial	Tetracycline	35–37
Cyclic AMP increase	Theophylline	12,13
Calcium ion channel disruption	Verapamil	14–16
Cyclic GMP decrease	Ipratropium bromide	38
Steroidal antiinflammatory	Methylprednisolone	39,40

inhibit chemotaxis at any concentration studied with either stimulant (data not shown).

Enzyme release: Several drugs were very effective inhibitors of enzyme release with both stimuli. Sulfinpyrazone, phenylbutazone, and colchicine inhibited the secretion of myeloperoxidase and β-D-glucuronidase when FMLP was used as a stimulus (FIG. 2). Sulfinpyrazone and phenylbutazone inhibited the release of myeloperoxidase when MCM was used as a stimulus (FIG. 3). These two drugs were much more effective inhibitors of enzyme release than were colchicine when FMLP was used as the stimulus. Auranofin first slightly stimulated and then depressed secretion of myeloperoxidase and β-D-glucuronidase when FMLP was

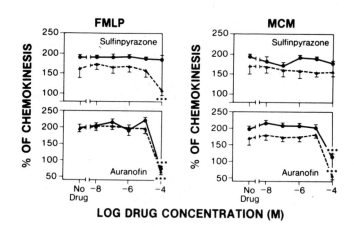

FIGURE 1. Effect of sulfinpyrazone and auranofin on chemotaxis. Chemokinesis measured in untreated neutrophils was considered 100%. Percent of chemokinesis was used to express the migration of the stimulated neutrophils over chemokinesis. In the *left panels,* FMLP was used as the stimulus at 1×10^{-8} M (*solid lines* and *closed circles*) or at 1×10^{-9} M (*dashed lines* and *closed triangles*) concentrations. In the *right panels,* MCM was used as the stimulus either undiluted (*solid lines* and *closed circles*) or at dilutions of 1:5 and 1:6 (*dashed lines* and *closed triangles*) in neutrophils treated with sulfinpyrazone and auranofin, respectively. The probability of a significant difference between measured chemotaxis and chemotaxis with no drug was (***) $p < 0.001$. (Reprinted with permission of Exp. Lung Res.[19])

used as the stimulant (FIG. 2) and stimulated and then depressed secretion of myeloperoxidase when MCM was used as the stimulant (FIG. 3). Like the other drugs, it had no effect on β-D-glucuronidase release when MCM was used as the stimulant (FIG. 3). Piroxicam, ibuprofen, tetracycline, theophylline, verapamil, ipratropium bromide, and methylprednisolone failed to inhibit enzyme release by either stimulant. As piroxicam, ibuprofen, tetracycline, and theophylline failed to inhibit enzyme release with either stimulant, other drugs that failed to release enzymes when FMLP was used as the stimulant were not tested using MCM.

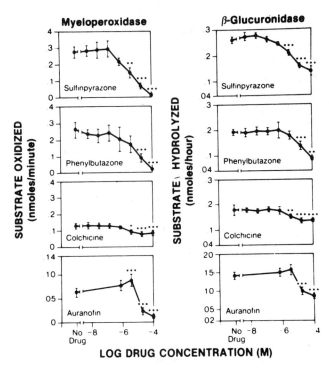

FIGURE 2. Effect of sulfinpyrazone, phenylbutazone, colchicine, and auranofin on the release of enzymes from neutrophils treated with FMLP. The *left panel* shows the measurements of myeloperoxidase released, and the *right panel* shows the measurements of β-D-glucuronidase released. The probability of a significant difference between measured enzyme release and enzyme release with no drug was (*) $p < 0.05$, (**) $p < 0.01$, and (***) $p < 0.001$. (Reprinted with permission of Exp. Lung. Res.[19])

Determination If Drugs That Impair Neutrophil Function or Inhibit Neutrophil Elastase Reduce Bacterial Killing by Neutrophils

Although most of the bacterial killing by neutrophils takes place inside the neutrophils, some of the antimicrobial actions of neutrophils takes place outside the neutrophil.[20-22] It is unclear whether the use of drugs that modify neutrophil function would have a detrimental effect on the killing of bacteria by neutrophils in patients in whom they are used. In these studies we carried out experiments *in*

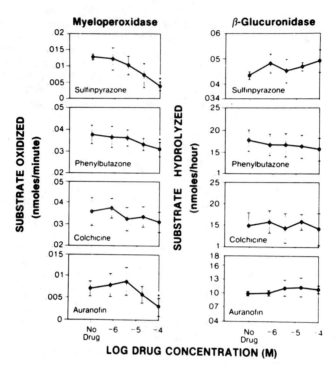

FIGURE 3. Effect of sulfinpyrazone, phenylbutazone, colchicine, and auranofin on the release of enzymes from neutrophils treated with MCM. The *left panel* shows the measurements of myeloperoxidase released, and the *right panel* shows the measurements of β-D-glucuronidase released. The probability of a significant difference between measured enzyme release and enzyme release with no drug was (*) $p < 0.05$. (Reprinted with permission of Exp. Lung Res.[19])

vitro to determine if neutrophil killing of *Pseudomonas aeruginosa* was decreased in the presence of inhibitors of neutrophil elastase or inhibitors of neutrophil azurophilic granule enzyme secretion.

Selection of Drugs

Three drugs were selected for study, the inhibitor of neutrophil elastase, U25651 (ICI Americas), colchicine, and auranofin. U25651 was chosen because neutrophil elastase has bactericidal actions.[23–26] Colchicine was chosen because we have already determined that it can reduce the neutrophil elastase concentration in bronchoalveolar lavage fluids from ex-smokers with emphysema, and it has a modest ability to prevent the secretions of enzymes by neutrophils.[19] In addition, it is known that colchicine at these doses does not predispose individuals to infections (unpublished results). Auranofin was chosen because at the doses used, it was the most effective inhibitor of degranulation of neutrophil enzymes. Its effect on chemotaxis would not be relevant in this test system.

Results

None of the three drugs tested had any effect on bacterial killing by neutrophils (FIG. 4). Control experiments showed that U25651, colchicine, and auranofin had no effect on bacterial growth. An experiment was performed to assure that these high concentrations of auranofin prevented enzyme release from the neutrophils. This experiment employed the same neutrophils and the same auranofin solution used in the bacterial experiments and was performed at the same time as the bacterial experiments (FIG. 5). Control samples for enzyme release contained only buffer. High concentrations of auranofin completely blocked enzyme release from the neutrophils.

Discussion

These studies were carried out to determine *in vitro* if agents that inhibit neutrophil elastase or prevent neutrophil degranulation reduce the ability of neutrophils to kill *P. aeruginosa*. Neutrophil elastase degrades *E. coli* proteins.[25,26] and works synergistically with myeloperoxidase and cathepsin G to kill *E. coli* and *Staphylococcus aureus*.[24]

In these studies an inhibitor of neutrophil elastase and inhibitors of neutrophil degranulation failed to modify the ability of neutrophils to kill *P. aeruginosa*. Therefore, if the *in vitro* conditions imitate the *in vivo* conditions, and if other

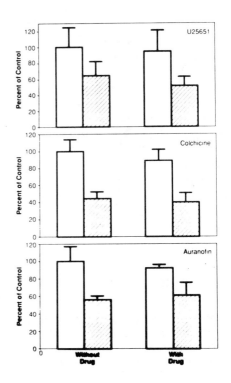

FIGURE 4. Effect of the neutrophil elastase inhibitor (U25651), colchicine, and auranofin on the killing of *P. aeruginosa* by neutrophils. The results were expressed as the percentage of control bacteria killed in order to compare the data from the three experiments. In each case, the percentage of colonies of bacteria killed was less in the presence of neutrophils than in their absence ($p < 0.01$), but the killing of the bacteria was the same in the presence and absence of the drug tested, $p > 0.05$.

bacteria are killed with similar mechanisms as those for *P. aeruginosa,* then treatment of patients with these kinds of therapy will not increase the likelihood of infection.

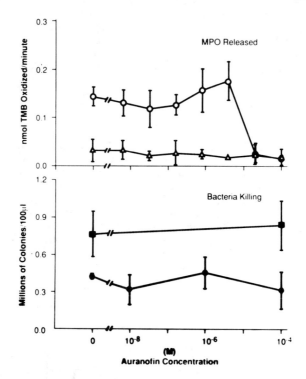

FIGURE 5. Myeloperoxidase released from human neutrophils (PMNs) and the killing of *P. aeruginosa* in the presence of varied concentrations of auranofin. Auranofin concentration was measured in moles. (*Top panel*) Vertical axis depicts the tetramethyl benzidine (TMB) oxidized per minute. *Open triangles* represent the myeloperoxidase (MPO) released by auranofin alone. *Open circles* represent the myeloperoxidase released by neutrophils treated with 5×10^{-7} M FMLP in the presence of varied concentrations of auranofin. Auranofin completely inhibited the secretion of myeloperoxidase at concentrations greater than 1×10^{-5} M. (*Bottom panel*) *Closed squares* represent bacterial growth with no neutrophils, whereas the *closed circles* show bacterial growth in the presence of neutrophils and varied concentrations of auranofin.

Clinical Trial of Colchicine to Reduce the Elastase Load in Cigarette Smokers with Emphysema[27]

Before drug analysis, colchicine was chosen as the first drug to try to reduce the elastase load in the lungs. We carried out a randomized double-blind placebo-controlled study to determine if colchicine could reduce several indices of elastase load or elastin breakdown in patients with COPD.[27] Indices of lung elastase load included bronchoalveolar lavage fluid neutrophil concentration, total neutrophils, and neutrophil elastase concentration. Elastase-generated fibrinopeptides were measured in plasma to determine overall neutrophil elastase exposure. Plasma elastin peptides and urinary desmosines were measured as indicators of lung elastin breakdown.

Design of the Study

The study was a pretreatment/posttreatment analysis of a group of patients. Subjects were placed on sufficient theodur to give a theophylline blood concentration between 10 and 20 $\mu g/ml$ and albuterol, two puffs three times per day, for the 21-day study period. At the end of the first week, bronchoalveolar lavage fluid, blood, and urine samples were obtained. The patients were then randomized (with blocking) to groups of approximately equal numbers of patients at any given time. Colchicine, 0.6 mg tablet, was placed into a capsule. A lactose filler was placed into capsules that were given to patients in the control group. The drugs were given to the patients at the pharmacy by a random selection code provided by the statistician. The physician and nurse who dealt with the patients and the technicians who performed the tests were unaware of the patient assignment group. Colchicine was not measured in the plasma because it rapidly disappears from the plasma in treated patients.[28] The nurse called the patients every second day to remind them to take the medication. Serum cotinine and thiocyanate were measured on the first day of colchicine treatment and 14 days later to evaluate smoking status. After 14 days of therapy with colchicine (21st day of the encounter), the bronchoalveolar lavage fluid, blood, and urine samples were obtained again.

Patient Selection Criteria

The protocol was approved by The Institutional Review Board, and all of the subjects signed appropriate consent forms before participating in these trials. Outpatients at the University of Texas Health Center at Tyler were recruited if they fulfilled the following criteria: (1) current cigarette smoking habit with at least 10 pack-years of smoking history, (2) irreversible airflow obstruction as defined by an $FEV_{1.0}$ less than 70% of predicted, and $FEV_{1.0}$/forced vital capacity (FVC) less than 0.7 and less than 20% reversibility, (3) $FEV_{1.0}$ greater than 1.2 liters, (4) age between 45 and 75 years, (5) no other significant medical illnesses such as heart failure or diabetes, and (6) no intake of steroids, antibiotics, or nonsteroidal antiinflammatory drugs.

Statistical Methods

Nonparametric tests of significance were used for analysis throughout. The Wilcoxon matched-pairs test was used to compare pre- and posttreatment values within drug groups, the Mann-Whitney test was used for between-group comparisons, and Spearman's rank correlation coefficient was used for correlations between variables. As no equivalent nonparametric method is available, analysis of covariance was used to adjust for differences in baseline variables for between-group comparisons. This was done after logarithmic transformations were performed on variables with highly skewed distributions. Unless otherwise noted, all p values reported were from nonparametric tests of significance.

Results

Measurements in plasma: The differences between plasma elastin peptides in control and treated groups or between pretreatment and posttreatment measurements in the same group were not significant.

Seventy-two percent of the 18 patients with emphysema examined had unmeasurable concentrations of elastase-generated fibrinopeptides in their plasmas. No statistically significant differences were noted between pre- and posttreatment measurements in the placebo-treated plasma ($p = 0.180$, $n = 8$) or between pre- and posttreatment measurements in the colchicine-treated plasma ($p = 0.180$, $n = 8$).

Urinary measurements: The differences between urinary desmosine and desmosine/creatinine ratios in control and treated groups or between pretreatment and posttreatment measurements in the same group were not statistically significant.

Measurements in bronchoalveolar lavage fluids: The differences between percentage of neutrophils, total neutrophils, and neutrophil elastase in control and treated groups or between pretreated and posttreated measurements in the same group were not statistically significant. A statistically significant correlation was noted between the concentrations of neutrophil elastase and the percentage of neutrophils ($r = 0.4$, $p = 0.003$).

Elastase-generated fibrinopeptides were unmeasurable in bronchoalveolar lavage fluids from 85% of the 46 patients with emphysema in whom they were measured. No statistically significant differences were found between pre- and posttreatment measurements in the placebo-treated lavage fluids ($p = 0.735$, $n = 23$) or between pre- and posttreatment measurements in the colchicine lavage fluids ($p = 0.484$, $n = 22$). Desmosine concentrations and elastin peptides were too low to measure in the bronchoalveolar lavage fluids.

Conclusions

We concluded that colchicine was not able to modify any of the variables related to elastase load in the lungs or of lung tissue destruction under the conditions used in this study.

Clinical Trial of Colchicine to Reduce the Elastase Load in Ex-Cigarette Smokers with Emphysema

After the clinical trial of colchicine was completed in current cigarette smokers, a similar trial was carried out in ex-smokers. The patients were similar to those in the study of current smokers except that they had at least a 10 pack-year history of cigarette smoking but no smoking for at least 12 months.

Statistical Analysis

Logarithms of raw data values were used for analysis of serum and lavage measurements. This calculation was carried out to change skewed distributions to symmetric ones and to minimize differences in sample variances. The transformed variables were then compared in two ways. First, the means of pretreatment values were compared with the means of posttreatment values within each group by the Student's t test for paired samples. Second, the magnitude of the mean change from pretreatment values was compared between groups using the usual independent-sample t test.

To account for possible effects of pretreatment differences between groups (e.g., regression to the mean), two additional methods were used. An analysis of covariance, as described by Fleiss,[29] was used to compare the posttreatment differences between groups (using transformed data), while adjusting for pretreatment values. Also, the percentage of change from pretreatment values (using the original, untransformed values for each individual) was tested using the nonparametric Mann-Whitney procedure.[30] Unless otherwise noted, all p values are from two-tailed tests.

Results

Measurements in bronchoalveolar lavage fluids: The levels of the neutrophil elastase concentration in the bronchoalveolar lavage fluids were statistically significantly smaller in the posttreatment samples (mean = 5.81, SD = 5.81) than in the pretreatment samples (mean = 12.13, SD = 7.26) in the colchicine-treated group ($p = 0.006$). The mean drop in the neutrophil elastase concentration in the colchicine-treated group was 51.5%. In addition, the magnitude of change in the neutrophil elastase concentration between pre- and posttreated groups was greater in the colchicine-treated than in the placebo-treated patients ($p = 0.005$). The group differences were still significant after adjusting for baseline levels by the nonparametric method ($p = 0.021$) and by the analysis of covariance method ($p = 0.029$). No statistically significant difference was found between pretreatment neutrophil elastase concentrations in the colchicine-treated group and placebo-treated group ($p = 0.55$). The differences between percentage of neutrophils and total neutrophils in control and treated groups or between pretreated and posttreated measurements in the same group were not statistically significant.

Urinary and blood measurements: There was no statistically significant improvement in either the urinary desmosine, urinary desmosine/creatine, or plasma elastin peptides in the patients treated with colchicine.

Discussion

The study demonstrated that colchicine can reduce the neutrophil elastase concentration in bronchoalveolar lavage fluids from exsmokers with COPD by an average of 51.5%. There was no statistically significant reduction in bronchoalveolar lavage fluid neutrophils, plasma elastin peptides, or urinary desmosines. The current studies showed that it is possible to modify neutrophil elastase concentration in the lungs of exsmokers with COPD. Although it is unclear why colchicine failed to reduce the elastase in the bronchoalveolar lavage fluids from current smokers,[27] it is most likely that this rather mild drug could not overcome the ongoing stimulus to neutrophil chemotaxis and degranulation provided by current cigarette smoking.

It is not clear whether reducing the neutrophil elastase in ex-smokers with COPD will slow the progress of their disease. Smokers who stop smoking reduce the rate of decline of $FEV_{1.0}$ from about 70 ml to a more "normal" rate of about 30 ml per year. It is not known why normal subjects sustain a decrease in $FEV_{1.0}$;

however, neutrophil elastase can be found in bronchoalveolar lavage fluids from normal subjects.[31] Consequently, it might be possible to reduce the rate of decline of $FEV_{1.0}$ in ex-smokers with COPD below that achievable with smoking cessation alone. Although "normal" rates of decline may not produce symptoms in otherwise normal people, a decline of 30 ml/year of $FEV_{1.0}$ could disable or kill patients with severe COPD. In addition, patients with a deficiency of α_1-protease inhibitor who do not smoke but who are developing emphysema and cannot afford treatment with Prolastin may be benefited by colchicine therapy.

These results demonstrate that reduction of the elastase load in the lungs of patients with COPD is an achievable goal and may be accomplished with drugs that are currently approved for use in other diseases. Although the obvious treatment for smoking-related COPD is smoking cessation, 75–80% of these patients cannot stop smoking, and any treatment that attacks the underlying cause of the disease would have enormous benefit in reduced morbidity and mortality of the millions of patients with COPD.

ANALYSIS AND CONCLUSIONS

The foregoing studies represent an attempt to determine if it is feasible to reduce the load of neutrophil elastase in the lungs of patients with emphysema. The comparison of drugs that modify neutrophil function demonstrates that several drugs are already available for use in man that significantly reduce the secretion of enzymes by neutrophils. None of them is perfect. Sulfinpyrazone was the safest and most effective drug at concentrations that are approved for use in man; however, auranofin was the best drug at concentrations slightly above those that are approved for use in man. The main side effect of auranofin in man is suppression of bone marrow. It would be worth determining if a drug could be designed that would keep the neutrophil suppressive effects and reduce the marrow suppressive effects. Suppression of the release of neutrophil enzymes or of neutrophil elastase did not modify the ability of the neutrophils to kill bacteria in vitro.

Colchicine reduced the neutrophil elastase in the lung washes of ex-smokers by more than 50%, but it did not suppress it in current smokers. The most obvious conclusion from these results is that the drug is not strong enough to suppress the load of neutrophil elastase when the ongoing inflammatory stimulus of cigarette smoking is occurring. It is unclear whether colchicine would be valuable in ex-smokers with emphysema, because their rate of decrease in lung function approaches that in nonsmokers after smoking cessation. Colchicine might also be considered for the treatment of patients with α_1-protease inhibitor deficiency who do not smoke, but are developing emphysema and cannot afford Prolastin treatment. Two additional kinds of studies should be performed before such treatment could be recommended. First, the study should be repeated with a larger group of patients who might benefit from the treatment, and second, a trial should be carried out to determine if the drug reduces the rate of decline of lung function or mortality. This would be the most useful test of the elastase-inhibitor hypothesis. Finally, if inhibitors of neutrophil elastase are approved for treatment of emphysema, they may prove to be synergistic in clinical effectiveness with drugs that suppress neutrophil enzyme secretion.

REFERENCES

1. JANOFF, A. 1985. Ann. Rev. Med. **36:** 206–216.
2. COHEN, A. B. & M. ROSSI. 1983. Am. Rev. Respir. Dis. **127:** 53–59.
3. COHEN, A., G. BATRA, R. PETERSEN, J. PODANY & E. NUGUEN. 1979. J. Appl. Physiol. **47:** 440–444.
4. COHEN, A. B., M. ROSSI, D. GECZY & L. KNIGHT. 1982. J. Clin. Invest. **69:** 794–798.
5. KAZMIEROWSKI, J. A., J. I. GALLIN & H. Y. REYNOLDS. 1977. J. Clin. Invest. **59:** 273–281.
6. COHEN, A. B., D. E. CHENOWETH & T. E. HUGLI. 1982. Am. Rev. Respir. Dis. **126:** 241–247.
7. MCGOWAN, S. E., P. J. STONE, G. L. SNIDER & C. FRANZBLAU. 1984. Am. Rev. Respir. Dis. **130:** 734–739.
8. CAMPBELL, E. J. & M. D. WALD. 1983. J. Lab. Clin. Med. **101:** 527–536.
9. BECKER, E. L. & J. H. SHOWELL. 1974. J. Immunol. **112:** 2055–2062.
10. DAHINDEN, C. & J. FEHR. 1980. J. Clin. Invest. **66:** 884–891.
11. RUDOLPH, S. A., P. GREENGARD & S. E. MALAWISTA. 1977. Proc. Natl. Acad. Sci. USA **74:** 3404–3408.
12. ZURIER, R. B., G. WEISSMAN, S. HOFFSTEIN, S. KAMMERMAN & H. H. TAI. 1974. J. Clin. Invest. **53:** 297–309.
13. ZURIER, R. B., S. HOFFSTEIN & G. WEISSMAN. 1973. J. Cell Biol. **58:** 27–41.
14. DELLA BIANCA, V., M. GRZESKOWIAK, P. DE TOGNI, M. CASSATELLA & F. ROSSI. 1985. Biochim. Biophys. Acta **845:** 223–236.
15. OSEAS, R. S., L. A. BOXER, C. BUTTERICK & R. L. BAEHNER. 1980. J. Lab. Clin. Med. **96:** 213–221.
16. O'FLAHERTY, J. T., C. L. SWENDSEN, C. J. LEESE & C. E. MCCALL. 1981. Am. J. Pathol. **105:** 107–113.
17. SOUTHWICK, F. S. & T. P. STOSSEL. 1983. Semin. Hematol. **20:** 305–321.
18. CARSON, M., A. WEBER & S. H. ZIGMOND. 1986. J. Cell Biol. **103:** 2707–2714.
19. STEVENS, M., E. MILLER & A. B. COHEN. 1989. Exp. Lung Res. **15:** 663–680.
20. COONROD, J. D. 1986. Semin. Respir. Infect. **1:** 118–129.
21. GANZ, T., M. E. SELSTED & R. I. LEHRER. 1986. Semin. Respir. Infect. **1:** 107–117.
22. WEISS, J., L. KAO, M. VICTOR & P. ELSBACH. 1985. J. Clin. Invest. **76:** 206–212.
23. SPITZNAGEL, J. K. & N. OKAMURA. 1983. Adv. Exp. Med. Biol. **162:** 5–17.
24. ODEBERG, H. & I. OLSSON. 1976. Infect. Immunol. **14:** 1276–1283.
25. JANOFF, A. & J. BLONDIN. 1974. Proc. Soc. Exp. Biol. Med. **145:** 1427–1430.
26. BLONDIN, J. & A. JANOFF. 1976. J. Clin. Invest. **58:** 971–979.
27. COHEN, A. B., W. GIRARD, J. MCLARTY, B. STARCHER, M. S. STEVENS, D. S. FAIR, D. DAVIS, H. JAMES, J. ROSENBLOOM & U. KUCICH. 1990. Am. J. Respir. Dis. **142:** 63–72.
28. WALLACE, S. L., B. OMOKOKU & N. H. ERTEL. 1970. Am. J. Med. **48:** 443–448.
29. FLEISS, J. L. 1986. *In* The Design and Analysis of Clinical Experiments. :186–219. John Wiley & Sons. New York.
30. DANIEL, W. W. 1978. *In* Applied Nonparametric Statistics. :82–86. Houghton Mifflin Company. Boston.
31. IDELL, S., U. KUCICH, A. FEIN, F. KUEPPERS, H. L. JAMES, P. N. WALSH, G. WEINBAUM, R. W. COLMAN & A. B. COHEN. 1985. Am. Rev. Respir. Dis. **132:** 1098–1105.
32. NEAL, T. M., M. C. VISSERS & C. C. WINTERBOURN. 1987. Biochem. Pharmacol. **36:** 2511–2517.
33. ABRAMSON, S., H. EDELSON, H. KAPLAN, R. LUDEWIG & G. WEISSMAN. 1984. Am. J. Med. **77**(4B): 3–6.
34. KAPLAN, H. B., H. S. EDELSON, H. M. KORCHAK, W. P. GIVEN, S. ABRAMSON & G. WEISSMAN. 1984. Biochem. Pharmacol. **33:** 371–378.
35. MARTIN, R., G. WARR, R. B. COUCH, H. TEAGER & V. KNIGHT. 1974. J. Infect. Dis. **129:** 110–116.

36. MARTIN, R., G. WARR, R. COUCH & V. KNIGHT. 1973. J. Lab. Clin. Med. **81:** 520–529.
37. STOCKLEY, R. A., S. L. HILL & H. M. MORRISON. 1984. Thorax **39:** 414–419.
38. CUGELL, D. 1986. Am. J. Med. **81:** 27–31.
39. DANNENBERG, A. M. 1979. Inflammation **3:** 329–343.
40. PETERS, W. P., J. F. HOLLAND & H. SENN. 1972. N. Engl. J. Med. **286:** 342–345.

Proteinases in Chronic Lung Infection[a]

R. A. STOCKLEY, S. L. HILL, AND D. BURNETT[b]

*Lung Immunobiochemical Research Laboratory
The Clinical Teaching Block
The General Hospital
Birmingham B4 6NH, UK*

The role of proteinases in the pathogenesis of chronic lung disease has been the subject of extensive investigation both *in vitro* and *in vivo*. The majority of studies have been concerned with factors that might alter the proteinase/antiproteinase balance and lead to the development of emphysema. This concentration of effort is understandable, because deficiency of one of the major antiproteinases (alpha$_1$-antitrypsin) is associated with the development of emphysema. The major proteinase inhibited by alpha$_1$-antitrypsin is neutrophil elastase which has been shown to produce emphysema in experimental animals through its ability to digest lung elastin.

Studies in man have been hampered by the inability to sample, directly, the lung interstitium where the destructive process is thought to occur. Furthermore, samples of harvested lung secretions are often contaminated by secretions from the major airways, particularly in the presence of bronchitis, thereby affecting the results.

In our early studies[1,2] in patients with chronic bronchitis and emphysema we were unable to detect any free elastase activity in the lung secretions while the patients were clinically well (free from acute exacerbations). However, the patients (like most with emphysema) had normal plasma and lung alpha$_1$-antitrypsin (α_1-AT). We therefore considered the possibility that subtle defects in the regulation of lung elastase and elastase inhibitors could exist, and this would be highlighted by the presence of an acute infection which would "stress" the system. Indeed, the studies[1,2] confirmed that the presence of infection resulted in lung inflammation and an increase in lung α_1-AT but also an excessive increase in lung neutrophil elastase (NE) and the presence of free elastase activity.

These data were obtained in bronchial secretions and their relevance to interstitial tissue destruction was uncertain. However, several observations have led us more recently to investigate the role of proteinases in the pathogenesis of chronic bronchial damage in the presence of acute and chronic lung infection. Indeed, previous work had shown that NE was capable of producing bronchial changes similar to those seen in patients with chronic bronchitis.[3] Furthermore, patients with α_1-AT deficiency also suffer from bronchial disease including bronchiectasis.[4] This suggests that the proteinase/antiproteinase balance in the bronchial tree may determine the degree of damage to larger airways. The ready access to the local environment of the bronchial tree and availability of secretions make studies of diseases of the airways easier than study of emphysema.

[a] R.A.S. is in receipt of a British Lung Foundation programme grant.
[b] British Lung Foundation Senior Research Fellow.

HOST RESPONSE TO INFECTION

The lung has developed a sophisticated defense system to deal with inhaled antigens and viable bacteria. When the bacterial load entering the lung is small, the organisms are removed by mechanisms including mucociliary clearance, local immunoglobulin binding, and ingestion by resident phagocytes. However, as the bacterial load rises, the organisms are able to overcome these defense mechanisms and a symptomatic infection ensues. This is characterized by an inflammatory response associated with recruitment and activation of circulating phagocytes (monocytes and neutrophils), increased transudation of plasma proteins, and often the expectoration of purulent secretions.

These infections are usually short-lived, and few studies have been performed of the effect of such episodes on the balance between proteinases and antiproteinases. Undoubtedly the recruitment and activation (with degranulation) of circulating neutrophils will increase the proteinase load in the lung as indicated by the observed rise in NE in lung secretions.[1,2] In addition, an increase occurs in plasma and secretion antiproteases, such as α_1-AT, as a result of the "acute phase" response and leakage into the lung because of inflammation.[1,5] This rise will tend to offset the increased proteinase burden. However, results from our laboratory demonstrated that acute infections in patients with chronic obstructive bronchitis resulted in the temporary release of neutrophil elastase in quantities sufficient to exceed the ability of lung antielastases to inhibit the enzyme.[1,2] The presence of free elastase activity in the lung could potentially lead to tissue damage. However, these short-lived episodes of infection do not appear to be associated with a demonstrable deterioration in lung function.[6]

CHRONIC INFECTIONS

The lung secretions from many patients with chronic lung disease often contain viable bacteria that can be isolated on bacterial culture. In some patients, particularly those with bronchiectasis (with and without cystic fibrosis), evidence of a chronic infective process is largely confined to the airways. This has been demonstrated recently in patients with cystic fibrosis and chronic Pseudomonas infection in which the organisms and neutrophil infiltrate are present predominantly within the airways.[7] The patients do not exhibit many of the features associated with acute lung infections (temperature, chest pain, and increased symptoms), but they appear to have a continuous low-grade inflammatory process characterized by the production of sputum.

Clinical observation of patients with bronchiectasis indicates that the type of secretions expectorated is usually constant, and this has been confirmed by analysis of data obtained from daily diary cards completed by the patients.[8] Thus, some patients continually produced mucoid (clear) secretions, whereas others usually produced mucopurulent (pale yellow) or frankly purulent (dark yellow/green) secretions.

Purulent lung secretions are thought to derive their green coloration from the presence of myeloperoxidase (MPO) within the neutrophil.[9] This suggests that such secretions contain more neutrophils and their products. Studies from our laboratories have confirmed that the number of neutrophils seen in lung secretions relates to the degree of purulence.[10] Furthermore, direct measurement of MPO in secretions using the method adapted from Bradley et al.[11], as described by Buttle and colleagues,[12] indicates a clear relation to the degree of purulence. Some preliminary results (expressed as units per milliliter) are summarized in FIGURE 1 for individual sputum samples obtained from patients with bronchiectasis. The values for purulent samples (mean = 1,314 units/ml; SE ± 186) were significantly higher (p <0.005) than those for mucopurulent samples (44.6; 122) which in turn were higher (p <0.005) than those for mucoid samples (15.1; 4.3). Thus, sputum contains neutrophils and their products, and the numbers are related to the purulent nature of the secretion.

To enter lung secretions, circulating neutrophils have to respond to chemoattractants within the lung. Indeed, lung secretions obtained from bronchiectatic patients contain factors that induce a chemotactic response in neutrophils. This chemoattraction is greatest for purulent rather than mucoid secretions,[13] and several peaks of activity have been identified after size exclusion high performance liquid chromatography.[14] The nature of these chemoattractants awaits

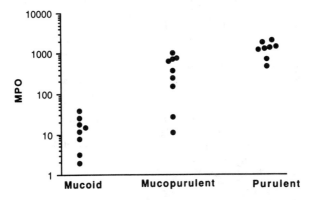

FIGURE 1. Myeloperoxidase activity of sputum is given (*vertical axis*) for individual patient samples. Results for mucoid (clear), mucopurulent (pale yellow/green), or purulent (dark yellow/green) samples are shown.

clarification, but they may be derived from a variety of sources including phagocytic cells (neutrophils and macrophages), complement activation, bacterial products, and tissue degradation products. Thus, the secretions contain factors that can attract neutrophils *in vitro*.

The importance of the chemoattractants *in vivo* can be inferred from the studies of Currie and colleagues,[15] who have shown that [111]Indium-labeled neutrophils demonstrate that the cells accumulate in bronchiectatic areas of the lung and are found subsequently in the secretions obtained from these patients. This continuing neutrophil traffic will increase the proteinase burden in the lung. In addition, neutrophils obtained from patients with chronic purulent bronchiectasis show evidence of increased activation and spontaneously degranulate, causing more digestion of connective tissue substrate (fibronectin) than normal neutrophils,[16] which will increase the potential proteinase burden further.

As just indicated, the purulent nature of secretions relates to the presence of a neutrophil product, MPO. In addition, elastolytic activity has also been shown to relate to the purulence of lung secretions.[10] Recent results indicate that mucoid samples rarely contain measurable enzyme activity, with only 7 of 27 samples being positive (mean value equivalent to 0.03 mg/ml of porcine pancreatic elastase; SD $\pm$ 0.01). Mucopurulent samples are usually positive (38 of 40) with a mean value of 2.03 mg/ml (SD $\pm$ 0.29) and purulent samples (n = 40) are always positive (mean 28.9 $\pm$ 2.8).

Several studies suggested that "elastase" activity is predominantly due to neutrophil elastase. Buttle et al.[12] showed a positive correlation (r = 0.83) of "elastase" activity (assessed by using the synthetic substrate succinyl-L-alanyl-L-alanyl-L-alanine paranitroanilide), with MPO activity (p <0.0001) in 25 samples of sputum of varying degrees of purulence, which suggests indirectly that both originate from the same source. Furthermore, studies in our own laboratory showed that the degradation of native elastin by bronchiectatic secretions can be reduced by the addition of serine proteinase inhibitors (such as α_1-AT) as well as the specific neutrophil elastase peptide chloromethyl ketone inhibitor methoxy-succinyl-alanyl-alanyl-prolyl-valyl-CMK, but not cysteine proteinase or metallo-proteinase inhibitors, confirming that most elastase activity is due to neutrophil elastase. Some of these results are summarized in FIGURE 2.

In patients with bronchiectasis the presence or absence of elastase activity appears to be a constant feature, whereas the clinical state and nature of the sputum remain stable. This point is summarized in FIGURE 3 in which the elastase activity of sputum samples is indicated for eight such patients on 12 consecutive days.

The effect of persistent neutrophil elastase activity in the bronchial secretions remains to be proven, but circumstantial and experimental evidence suggests it has a major role in the pathogenic changes seen in the airways. Early animal studies on the effect of neutrophil elastase instilled into the lung via the trachea concentrated on its effect on connective tissue and its role in the pathogenesis of emphysema, and little attention was paid to the airways. Subsequent studies have

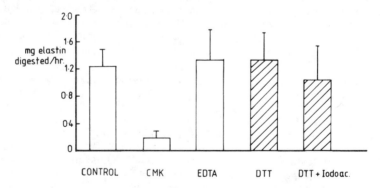

FIGURE 2. Elastase activity of sputum is shown as milligrams of particulate elastin degraded/hour (*vertical axis*). Results are expressed as mean $\pm$ SE (n = 6). The control results are for sputum alone, and the remaining histograms are in the presence of several proteinase inhibitors. CMK = methoxysuccinyl-alanyl-alanyl-prolyl-valyl-chlormethylketone (1 μM); EDTA = ethylenediamene tetraacetate (10 mM); DTT = dithiothreitol (1 mM); Iodoac = iodoacetic acid (100 mM).

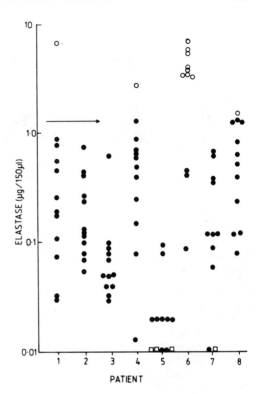

FIGURE 3. Elastase activity (determined by degradation of fluoresceinated elastin) collected from eight bronchiectatic patients on 12 consecutive days. Results are expressed as micrograms per 150 μl sputum (*vertical axis*). *Open squares* are samples with no detectable activity (<0.01). *Open circles* are results that would have been detectable by the elastin agarose diffusion method utilized previously.[11]

shown that the enzyme causes epithelial damage and mucus gland hyperplasia.[3] Furthermore, the enzyme causes goblet cell secretion,[17] the release of glycoconjugates from tracheal epithelium,[18] and induces secretion from subepithelial airway serous glands.[19] In addition, neutrophil elastase also affects the beating of ciliated epithelial cells *in vitro*,[20] and this can be reproduced by elastase-positive secretions from patients with bronchiectasis.[21] More recent preliminary studies have investigated the role of neutrophils and neutrophil elastase in causing direct damage to bronchial epithelial cells using *in vitro* experimentation.[22,23] Results confirmed the potential of NE in detaching bronchial epithelium from extracellular matrix and causing epithelial disruption.

These observations suggest that neutrophil elastase may be responsible (at least in part) for many of the features of bronchiectasis. The patients' lungs show evidence of loss of connective tissue from the bronchi, mucus gland hyperplasia, mucus hypersecretion, reduced mucociliary clearance,[24] and epithelial damage. Thus, it has been proposed that progression of airways damage in some patients with bronchiectasis may be a direct result of free elastase activity in the airways.[25] Pathologic evidence of the relationship of elastase activity to the features of bronchiectasis may prove difficult to obtain, because it will be necessary to demonstrate either improvement or deterioration of morphologic changes in the absence or presence of continuing elastase activity.

However, it is possible to obtain some indirect evidence for the potential

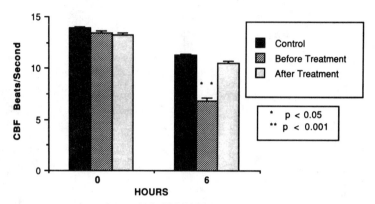

FIGURE 4. Ciliary beat frequency of normal nasal epithelium in medium alone (control); in the presence of patients' sputum sol phase before and after treatment. The significance of the reduction below control after 6 hours is shown. The 6-hour results of sputum sol before treatment are also significantly lower than are those of sputum after treatment (p <0.0001). Histograms are mean ± SE (n = 6, all results).

effects of elastase on bronchial epithelium by studies that alter the elastase activity in the lung. As mentioned above, elastase may damage epithelial cells or lead to secretion from epithelial cells or serous glands. Thus, elastase activity in the airways' secretions should relate to the presence of released epithelial cell contents, secretion from glands, and damage to epithelial cells. Clearly there could be several potential markers of these processes, but few have been studied.

Our own studies have included some potential markers. Firstly, a 51-kD cytoplasmic protein L1 has been described in bronchial epithelium.[26] Damage to these cells could result in its release, and indeed secretion levels of L1 increase with the degree of sputum purulence[27] as does the elastase level. Unfortunately L1 is also found in the cytoplasm of neutrophils, and thus the secretion concentration of L1 may largely reflect the presence and number of these cells alone.

An alternative marker is the cysteine proteinase Cathepsin B. This enzyme is not present in neutrophils but is present in bronchial epithelium, serous cells, and ducts of submucosal glands.[28] The concentrations of Cathepsin B activity relate to the purulent nature of the lung secretions.[29] Furthermore, recent work has shown a direct correlation (r = 0.85; p <0.0001) between elastase activity and Cathepsin B activity in secretions (n = 25) from patients with bronchiectasis.[12] It is uncertain at present if the release of Cathepsin B activity is the result of active secretion due to stimulation of the glands by NE or leakage from epithelial cells damaged by NE, and further studies will be necessary to clarify the mechanism.

A further marker providing indirect evidence for the potential role of neutrophil elastase in epithelial damage has been obtained from the studies of ciliary function *in vitro*. Elastase-positive secretions from patients with bronchiectasis have been shown to decrease ciliary beat frequency of normal ciliated nasal epithelium over 6 hours when compared to a control sample (FIG. 4). This effect is reversed by the addition of α_1-AT in sufficient quantities to inhibit the enzyme activity, suggesting that the effect is due (at least in part) to the presence of NE.[21] Three studies[21,27,29] indicate potential markers for the study of the deleterious effects of NE on the airways and provide a mechanism for monitoring the destructive process and the effects of intervention therapy (see below).

Most studies implicating NE in chronic infection have concentrated on its role in directly causing damage to the bronchial tree and lung interstitium. However, this enzyme has also been considered to play a role in several other aspects of the infective process. Studies have suggested that NE may enhance the inflammatory process by the activation of complement[30] or reduce inflammation by the removal of immune complexes.[31] Also NE has been implicated in bacterial killing,[32] while at the same time promoting their colonization of the lung by the destruction of secretory immunoglobulins.[33] The net result and importance of these conflicting roles are unknown.

EFFECT OF TREATMENT

The presence of active proteolytic enzymes within the airways likely contributes to lung damage in the presence of chronic infection. The presence of elastase relates to neutrophil recruitment as part of the host response to bacteria in the airways. This process can be reversed by effective antimicrobial therapy. After successful therapy the sputum changes from purulent to mucoid. The sputum becomes less chemoattractant for neutrophils,[13] the neutrophils become less activated,[34] and elastase activity disappears.[8,10] The secretions become less ciliotoxic[21] as shown in FIGURE 4, and sputum volume decreases.[8] Direct measurements of epithelial damage have not been assessed, but two of the potential markers, L1 (FIG. 5) and Cathepsin B activity, decrease.[29] These studies indicate that many of the potential effects of NE can be reversed, although in such a complex system it is difficult to attribute the changes to removal of NE alone. Other proteinases and products, such as oxygen-free radicals from phagocytes, in addition to many cytokines and bacterial products, will also decrease.

The effect of antibiotic therapy, however, is often short-lived,[10] and the secre-

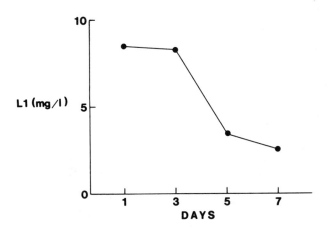

FIGURE 5. The L1 concentration in sputum is shown over consecutive days after the commencement of successful antibiotic therapy for an individual patient.

tions return to their purulent nature within a few weeks.[35] Thus, some patients require long-term or continuous antibiotic therapy to maintain the beneficial response. The hazards of long-term antibiotic therapy in these patients remains to be determined, but concern about superinfections has led to consideration of alternative ways of decreasing the lung proteases and their potential effects.

Clearly one approach would be the use of specific antiproteases, including those that inhibit neutrophil elastase. The aim of such therapy, whether administered locally or systemically, would be to reduce the activity of elastase released into the airways rather than that within or in close proximity to the cell where it may perform a critical role in bacterial destruction.[32] Such studies will clarify the role of NE in the pathogenesis of airways damage.

Another approach would be to limit the degree of neutrophil recruitment and activation by the use of alternative therapeutic agents. Many drugs including mucolytic agents,[36] nonsteroidal antiinflammatory agents,[37] and corticosteroids[38] are known to have an effect, but few studies have been completed. Whether such approaches will prove harmful remains to be determined. Certainly factors that reduce the host response could result in further bacterial proliferation. In a short study of the effect of corticosteroids on chronic lung infection in cystic fibrosis, no major adverse effects were seen, and the patients' health improved.[39] The success of this study indicates that the role of antiinflammatory agents in chronic lung infections is worthy of further careful study.

OTHER PROTEINASES

For historical reasons most of the emphasis has been confined to studies concerning the role of neutrophil elastase. However, other enzymes have received some attention particularly in bronchiectasis associated with cystic fibrosis. In these patients another neutrophil enzyme (Cathepsin G) has been identified.[40] The effect of this enzyme is less well studied, although it can cause serous cell secretion[19] and may assist in the destruction of lung elastin.[41] In addition, Cathepsin G (as well as NE) has been implicated in bacterial killing.[42]

More recently two further neutrophil serine proteases that show homology with NE and Cathepsin G have been identified with antibacterial properties,[43] and an elastolytic enzyme (proteinase 3) that is distinct from NE but capable of causing emphysema[44] has also been isolated. The role of these enzymes in lung defense and damage remains to be determined.

More attention has been paid to the role of proteinases derived from *Pseudomonas aeruginosa* which commonly colonize the airways. In such patients Pseudomonas elastase has been shown to account for some of the sputum "elastase" activity[45] and is even more potent than NE in causing epithelial disruption *in vitro*.[23] The role of this and other bacterial proteinases remains to be determined, although they may play a more indirect role of the cleavage and inactivation of lung antiproteinases,[46] thereby contributing to a proteinase/antiproteinase imbalance and cleavage of immunoglobulins,[47] thereby promoting bacterial colonization and maintaining the chronic infective process.

The presence of Cathepsin B has been discussed previously. It was initially thought that this enzyme could be elastolytic and therefore contribute to tissue damage, but its true role remains to be determined. Undoubtedly other enzymes derived from phagocytes, including Cathepsin L[48] and macrophage metalloelastase,[49] as well as those from bacteria and epithelial cells will be present in the secretions of patients with chronic lung infections. Further studies are needed to detect these enzymes and determine their role.

CONCLUSIONS

Chronic lung infections are associated with an exuberant and persistent host response, resulting in lung inflammation and phagocyte recruitment and activation. The release of neutrophil proteases is a constant feature and likely to be the cause of some of the clinical and pathologic features of the disease. Therapeutic strategies aimed at controlling the host response or its consequences should be explored and their effects on disease progression monitored.

ACKNOWLEDGMENTS

The authors thank E. Ford for typing the manuscript. L1 protein was measured by Dr. I. Dale (Oslo, Norway).

REFERENCES

1. STOCKLEY, R. A. & D. BURNETT. 1979. Am. Rev. Respir. Dis. **120:** 1081–1086.
2. STOCKLEY, R. A. & D. BURNETT. 1980. Bull. Eurp. Physiopath. Resp. **16** (suppl): 261–271.
3. SNIDER, G. L., P. J. STONE, E. C. LUCEY *et al.* 1985. Am. Rev. Respir. Dis. **132:** 703–708.
4. LONGSTRETCH, G. F., S. A. WEITZMAN, R. J. BROWNING *et al.* 1975. Chest **67:** 233–235.
5. STOCKLEY, R. A., M. MISTRY, A. R. BRADWELL & D. BURNETT. 1979. Thorax **34:** 777–782.
6. FLETCHER, C., R. PETO, T. CECILY & F. E. SPEIZER. 1976. The Natural History of Chronic Bronchitis and Asthma. An Eight Year Study of Early Chronic Obstructive Lung Disease in Working Men in London. Oxford University Press. Oxford.
7. BALTIMORE, R. S., C. D. C. CHRISTIE & G. J. WALKER SMITH. 1989. Am. Rev. Respir. Dis. **140:** 1650–1661.
8. HILL, S. L., H. M. MORRISON, D. BURNETT & R. A. STOCKLEY. 1986. Thorax **41:** 559–565.
9. ROBERTSON, A. J. 1952. Lancet **i:** 12–15.
10. STOCKLEY, R. A., S. L. HILL, H. M. MORRISON & C. M. STARKIE. 1984. Thorax **39:** 408–413.
11. BRADLEY, P. P., D. A. PRIEBAT, R. D. CHRISTENSEN & G. ROTHSTEIN. 1982. J. Invest. Dermatol. **78:** 206–209.
12. BUTTLE, D. J., D. BURNETT & M. ABRAHAMSON. 1990. Scand. J. Clin. Lab. Invest. **50:** 509–516.
13. STOCKLEY, R. A., J. SHAW, S. L. HILL & D. BURNETT. 1988. Clin. Sci. **74:** 645–650.
14. STOCKLEY, R. A., S. L. HILL, P. DRAGICEVIC *et al.* 1988. Royal Society of Medicine Services International Congress and Symposium Series. A. J. Howard, ed. No. 139: 19–30. London.
15. CURRIE, D. C., S. H. SAVERYMUTTU, S. NEEDHAM *et al.* 1987. Lancet **i:** 1335–1339.
16. BURNETT, D., A. CHAMBA & R. A. STOCKLEY. 1987. Lancet **ii:** 1043–1046.
17. BREUER, R., T. G. CHRISTENSEN, E. C. LUCEY *et al.* 1987. Am. Rev. Respir. Dis. **136:** 698–703.
18. BREUER, R., T. G. CHRISTENSEN, R. M. NILES *et al.* 1989. Am. Rev. Respir. Dis. **139:** 779–782.

19. SOMMERHOFF, C. P., J. A. NADEL, C. B. BASBAUM & G. H. CAUGHEY. 1990. J. Clin. Invest. **85:** 682–689.
20. TEGNER, H., K. OHLSSON, N. G. TOREMALM & C. VON MECKLENBURG. 1979. Rhinology **17:** 199–206.
21. SMALLMAN, L. A., S. L. HILL & R. A. STOCKLEY. 1984. Thorax **39:** 663–667.
22. RICKARD, K. & S. RENNARD. 1989. Am. Rev. Respir. Dis. **139** (no. 4, part 2): A406.
23. AMITANI, R., R. WILSON, R. READ et al. 1989. Am. Rev. Respir. Dis. **139** (no. 4, part 2): A406.
24. CURRIE, D. C., D. PAVIA, J. E. AGNEW et al. 1987. Thorax **42:** 126–130.
25. STOCKLEY, R. A. 1987. Clin. Chest Med. **8** (no. 3): 481–494.
26. FAGERHOL, M. K., I. DALE & T. ANDERSSON. 1979. Bull. Eur. Physiopathol. Respir. **16** (suppl): 273–282.
27. STOCKLEY, R. A., I. DALE, S. L. HILL & M. K. FAGERHOL. 1984. Scand. J. Clin. Lab. Invest. **44:** 629–634.
28. HOWIE, A. J., D. BURNETT & J. CROCKER. 1985. J. Pathol. **145:** 307–314.
29. BURNETT, D. & R. A. STOCKLEY. 1985. Clin. Sci. **68:** 469–474.
30. VENGE, P. & I. OLSSON. 1975. J. Immunol. **115:** 1505–1508.
31. DORING, G., W. GOLDSTEIN, K. BOTZENHART et al. 1986. Clin. Exp. Immunol. **64:** 597–605.
32. OLSSON, I., H. ODEBERG, J. WEISS & P. ELSBACH. 1978. In Neutral Proteases of Human Polymorphonuclear Leukocytes.: 18–32. Urban and Schwarzenberg. Baltimore-Munich.
33. NIEDERMAN, M. S., W. W. MERRILL, L. M. POLOMSKI et al. 1986. Am. Rev. Respir. Dis. **133:** 255–260.
34. HILL, S. L., A. CHAMBA, D. BURNETT & R. A. STOCKLEY. 1989. Am. Rev. Respir. Dis. **139** (no. 4, part 2): A36.
35. STOCKLEY, R. A. 1988. Br. J. Dis. Chest. **82:** 209–219.
36. STOCKLEY, R. A., J. SHAW & D. BURNETT. 1988. Agents & Actions **24:** 292–296.
37. MATZNER, V., R. DREXLER & M. LEVY. 1984. Eur. J. Clin. Invest. **14:** 440–443.
38. PERPER, R. J., M. SANDA, G. CHINEA & A. L. ORONSKY. 1974. J. Lab. Clin. Med. **84:** 394–406.
39. AUERBACH, H. S., M. WILLIAMS, J. A. KIRKPATRICK & H. R. COLTEN. 1985. Lancet **ii:** 686–688.
40. GOLDSTEIN, W. & G. DORING. 1986. Am. Rev. Respir. Dis. **134:** 49–56.
41. BOUDIER, C., P. LAURENT & J. G. BIETH. 1984. Adv. Exp. Med. Biol. **167:** 313–317.
42. ODEBERG, H. & I. OLSSON. 1975. J. Clin. Invest. **56:** 1118–1124.
43. CAMPANELLI, D., P. A. DETMERS, C. F. NATHAN & J. E. GABAY. 1990. J. Clin. Invest. **85:** 904–915.
44. KAO, R. C., N. G. WEHNER, K. M. SKUBITZ et al. 1988. J. Clin. Invest. **82:** 1963–1973.
45. TOURNIER, J.-M., J. JAQUOT, E. PUCHELLE & J. G. BIETH. 1985. Am. Rev. Respir. Dis. **132:** 521–528.
46. MORIHARA, K., H. TSUZUKI & K. ODA. 1979. Infect. Immun. **24:** 188–193.
47. FICK, R. B., R. S. BALTIMORE, S. U. SQUIER et al. 1985. J. Infect. Dis. **151:** 589–598.
48. MASON, R. W., D. A. JOHNSON, A. J. BARRETT & H. A. CHAPMAN. 1986. Biochem. J. **233:** 925–927.
49. SENIOR, R. M., N. L. CONNOLLY, J. D. CURY et al. 1989. Am. Rev. Respir. Dis. **139:** 1251–1256.

Protease Mechanisms in the Pathogenesis of Acute Lung Injury

THOMAS L. PETTY

*Presbyterian/St. Luke's Center For Health Sciences Education
and Department of Medicine
University of Colorado School of Medicine
Denver, Colorado 80218*

The problem of acute lung injury resulting in the acute respiratory distress syndrome (ARDS) continues to fascinate and frustrate basic researchers and clinicians alike. Numerous purported cellular and chemical mediators appear to be involved, but the exact sequence of interrelated events that ultimately damage the air-blood interface and result in inflammatory pulmonary edema remains to be understood. Both humoral agents and cell products damage the lungs' epithelium and endothelium; many mediators may be involved. It remains a reasonable hypothesis that an attack on the lung's surfactant system is the final common pathway into the clinical state known as ARDS.

PATHOGENESIS AND PATHOPHYSIOLOGY OF ARDS

Much of the following has been published elsewhere and is modified for present emphasis.[1]

ARDS is a protein-rich inflammatory pulmonary edema. It may begin as damage to endothelial cells and with large numbers of neutrophils in the interstitium of the lungs. Electron microscopic studies of lung tissues from patients who died during the initial phases of ARDS always show acute endothelial injury, which presumably is associated with increased permeability and interstitial pulmonary edema.[2,3] However, recent experimental evidence indicates that exclusive damage to endothelial cells by endotoxin does not lead to the full-blown syndrome of inflammatory *alveolar edema* or the structural damage of alveolar epithelial cells in experimental animals.[4] It is well established that major *epithelial injury* is present along with protein-rich edema in patients who die from ARDS.[5] This extracellular material forms hyaline membranes; cellular debris is also present. A typical, light microscopic example of the pathology of ARDS is illustrated in FIGURE 1A and B. These findings are found focally throughout human lungs.[2,3] Small amounts of protein normally reach the alveolar surface by a process that selectively limits the size of lavage molecules. In ARDS lung injury, this size selectivity is lost.[6] By contrast, selectivity by the size of molecules that traverse the alveolar barrier in cardiogenic pulmonary edema is preserved.[6] Consequently, epithelial injury may result in edema flooding the alveolar spaces of the lung and the disruption and destruction of the air-blood interface. A recent clinical study suggests that sequential measurements of pulmonary edema protein concentrations may provide a clinically useful index of alveolar epithelial function.[7] Thus, bronchoalveolar lavage (BAL) may offer additional insight into whether or not ARDS will occur and what the patient's prognosis might be.[7]

The mechanisms that create this inflammatory edema and necrosis of the

267

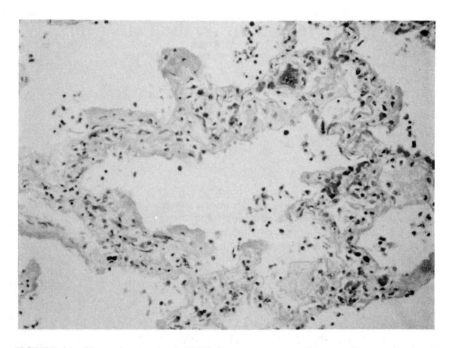

FIGURE 1A. Photomicrograph of ARDS disease processes 48 hours after massive head trauma and hemorrhagic shock with possible aspiration in a 24-year-old man. Note the cellular infiltrate and the hyaline membrane formation (magnification ×40). (Reprinted with permission from Year Book Publishers, Inc.[1])

alveolar epithelium are undoubtedly extremely complex. The clinical features that accompany the sequence of injurious events that result in ARDS following the original injury have been compared to an avalanche that cannot be controlled once it has begun. Many triggering factors have been proposed, including activation of the complement system, activation and aggregation of neutrophils, endothelial factors, an enhancing role of endotoxin, tumor necrosis factor, and platelet activation factor; these are all possibilities based on experimental models. Chemotaxins from monocytes and alveolar macrophages may cause adhesion of neutrophils to endothelial cells in the lung as part of the process. Any or all of these events may follow activation of the complement system, or during sepsis, neutrophils and macrophages may be activated by other mechanisms or by themselves. Evidence of selective leukosequestration in the lung after complement activation and not simple sieving of leukocyte aggregates adds credence to the complement-leukocyte activation and adhesion hypothesis in the pathogenesis of ARDS.[8] These activated and sequestered leukocytes would certainly be the source of vasoactive materials as could materials that damage endothelial and epithelial cells such as eicosanoids, proteases, and toxic oxygen species.[8-12] Together these studies suggest a specific receptor site for activated complement fragments in the lung.[7-11]

Earlier the Denver group became interested in the possibility that a major mechanism in acute lung injury might be microthrombi and fibrin deposition which are histopathologic findings in ARDS as well as in various animal models.[13-15] An elevated fibrinogen degradation product (D-dimer) was found within 60 hours after the start of ARDS in a small clinical series.[13] But further evidence has downgraded the importance of the coagulation system in the development of acute lung injury.[14-16] Therefore, although the coagulation system may be activated and involved, it is probably not a central pathogenetic mechanism resulting in acute lung injury.

The Activated Neutrophil and the Pathogenesis of Acute Lung Injury in ARDS

Numerous clinical observations in experimental models have focused on the neutrophil as the central mediator of acute lung injury in at least most patients with ARDS. It is a fact that large numbers of neutrophils collect in the lung during the early phases of acute lung injury.[17-18] In human studies large numbers of neutrophils were recovered from the BAL of patients with ARDS compared with few neutrophils in mechanically ventilated control patients and normal volunteers.[18] The number of neutrophils recovered correlated with the magnitude of abnormality in gas exchange and the degree of lung protein permeability. These relationships were statistically significant. In addition, serial bronchoalveolar lavages at 72-hour intervals showed decreasing percentages of neutrophils in six patients with ARDS, which was also statistically significant. These observations

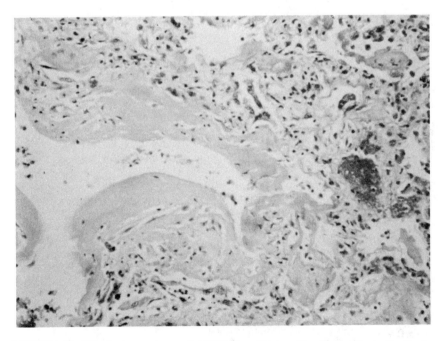

FIGURE 1B. Histologic specimen of ARDS in a 58-year-old patient with septic shock. Notice the hyaline membrane formation and cellular infiltration (magnification ×100). (Reprinted with permission from Year Book Publishers, Inc.[1])

in humans may be the alveolar mirror of the emerging blood leukopenia which heralds the onset of ARDS in certain patients at risk. It has been found that when ARDS risk factors are present, a falling white blood cell count to less than 4,000 per mm^3 is a fairly reliable predictor of the subsequent development of the full clinical syndrome.[19] Evidence of abnormal neutrophil-pulmonary interaction in ARDS has also been offered in various stages of ARDS, including the resolution and recovery phase.[20]

If the neutrophil is key to acute lung injury, what mechanisms are involved? Release of neutrophil elastase is a possibility, but increased elastase activity was not detected in one study of patients with ARDS.[18] This may be explained by the finding that elastase appeared to be complexed with alpha$_1$-antitrypsin.[18] By contrast, increased neutrophil collagenase activity was found in bronchoalveolar lavage ARDS fluid.[18] Other investigators have found increased elastase activity in the pulmonary lavage fluid of patients with ARDS.[21-23] These elastases are most likely derived from neutrophils. Also, high levels of myeloperoxidase activity were found in patients with ARDS.[18] The myeloperoxidase was cytotoxic for normal lung parenchymal cells when incubated in the presence of the myeloperoxidase system, co-factors, H_2O_2, and halide anion.[18] These and other investigators concluded that the neutrophil and its products, that is, proteinases and toxic oxygen species, play a central role in the pathogenesis of ARDS.[18,21,23] After all, literally billions of granulocytes traffic the lung every minute. Many remain in residence in a sizable "marginated pool" which is constantly changing. It is perfectly reasonable to believe that some stimulus, such as sepsis with complement activation, activates the neutrophils,[24] causing them to migrate through the endothelium and break through the epithelial barrier, during which time they release toxic oxygen species, collagenases, other proteinases, eicosanoids, or most likely a combination of active factors.[23,25] In further support of the primary role of neutrophils in ARDS is the deterioration of oxygenation which occurs during the resolution of neutropenia after chemotherapy in patients with acute diffuse lung injury.[25] However, it is also a fact that ARDS can occur in neutropenic patients;[26] thus, the neutrophil must not be absolutely crucial to the development of acute lung injury.[27]

In support of this notion is the observation that intrapulmonary sequestration of aggregated neutrophils does not necessarily result in lung injury.[28] It is probably fair to state that all of the foregoing studies and observations, taken *in toto,* leave some uncertainty about the role of the neutrophil in the pathogenesis of ARDS.[27,28]

No matter what the mechanism of initiation and amplification of the original lung injury may be, protein-rich fluid that carries inflammatory mediators ultimately moves from the interstitial space beyond the alveolar barrier and into the alveolar spaces. When this happens, alveolar units become flooded with these injurious materials. It is an interesting observation that increased numbers of neutrophils were found in the alveolar lavage fluid in patients at risk of ARDS as well as in ARDS itself, indicating that the neutrophils get into the alveoli *before* increased permeability pulmonary edema occurs. These neutrophils have a greater than normal adherence to plastic surfaces. The factor(s) that increase adherence includes both lipid and protein characteristics.[29] The structural basis for increased alveolar epithelial permeability is uncertain, nor is it known why a rapid progression from interstitial to alveolar edema occurs, a phenomenon that is so often observed clinically.[29] It is likely that the epithelial lining breaks occur, possibly because of locally increased microvascular pressure that also occurs in the early stages of acute lung injury.[30] Direct epithelial damage which follows aspiration or inhalation of toxins appears, from my bedside observations, to produce the most rapidly progressive pulmonary edema, again suggesting that the epithelial barrier to fluid formation is most important. These clinical observations are supported by experimental work.[31,32]

In any case, massive amounts of inflammatory mediators are released that may include several possible amplifying factors. The release of inflammatory mediators and amplifiers may require a certain sequence or combination that result in ARDS. Other inflammatory mediators, such as leukotriene B_4, are chemotactic for neutrophils and mononuclear cells.[32] Still other arachidonic acid lipoxygenase pathway products also contribute to increased permeability.[33] Increased microvascular pressure responses within the pulmonary microcirculation are mediated by cycloxygenase products.[34,35] In addition, platelet activating factor (PAF), if involved, may act also as a powerful agent in causing constriction on the venous side of the lung's microcirculation.

The research for a circulating marker that could predict the onset of ARDS continues. Some early evidence suggested that increased levels of complement fragments could reflect neutrophil activation as an event that could herald forthcoming acute lung injury.[24] More recent studies have found no correlation between elevated levels of C_5 or C_3 and the development of ARDS, but suggest a possible role of endotoxin in combination with complement activation as one mechanism in the development of ARDS.[36] Further studies by the same investigators found no relation between the terminal complement complex (sC5b-9) and the development of ARDS.[37]

Thus, no matter the sequence of events, it is likely that a vicious cycle which includes increased permeability and increased microvascular pressure conspires to flood the lung with inflammatory mediators, which damage the alveolar epithelium.[38-40] Some of the possible pathways that result in increased permeability as well as the vascular and bronchial responses of ARDS are depicted in FIGURE 2.[39] The proposed interplay between membrane damage and leak, mediator production, and microvascular response is shown in FIGURE 3.[41]

In summary, an almost endless number of mediators, both cellular and humoral, and interrelated pathways that can alone or synergistically cause acute inflammatory lung injury, seem possible. These are: complement products, endotoxin, proteases, toxic oxygen species, eicosanoids (cyclooxygenase products and lipoxygenase products), coagulation factors, platelet activating factor, vasoactive substances, and combinations. It must be stated that some or all of these mechanisms may be present, at least to a degree, in patients at risk who develop only mild to moderate abnormalities in oxygenation. Thus, the full-blown clinical state may be only the "tip of the iceberg".[42] The numerous possibilities appear almost like a spider web, creating intriguing fantasies concerning the actual sequence of events and possible therapeutic interventions.

Surfactant Deficiency as the Final Common Pathway in ARDS

We observed an abnormally elevated minimum surface tension in material obtained from specimens of minced human lungs obtained at autopsy in our first series of patients.[43] Later we found both qualitative and quantitative surfactant abnormalities in lavage fluid of patients with ARDS obtained from fresh whole human lungs at autopsy, compared with normal controls.[44,45] These unique observations were possible because of the availability of living cadavers that had suffered brain death after major body trauma which, coincidentally, resulted in

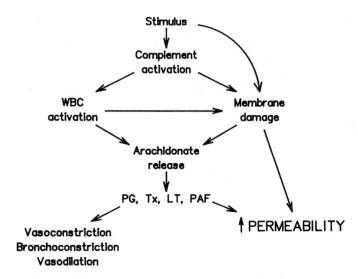

FIGURE 2. Illustration of some possible pathways that result in increased permeability as well as vascular and bronchial responses in ARDS. (Modified with permission from Mc-Donald, T. J., N. F. Voelkel, T. L. Petty & J. D. Wilson. 1987. The Adult Respiratory Distress Syndrome. Drugs Used in Respiratory Diseases. Williams & Wilkins. Baltimore.)

ARDS. Some cadavers had other organs suitable for transplantation. Fresh whole human lungs were obtained only a few hours after the onset of ARDS or in the absence of ARDS (controls) and before infection or fibrosis was present. These studies offered additional evidence that an attack on the surfactant system of the lung may well be the final common pathway to ARDS. It is a reasonable hypothesis that endothelial and epithelial damage can destroy the surfactant system possibly via oxidative or proteolytic mechanisms. Certainly surfactant is vulnerable to oxidative and proteolytic damage. Substantial evidence indicates that oxidants are abundant in human ARDS.[46,47] It is also a fact that excessive proteases have been found in some patients with ARDS.[18,22,23] Another possibility would be dam-

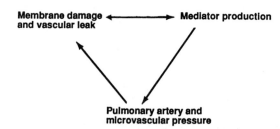

FIGURE 3. Simplified scheme showing interrelationship between membrane damage, mediator production, and pulmonary microvascular pressure. (Courtesy of Norbert Voelkel. Reprinted with permission from Voelkel, N. F.[41])

age of the type II alveolar lining cells, the source of surfactant. Finally, surfactant could simply be "washed" into the circulating blood. If surfactant deficiency causes alveolar instability and collapse, another vicious cycle would be set in motion. This is conceptualized in FIGURE 4.[48]

Increased elastic recoil, due to unopposed surface forces at the air-blood interface, could produce local hydrostatic forces (stresses) that would, in turn, create more leakage from extra alveolar vessels which could create more alveolar flooding with inflammatory and protease mediators.[49] Studies in intact humans give further credence to the surfactant hypothesis as a key "player," although clearly a late one in the cascade of events resulting in ARDS.[50]

If surfactant damage or insufficiency is truly the final common pathway and

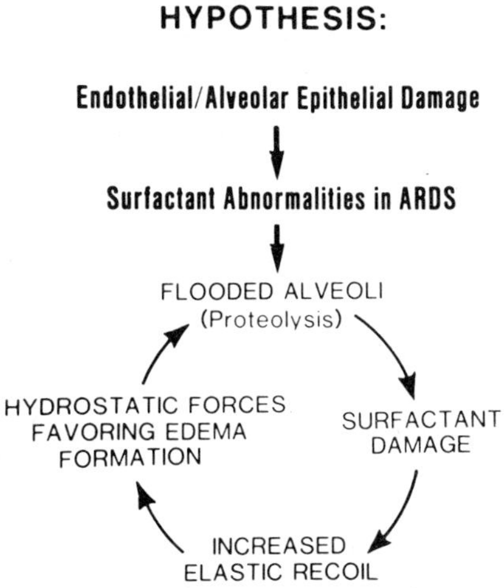

HYPOTHESIS:

Endothelial/Alveolar Epithelial Damage

↓

Surfactant Abnormalities in ARDS

↓

FLOODED ALVEOLI
(Proteolysis)

HYDROSTATIC FORCES
FAVORING EDEMA
FORMATION

SURFACTANT
DAMAGE

INCREASED
ELASTIC RECOIL

FIGURE 4. Hypothesis to explain possible mechanisms of endothelial/alveolar epithelial damage which in turn injures the surfactant producing Type II pneumocytes, resulting in reduced surfactant production. Once alveoli are flooded with proteases and/or free radicals following alveolar capillary injury and leak, further destruction of surfactant probably occurs. In this situation, a vicious cycle of increased elastic recoil, because of surfactant abnormalities, favors further alveolar flooding with edema fluid due to hydrostatic forces (stresses). (Reprinted with permission from Petty, T. L.[48])

thus a key step in the development of ARDS, therapeutic opportunities in surfactant replacement offer exciting possibilities in early intervention. In addition, the use of antioxidants with antiproteases would seem appropriate in incipient stages of disease (see below).

CURRENT STATUS OF ARDS: 1991

Today there are an estimated 150,000 cases of ARDS from all causes each year in the United States, but this figure is only an estimate.[51] Survival rates approximate only 40%.[52,53] Methods of supportive care are being continuously revised, or developed, and studied in controlled clinical trials, but it is doubtful that any form of life support will greatly increase survival into even a range of 70–80% or more. The reason is that the injured lung requires a long time for repair, and during this interval the risk of sepsis and multiple organ system failure makes a successful outcome unlikely in all too many patients, often young people who were previously healthy. If we are to make a major breakthrough in prevention and management, we will have to be disciplined about the definition of ARDS and its complications and ultimately find the responsible mechanisms and the blockers of these mechanisms in incipient stages of disease or in patients at risk of ARDS.[52,54]

Current understandings concerning the pathogenesis of ARDS suggest a new approach to preventive management. If we focus on patients at risk of ARDS, based on epidemiologic considerations,[51–53] and particularly if we can find accurate markers that accurately herald the onset of ARDS in tracheobronchial secretions, blood, or even urine, we will be in a position to embark on controlled clinical trials aimed at improving survival. I can envision a multicenter protocol in which patients at risk receive bronchoalveolar lavage in order to identify excess polymorphonuclear leukocytes in alveolar spaces, as well as the presence of excess proteases and evidence of toxic oxygen species excess and possibly related surfactant abnormalities. When reliable factors that can predict the onset of refractory hypoxemia and the reduced lung compliance of ARDS are identified, antiproteases, oxidant scavengers, and surfactant replacement will become an exciting therapeutic approach. Surfactant replacement itself might alter the course and prognosis of ARDS if it produced its physiologic mechanism of opening closed lung units, which could also mitigate the local hydrostatic factors in edema formation. A systematized and strategic attack on the protease-oxidant-surfactant system seems most likely to be successful, particularly if instituted early in the course of emerging lung injury *before* irreversible alveolar injury and cell necrosis occur. When the air-blood interface is damaged and replaced by hyaline membrane formation, recovery will be effected only if lung regeneration and repair occur by virtue of proliferation of Type II pneumocites with associated repair of a damaged basement membrane and subsequent differentiation into Type I gas exchange cells.

Today it is time to redefine and classify ARDS based on new knowledge of the basic biochemical and cellular mechanisms involved with acute lung injury.[55–57] At least one of these mechanisms appears to be the release of damaging proteases derived from polymorphonuclear leukocytes.

ACKNOWLEDGMENT

Appreciation is expressed to Kay Bowen for manuscript assistance.

REFERENCES

1. PETTY, T. L. 1990. Acute respiratory distress syndrome (ARDS). Disease-a-Month **36:** 7–58.
2. BACHOFEN, M. & E. R. WEIBEL. 1977. Alterations of the gas exchange apparatus in adult respiratory insufficiency associated with septicemia. Am. Rev. Respir. Dis. **116:** 589–615.

3. BACHOFEN, M. & E. R. WEIBEL. 1982. Structural alterations of lung parenchyma in the adult respiratory distress syndrome. Clin. Chest Med. 3: 35–56.
4. WIENER-KRONISH, J. P., L. GATES, K. H. ALBERTINE et al. 1988. The effects of alveolar E.-coli endotoxin in lung endothelial and epithelial permeability in anesthetized sheep. Am. Rev. Respir. Dis. 137: 225A.
5. LAMY, M., R. FALLAT, E. KOENIGER et al. 1976. Pathologic features and mechanisms of hypoxemia in adult respiratory distress syndromes. Am. Rev. Respir. Dis. 114: 267–284.
6. HOLTER, J. F., J. E. WEILAND, E. R. PACH et al. 1986. Protein permeability in the adult respiratory distress syndromes. J. Clin. Invest. 78: 1513–1522.
7. WIENER-KRONISH, J. P. & M. A. MATTHAY. 1989. Sequential measurements of pulmonary edema protein concentrations provide a reliable index of alveolar epithelial function and prognosis in patients with the adult respiratory distress syndrome. Clin. Res. 37: 165A.
8. KLAUSNER, J. M., L. KOBZIK & C. R. VALERI. 1988. Selective lung leukosequestration after complement activation. J. Appl. Physiol. 65: 80–88.
9. HENSON, P. M., G. L. LARSEN, R. O. WEBSTER et al. 1982. Pulmonary microvascular alterations and injury induced by complement fragments: Synergistic effort of complement active neutrophil sequestration and prostaglandins. Ann. N.Y. Acad. Sci. 384: 287–300.
10. LARSEN, G. L., R. O. WEBSTER, G. S. WORTHEN et al. 1985. Additive effect of intravascular complement activation and brief episodes of hypoxia in producing increased permeability in the rabbit lung. J. Clin. Invest. 75: 902–910.
11. HARLAN, J. M. 1985. Leukocyte-endothelial interactions. Blood 65: 513–525.
12. WEBSTER, R. O., G. L. LARSEN & P. M. HENSON. 1982. In vivo clearance and tissue distribution of C5a and C5a des arginine complement fragments in rabbits. J. Clin. Invest. 70: 1177–1183.
13. HAYNES, J. B., T. M. HYERS, P. C. GICLAS et al. 1980. Elevated fibrin (ogen) degradation products in the adult respiratory distress syndrome. Am. Rev. Respir. Dis. 122: 841–847.
14. MANWARING, D., D. THORNING & P. CURRERI. 1978. Mechanisms of acute pulmonary dysfunction induced by fibrinogen degradation product D. Surgery 84: 45–54.
15. SALDEEN, T. 1976. Trends in microvascular research—the microembolism syndrome. Microvasc. Res. 11: 227–259.
16. BINDER, A. S., K. NAKAHARA, K. OHKUDA et al. 1979. Effect of heparin or fibrinogen depletion on lung fluid balance in sheep after emboli. J. Appl. Physiol. 47: 213–219.
17. BRIGHAM, K. L. 1982. Mechanisms of lung injury. Clin. Chest Med. 3: 9–24.
18. WEILAND, J. E., W. B. DAVID, J. F. HOLTER et al. 1986. Lung neutrophils in the adult respiratory distress syndrome: Clinical and pathophysiological significance. Am. Rev. Respir. Dis. 133: 218–225.
19. THOMMASEN, H. V., J. A. RUSSELL, W. J. BOYKO et al. 1984. Transient leucopenia associated with adult respiratory distress syndrome. Lancet 1: 809–812.
20. WARSHAWSKI, F. J., W. J. SIHHOLD, A. A. DREIDGEN et al. 1986. Abnormal neutrophil—pulmonary interaction in the adult respiratory distress syndrome. Am. Rev. Respir. Dis. 133: 787–804.
21. BRIGHAM, K. L. & B. MEYRICK. 1984. Interactions of granulocytes with the lungs. Circ. Res. 54: 623–635.
22. ZHEATLIN, L. M., E. R. JACOBS, M. E. HANLEY et al. 1986. Plasma elastase levels in the adult respiratory distress syndrome. J. Crit. Care 1: 39–47.
23. LEE, C. T., A. M. FEIN, M. LIPPMAN et al. 1981. Elastolytic activity in pulmonary lavage fluid from patients with adult respiratory distress syndrome. N. Engl. J. Med. 304: 192–196.
24. TIL, G. D., K. J. JOHNSON & R. KUNKEL. 1982. Intravascular activation of complement and acute lung injury: Dependency on neutrophils and toxic oxygen metabolites. J. Clin. Invest. 69: 1126–1135.
25. RINALDO, J. E. & H. BOROVETZ. 1985. Deterioration of oxygenation and abnormal lung microvascular permeability during resolution of leukopenia in patients with diffuse lung injury. Am. Rev. Respir. Dis. 131: 579–583.

26. MAUNDER, R. J., R. C. HACKMAN, E. RIFFI et al. 1986. Occurrence of the adult respiratory distress syndrome in neutropenic patients. Am. Rev. Respir. Dis. 133: 313–316.
27. GLAUSER, F. L. & R. P. FAIRMAN. 1985. The uncertain role of the neutrophil in increased permeability edema. Chest 88: 601–607.
28. SHAW, J. & P. HENSON. 1982. Pulmonary sequestration of activated neutrophils: Failure to induce light-microscopic evidence of lung injury in rabbits. Am. J. Pathol. 108: 17–23.
29. FOWLER, A. A., T. M. HYERS, B. J. FISHER et al. 1987. The adult respiratory distress syndrome: Cell populations and soluble mediators in the air spaces of patients at high risk. Am. Rev. Respir. Dis. 136: 1225–1231.
30. STAUB, N. C. 1984. Pathophysiology of pulmonary edema. In Edema. N. C. Staub & A. E. Taylor, eds.: 719–747. Raven Press. New York.
31. EAGEN, E. A., R. M. NELSON & I. H. GESSNER. 1977. Solute permeability of the alveolar epithelium: Acute hemodynamic pulmonary edema in dogs. Am. J. Physiol. 233: H80–H86.
32. COTTRELL, T. S., O. R. LEVINE, R. M. SENIOR et al. 1967. Electron microscopic alterations at the alveolar level in pulmonary edema. Circ. Res. 21: 783–797.
33. MARTIN, T. R., L. C. ALTMAN, R. K. ALBERT et al. 1984. Leukotriene B4 production by the human alveolar macrophage: A potential mechanism for amplifying inflammation in the lung. Am. Rev. Respir. Dis. 129: 106–111.
34. SEEGER, W., M. MENGER, D. WALMRATH et al. 1987. Arachidonic acid lipoxygenase pathways and increased vascular permeability in isolated rabbit lungs. Am. Rev. Respir. Dis. 136: 964–972.
35. MALIK, A. B., M. B. PERLMAN, J. A. COOPER et al. 1985. Pulmonary microvascular effects of arachidonic acid metabolites and their role in lung vascular injury. Fed. Proc. 44: 36–42.
36. PARSONS, P. E., G. S. WORTHEN, R. M. TATE et al. 1989. Endotoxin and complement fragments (CSF) in plasma, but not elevated CSF levels alone correlate with the development of ARDS. Am. Rev. Respir. Dis. 140: 294–300.
37. PARSONS, P. E. & P. G. GICLAS. 1990. The terminal complement complex (sC5b-9) is not specifically associated with the development of the Adult Respiratory Distress Syndrome. Am. Rev. Respir. Dis. 141: 98–103.
38. SEEGER, W., H. WOLF, G. STAHLER et al. 1982. Increased pulmonary vascular resistance and permeability due to arachidonate metabolism in isolated rabbit lungs. Prostaglandins 23: 157–173.
39. MCDONNELL, T. J., N. F. VOELKEL & T. L. PETTY. 1987. The adult respiratory distress syndrome. In Drug Use in Respiratory Disease. J. D. Wilson, ed.: 208–222. Williams & Wilkins. Baltimore.
40. SAKAI, A., S. CHANG & N. F. VOELKEL. 1989. The importance of vasoconstriction in lipid mediator induced pulmonary edema. J. Appl. Physiol. 66: 2667–2674.
41. VOELKEL, N. F. 1989. The Adult Respiratory Distress Syndrome. Klin. Wochenschr. 67: 559–562.
42. RINALDO, J. E. & R. M. ROGERS. 1982. Adult Respiratory Distress Syndrome: Changing concepts of lung injury and repair. N. Engl. J. Med. 306: 900–909.
43. ASHBAUGH, D. G., D. B. BIGELOW, T. L. PETTY et al. 1967. Acute respiratory distress in adults. Lancet 2: 319–323.
44. PETTY, T. L., O. K. REISS, G. W. PAUL et al. 1977. Characteristics of pulmonary surfactant in adult respiratory distress syndrome associated with trauma and shock. Am. Rev. Respir. Dis. 115: 531–536.
45. PETTY, T. L., G. W. SILVERS, G. W. PAUL et al. 1979. Abnormalities in lung elastic properties and surfactant function in adult respiratory distress syndrome. Chest 75: 571–574.
46. BALDWIN, S. R., R. H. SIMON, C. M. GRUM et al. 1986. Oxidant activity in expired breath of patients with adult respiratory distress syndrome. Lancet 1: 11–14.
47. COCHRANE, C. G., R. SPRAGG & S. D. REVAK. 1983. Pathogenesis of the adult respiratory distress syndrome. Evidence of oxidant activity in bronchoalveolar lavage fluid. J. Clin. Invest. 71: 754–761.

48. PETTY, T. L. 1981. Adult respiratory distress syndrome: Historical perspective and definition. Semin. Respir. Med. **2:** 99–103.
49. ALBERT, R. K. 1981. Factors affecting transvascular fluid and protein movement in pulmonary edema and ARDS. Semin. Respir. Med. **2:** 109–113.
50. HALLMAN, W., R. SPRAGG, J. H. HARRELL *et al.* 1982. Evidence of lung surfactant abnormality in respiratory failure. J. Clin. Invest. **70:** 673–683.
51. MURRAY, J. F. & THE STAFF OF THE DIVISION OF LUNG DISEASES. 1977. National Heart, Lung & Blood Institute: Mechanisms of Acute Respiratory Failure. Am. Rev. Respir. Dis. **115:** 1071–1078.
52. FOWLER, A. A., R. F. HAMMAM, J. T. GOOD *et al.* 1983. Adult Respiratory Distress Syndrome: Risk with common predispositions. Ann. Intern. Med. **98:** 593–595.
53. BAUMANN, W. E., R. C. SAVY, M. KOSS *et al.* 1986. Incidence and mortality of adult respiratory distress syndrome: A prospective analysis from a large metropolitan hospital. Crit. Care Med. **14:** 1–4.
54. PEPE, P. E., P. T. POTKINS P. H. RENS *et al.* 1982. Clinical predictors of the adult respiratory distress syndrome. Am. J. Surg. **144:** 124–130.
55. MURRAY, J. F., M. A. MATTHAY, J. LUCE *et al.* 1988. An expanded definition of the adult respiratory distress syndrome. Am. Rev. Respir. Dis. **138:** 720–723.
56. PETTY, T. L. 1988. ARDS: Refinement of Concept and Redefinition. Am Rev Resp Dis **138:** 724.
57. MATTHAY, M. A. 1989. The Adult Respiratory Distress Syndrome. New insights into diagnosis, pathophysiology and treatment. West. J. Med. **150:** 187–194.

Protease Injury in Airways Disease

STEPHEN I. RENNARD,[a] KATHLEEN RICKARD,
JOE D. BECKMANN, GUILLERMO HUERTA,
SEKIYA KOYAMA, RICHARD A. ROBBINS,
DEBRA ROMBERGER, JOHN SPURZEM, AND
AUSTIN B. THOMPSON

University of Nebraska
Department of Internal Medicine
Pulmonary and Critical Care Medicine Section
Omaha, Nebraska 68198–2465

Proteases are widely recognized as a cause of injury to the alveolar structures of the lower respiratory tract. A major role for proteases, particularly neutrophil elastase, is suggested in the pathogenesis of tissue destruction accompanying emphysema.[1,2] This concept originated in the observation that patients with a genetic deficiency of alpha$_1$-protease inhibitor (α_1-PI) had a markedly increased incidence of emphysema.[3] As α_1-PI is the major serum inhibitor of neutrophil elastase,[4] and as elastase instilled into the lungs of experimental animals can lead to the development of emphysema,[5,6] the theory was proposed that emphysema resulted from an excess of protease overcoming the antiprotease screen in the lower respiratory tract. This "protease-antiprotease" theory has been extended to cigarette smoke-induced emphysema in which excess protease burden is thought to result from chronic smoke-induced inflammation,[7–9] and loss of antiprotease protection is thought to result from chemical inactivation of antiproteases.[10–12] Most attention on the balance between proteases and antiproteases has focused on the alveolar structure and emphysema. These studies are reviewed in detail elsewhere in this volume. Proteases and antiproteases also play a role in airways disease. Airways dysfunction, for example, as evidenced by hyperreactivity and reversible obstruction is a feature of α_1-PI deficiency separate from emphysema.[4,13] Similarly, instillation of elastase into the lower respiratory tract can lead not only to alveolar emphysema, but also to long-term alterations of airways architecture.[14,15] Consequently, the ability of proteases to damage airway epithelium may play a role in the pathogenesis of chronic airways disease.

PROTEASES IN DISEASE

Proteases are present in the airways in airways disease. Sputum from patients with both chronic bronchitis[16–18] and cystic fibrosis[19] contains very large accumulations of inflammatory cells, predominantly neutrophils. Not surprisingly, large concentrations of proteases have been detected in both sputum[20–22] and, more recently, in bronchoalveolar lavage fluid[23,24] from these patients. Active neutro-

[a] Address for correspondence: Stephen I. Rennard, M.D., Chief, Pulmonary and Critical Care Medicine Section, Department of Internal Medicine, University of Nebraska Medical Center, 600 South 42nd Street, Omaha, Nebraska 68198–2465.

phil elastase has been reported in both patients with cystic fibrosis[20,21] and selected individuals with chronic bronchitis,[24] suggesting that the antiprotease screen can be overcome by either the quantity of elastase present or inactivation of the antiprotease screen or both. In most patients with chronic bronchitis at times of clinical quiescence, however, neutrophil elastase can be detected, but it apparently is quantitatively inhibited by antiproteases.[24] Despite the lack of active neutrophil elastase in pooled bronchoalveolar lavage fluid, it is likely that after secretion the elastase is transiently active, at least within the local environment. Thus, considerable evidence indicates that the airway epithelium is exposed to active neutrophil elastase in inflammatory airways disease.

Proteases other than neutrophil elastase also likely contribute to tissue injury in airways disease. Macrophages can release a spectrum of proteases;[25] however, the role of these enzymes in airways disease is largely unexplored. In asthma, in addition to neutrophils that can be present in the airways, eosinophils and mast cells are present.[26–28] Both cells are capable of releasing proteases with a spectrum of substrate specificities.[29,30] In addition, both eosinophil and mast cell products have been reported in bronchoalveolar lavage fluid of persons with asthma.[28] Bacterial colonization can lead to the presence of bacterial proteases in airways disease. This has been studied best in cystic fibrosis in which colonization with *Pseudomonas aeruginosa* leads to release of pseudomonas elastase.[31–33] Whereas the pathogenetic role of pseudomonas elastase is controversial, it is not inhibited by α_1-PI.[34] As a result, active pseudomonas elastase can routinely be found in respiratory tract samples. It is likely that bacterial proteases are also present in other patients with bronchiectasis or chronic bronchitis who have chronic airways colonization. The parenchymal cells of the airway, epithelial cells, and fibroblasts also likely release proteases that might cause damage in certain pathologic states. Fibroblasts from a number of sources have been reported to produce collagenases capable of degrading interstitial collagens.[35,36] Tumor cells derived from epithelial surfaces can release proteases capable of degrading basement membrane.[37,38] It is likely that normal cells, including normal airway epithelial cells, will have a similar capability. Therefore, it is likely that a spectrum of proteases with differing substrate specificities could contribute to the pathogenesis of airways disease in various pathologic settings.

The best studied of the proteases that might contribute to airways disease is neutrophil elastase. This enzyme has a broad spectrum of substrates and likely can affect airway function through several pathways.[39] Destruction of complement receptors by neutrophil elastase has been suggested to contribute to the bactericidal defect present in cystic fibrosis.[40] Snider and colleagues have demonstrated that direct instillation of this enzyme into airways of hamsters leads to changes that resemble those of chronic bronchitis.[14,15] Acute instillation is followed by a rapid loss of recognizable granulated secretory cells, presumably due to degranulation.[14] Although the mechanism that leads to acute degranulation is unclear, a direct effect of the elastase on the epithelial cells is possible. In support of this, neutrophil elastase[41] and bacterial elastase[42] have been reported to be potent stimuli for airway secretion.

EFFECTS OF PROTEASES ON AIRWAY EPITHELIAL CELLS

Following the initial exposure to elastase, the number of granulated secretory cells is rapidly recovered. By 3 days, the number has been restored to normal. By 16 days, "overshoot" occurs with a significant increase in the percentage of granulated secretory cells compared to baseline conditions or to control animals exposed to saline solution alone.[15] The secretory cells that accumulate under these conditions have been noted to have an increased number and density of secretory granules. Importantly, the increase in the number of these altered granulated secretory cells appears to be persistent. Christensen et al.[15] reported that no decrease in these metaplastic changes occurred over a period of 12 months, as long as the animals were studied. Taken together, these studies suggest that intermittent exposures to excess elastase may account for the marked goblet cell metaplasia that characterizes both asthma and chronic bronchitis.[43] Interestingly, the persistence of the changes in the hamster model system resembles the persistence of similar anatomic features in bronchitis and asthma, even after long periods of disease quiescence.[44] Persistence of anatomic changes after limited exposure to inflammatory cell proteases could contribute to the increased development of airways disease in adults who experienced bronchiolitis in childhood.[45,46]

The mechanisms by which neutrophil elastase damages the airway epithelium and causes goblet cell metaplasia are, as yet, unclear. One possible explanation, however, is that elastase may disrupt the adhesion of the airway epithelial cells with the underlying basement membrane. Rickard and Rennard[47] have demonstrated that both neutrophils and purified human neutrophil elastase can cause the detachment of cultured airway epithelial cells.[47] This detachment can be blocked by the addition of a variety of serine protease inhibitors including α_1-PI, suggesting that neutrophil elastase plays a major role.

The condition of the epithelial cells, moreover, may affect the sensitivity to proteases. Airway epithelial cells possess tight junctions[48] and cultured airway epithelial cells form tight junctions when maintained in culture for 10 days to 2 weeks.[49] These tight junctions form a barrier, preventing the free diffusion of various molecules across the epithelial cell monolayer. The electrical resistance of epithelial cell monolayers, for example, can be shown to increase dramatically with the development of this barrier.[49] Concurrent with the development of electrical resistance, the airway epithelial cell monolayer becomes resistant to detachment induced by elastase added on the apical surface.[50] In contrast, when elastase is added to the basal surface, detachment is rapid. Although the exact mechanism for the development of this polarity is unclear, it is possible that an epithelium with intact tight junctions could function to confine the inflammatory protease-rich milieu to the intraluminal space. Interestingly, if airway epithelial cell monolayers that are resistant to apically exposed elastase-induced detachment are exposed to cigarette smoke extract, their ability to resist elastase-induced detachment is lost.[50] Such a combined injury might play a role in the development and the exacerbation of airways disease by cigarette smoke.

The loss of epithelial cells in asthma is well recognized. Creola bodies, representing clusters of desquamated airway epithelial cells, were reported in 1962[51] in the sputum of asthmatic persons with acute exacerbations. Although a variety of mechanisms, including mechanical disruption due to the hydrostatic pressure of edema fluid, are possible, it seems reasonable that the proteases could play a role in the desquamation of airway epithelial cells. The loss of airway epithelial cells could affect airway function in a number of ways. First, loss of barrier function could both provide relatively easy access to the subepithelium of substances normally confined to the airway lumen and allow plasma constituents to more easily reach the air-liquid interfaces. Second, the airway epithelium can produce smooth muscle relaxing factor(s).[52,53] Loss of epithelium might be expected to lead to an increase in airways tone. Third, the airway epithelial cells are a major source of neutral endopeptidase,[54] the cell surface protease responsible for degrading neuropeptide bronchoconstrictor substance P. Loss of airway epithelium is associated with increased bronchial reactivity in the presence of a variety of stimuli,[55,56] including substance P.[57] The increased sensitivity to neuropeptidase is presumably due to decreased degradation of the active neuropeptide.[57,58] Taken together, it is likely that loss of airway epithelium could be associated with increased bronchial reactivity. Beasley *et al.*[59] have provided direct data supporting this concept. Bronchoalveolar lavage epithelial cell recovery, presumably a reflection of epithelial cell desquamation, is increased in asthmatic persons in direct relationship to increased airways inactivity.

Spurzem and colleagues[60] evaluated epithelial cells recovered by bronchoalveolar lavage in patients with chronic bronchitis. No increases in epithelial cell recovery were reported. However, an increase was noted in the percentage of epithelial cells that were goblet cells. Moreover, a striking correlation existed between the goblet cell percentages and the airflow as reflected in the FEV_1 as percent predicted. Although airways reactivity was not assessed by Spurzem *et al.*, patients with reversible airflow obstruction or clinical history consistent with "asthma" were excluded. Perhaps the loss of epithelial cell integrity correlates with reactivity and the goblet cell metaplasia correlates with progression to fixed airflow obstruction.

The cellular and biochemical pathways that lead to epithelial cell metaplasia and fixed airflow obstruction are also unclear. Airway epithelial cells, however, appear to respond to injury as do other cells. Transforming growth factor-β, for example, has been suggested to modulate wound healing in certain model systems.[61] Airway epithelial cells in culture undergo squamous metaplasia[62,63] and increase fibronectin production[64] in response to TGF-β. Fibronectin, a large multifunctional glycoprotein, is known to be deposited at sites of injury and repair when it is thought to contribute to morphogenesis.[65,66] It is unclear if protease injury in the airway is associated with increased TGF-β and fibronectin production. Fibronectin, however, is a substrate for many proteases including mast cell protease[67] and neutrophil elastase,[68] both of which can release biologically active products from fibronectin.[69] Thus, it appears likely that proteases released during airways inflammation not only can injure the airways, but also can affect airway repair processes.

SUMMARY

In summary, proteases are present in the airway in inflammatory airways disease. These enzymes can damage the airway epithelium. As a consequence, airway function can be altered, and long-term changes in airway anatomy can result. Although the exact cellular and biochemical mechanisms that lead to these changes are incompletely described, it seems likely that they will play important roles in clinical airways disease. As such, these pathways may represent novel opportunities for therapeutic intervention.

REFERENCES

1. SNIDER, G. L. 1981. The pathogenesis of emphysema: Twenty years of progress. Am. Rev. Respir. Dis. **124:** 321–324.
2. JANOFF, A. 1985. Elastases and emphysema. Am. Rev. Respir. Dis. **132:** 417–433.

3. ERIKSON, S. 1964. Pulmonary and emphysema and alpha, anti-trypsin deficiency. Acta Med. Scand. 177: 175.
4. GADEK, J. E. & R. G. CRYSTAL. 1983. α_1-antitrypsin deficiency. In The Metabolic Basis of Inherited Disease. Stanbury et al., eds.: 1450–1467. McGraw-Hill. New York.
5. JANOFF, A., B. SLOAN, G. WEINBAUM, V. DAMIANO, R. A. SANDHAUS, J. ELIAS & P. KIMBEL. 1977. Experimental emphysema induced with purified human neutrophil elastase: Tissue localization of instilled protease. Am. Rev. Respir. Dis. 115: 461–478.
6. SENIOR, R. M., H. TEGNER, C. CUHN, K. OHLSSON, B. C. STARCHER & J. A. PIERCE. 1977. The induction of pulmonary emphysema with human leukocyte elastase. Am. Rev. Respir. Dis. 116: 469–475.
7. ROBBINS, R. A., A. B. THOMPSON, S. KOYAMA, J. P. METCALF, K. A. RICKARD, J. R. SPURZEM & S. I. RENNARD. Cigarette smoking and neutrophils. Submitted.
8. NIEWOEHNER, D. E., J. KLEINERMAN & D. B. RICE. 1974. Pathologic changes in the peripheral airways of young cigarette smokers. N. Engl. J. Med. 291: 755.
9. HUNNINGHAKE, G. W. & R. G. CRYSTAL. 1983. Cigarette smoking and lung destruction. Accumulation of neutrophils in the lungs of cigarette smokers. Am. Rev. Respir. Dis. 128: 833–838.
10. JANOFF, A. & R. DEARING. 1982. Alpha-1-proteinase inhibitor is more sensitive to inactivation by cigarette smoke than leukocyte elastase. Am. Rev. Respir. Dis. 126: 691–694.
11. COHEN, A. B. & H. L. JAMES. 1982. Reduction of the elastase inhibitory capacity of alpha-1-antitrypsin by peroxides in cigarette smoke. An analysis of brands and filters. Am. Rev. Respir. Dis. 126: 25–30.
12. GADEK, J. E., G. A. FELLS & R. G. CRYSTAL. 1979. Cigarette smoking induces functional antiprotease deficiency in the lower respiratory tract of humans. Science 206: 1315–1316.
13. SCHWARTZ, R. H., J. D. VAN ESS, D. E. JOHNSTONE, E. M. DREYFUSS, M. A. ABRISHARMI & H. CHAI. 1977. Alpha-1-antitrypsin in childhood asthma. J. Allerg. Clin. Immunol. 59: 31.
14. BREUER, R., E. C. LEUCY, P. J. STONE, T. G. CHRISTENSEN & G. L. SNIDER. 1985. Proteolytic activity of human neutrophil elastase and porcine pancreatic trypsin causes bronchial secretory cell metaplasia in hamsters. Exp. Lung Res. 9: 167–175.
15. CHRISTENSEN, T. G., A. L. KORTHY, G. L. SNIDER & J. A. HAYES. 1977. Irreversible bronchial goblet cell metaplasia in hamsters with elastase-induced panacinar emphysema. J. Clin. Invest. 59: 397–404.
16. CHODOSH, S. & T. C. MEDICI. 1971. The bronchial epithelium in chronic bronchitis. I. Exfoliative cytology during stable, acute bacterial infection and recovery phases. Am. Rev. Respir. Dis. 104: 888–898.
17. CARILLI, A. D., R. S. GOHD & D. BROWN. 1970. A cytologic study of chronic bronchitis. Am. Rev. Respir. Dis. 101: 696–699.
18. GHAFOURI, M. A., K. D. PATIL & I. KASS. 1984. Sputum changes associated with the use of ipratropium bromide. Chest 86: 387–393.
19. BOAT, T. F. 1988. Cystic fibrosis. In Textbook of Respiratory Medicine. J. F. Murray & J. A. Nadel, eds.: 1126–1152. W. B. Saunders. Baltimore, MD.
20. SUTER, S., U. B. SCHAAD, H. TEGNER, K. OHLSSON, D. DESGRANDCHAMPS & F. A. WALDVOGEL. 1986. Levels of free granulocyte elastase in bronchial secretions from patients with cystic fibrosis: Effects of antimicrobial treatment against Pseudomonas aeruginosa. J. Infect. Dis. 153: 902–909.
21. GOLDSTEIN, W. & G. DORING. 1986. Lysosomal enzymes from polymorphonuclear leukocytes and proteinase inhibitors in patients with cystic fibrosis. Am. Rev. Respir. Dis. 134: 49–56.
22. BURNETT, D. & R. A. STOCKLEY. 1985. Cathepsin B-like cysteine proteinase activity in sputum and bronchoalveolar lavage samples: Relationship to inflammatory cells and effects of corticosteroids and antibiotic treatment. Clin. Sci. 68: 469–474.
23. STOCKLEY, R. A. 1988. Chronic bronchitis: The antiproteinase/proteinase balance and the effect of infection and corticosteroids. Clin. Chest Med. 9: 643–656.

24. Fujita, J., N. Nelson, D. Daughton, C. Dobry, J. R. Spurzem, S. Irino & S. Rennard. 1990. Evaluation of elastase and antielastase balance in patients with pulmonary emphysema. Am. Rev. Respir. Dis. **142:** 57–62.

25. Sibille, Y. & H. Y. Reynolds. 1990. Macrophages and polymorphonuclear neutrophils in lung defense and injury. Am. Rev. Respir. Dis. **141:** 471–501.

26. Reid, L. M., G. J. Gleich, J. Hogg, J. Kleinerman & L. A. Laitinen. 1989. Pathology. *In* The Roles of Inflammatory Processes in Airway Hyperresponsiveness. Holgate, ed.: 36–79. Blackwell Scientific Publications. London.

27. Laitinen, L. A. & A. Laitinen. 1988. Mucosal inflammation and bronchial hyperreactivity. Eur. Respir. J. **5:** 488–489.

28. Wardlaw, A. J., S. Dunnette, G. J. Gleich, J. V. Collins & A. B. Kay. 1988. Eosinophils and mast cells in bronchoalveolar lavage in mild asthma; relationship to bronchial hyperreactivity. Am. Rev. Respir. Dis. **137:** 62–69.

29. Butterfield, J. H., D. E. Maddox & G. J. Gleich. 1983. The eosinophil leukocyte: Maturation and function. Clin. Immunol. Rev. **1:** 187–306.

30. Schwartz, L. B., T. R. Bradford, A. M. Irani, G. Deblois & S. S. Craig. 1987. The major enzymes of human mast cell secretory granules. Am. Rev. Respir. Dis. **135:** 1186–1189.

31. Suter, S., U. B. Schaad, L. Roux, U. E. Nydegger & F. A. Waldvogel. 1984. Granulocyte neutral proteases and pseudomonas elastase as possible causes of airway damage in patients with cystic fibrosis. J. Infect. Dis. **149:** 523–531.

32. Tournier, J. M., J. Jacquot, E. Puchelle & J. G. Bieth. 1985. Evidence that Pseudomonas aeruginosa elastase does not inactivate the bronchial inhibitor in the presence of leukocyte elastase. Studies with cystic fibrosis sputum and with pure proteins. Am. Rev. Respir. Dis. **132:** 524–528.

33. Doring, G., W. Goldstein, A. Roll, P. O. Schitz, N. Hiby & K. Botzenhart. 1985. Role of Pseudomonas aeruginosa exoenzymes in lung infections of patients with cystic fibrosis. Infect. Immunol. **49:** 557–562.

34. Patrines, M. & J. G. Bieth. 1989. Pseudomonas aeruginosa elastase does not inactivate alpha-1-proteinase inhibitor in the presence of leukocyte elastase. Infect. Immunol. **57:** 3793–3797.

35. Harris, E. D., Jr., H. G. Welgus & S. M. Krane. 1984. Regulation of the mammalian collagenases. Collagen Relat. Res. **4:** 493–512.

36. Krane, S. M. 1984. Collagen degradation. Prog. Clin. Biol. Res. **154:** 89–102.

37. Goldfarb, R. H. & L. A. Liotta. 1986. Proteolytic enzymes in cancer invasion and metastasis. Semin. Thromb. Hemostasis **12:** 294–307.

38. Varani, J. 1987. Interaction of tumor cells with the extracellular matrix. Rev. Biol. Cellular **12:** 1–113.

39. Travis, J. 1988. Structure, function and control of neutrophil proteinases. Am. J. Med. **84:** 37–42.

40. Berger, M., R. U. Sorensen, M. F. Tosi, D. G. Dearborn & G. Doring. 1989. Complement receptor expression on neutrophils at an inflammatory site, the pseudomonas-infected lung in cystic fibrosis. J. Clin. Invest. **84:** 1302–1313.

41. Kim, K. C., J. Nassiri & J. S. Brody. 1989. Mechanisms of airway goblet cell mucin release. Studies with cultured tracheal surface epithelial cells. Am. J. Respir. Cell. Mol. Biol. **1:** 137–143.

42. Boat, T. F., P. I. Cheng, J. D. Klinger, C. M. Liedtke & B. Tandler. 1984. Proteinases release mucin from airways goblet cells. Ciba Found. Symp. **109:** 72–88.

43. McDowell, E. M. & F. F. Beals (Eds). 1986. Bronchial inflammatory reactions. *In* Biopsy Pathology of the Bronchi.: 140–191. W. B. Saunders. Baltimore, MD.

44. Lundgren, R., M. Soderberg, P. Horstedt & R. Stenling. 1988. Morphological studies of bronchial mucosal biopsies from asthmatics before and after ten years of treatment with inhaled steroids. Eur. Respir. J. **1:** 883–889.

45. Gurwitz, D., C. Mindorff & H. Levison. 1981. Increased incidence of bronchial reactivity in children with a history of bronchiolitis. J. Pediatr. **98:** 551–555.

46. Pagtakhan, R. D., M. H. Reed & V. Chernick. 1986. Is bronchiolitis in infancy an

antecedent of chronic lung disease in adolescence and adulthood? J. Thorac. Imaging **1:** 34–40.

47. RICKARD, K. & S. RENNARD. 1989. Neutrophil elastase causes detachment of bronchial epithelial cells from extracellular matrix. Am. Rev. Respir. Dis. **139:** 406.
48. BREEZE, R. G. & E. B. WHEELDON. 1977. The cells of the pulmonary airways. Am. Rev. Respir. Dis. **116:** 705–777.
49. WIDDICOMBE, J. H. 1988. Electrical methods for studying ion and fluid transport across airway epithelia. In Methods in Bronchial Mucology. P. C. Braga & L. Allegra, eds.: 335–345. Raven Press. New York.
50. RICKARD, K., G. PROROK, G. HUERTA, A. B. THOMPSON, R. A. ROBBINS & S. I. RENNARD. 1990. Bronchial epithelial cell (BEC) monolayers acquire resistance to protease mediated detachment: A means to compartmentalize inflammatory mediators in the airway. Am. Rev. Respir. Dis. **141:** A110.
51. NAYLOR, B. 1962. The shedding of the mucosa of the bronchial tree in asthma. Thorax **17:** 69–72.
52. FLAVAHAN, N. A., L. L. AARHUS, T. J. RIMELE & P. M. VANHOUTTE. 1985. Respiratory epithelium inhibits bronchial smooth muscle tone. J. Appl. Physiol. **58:** 834–838.
53. TSCHIRHART, E. & Y. LANDRY. 1986. Airway epithelium releases a relaxant factor: Demonstration with substance P. Eur. J. Pharmacol. **132:** 103–104.
54. RONCO, P., H. POLLARD, M. GALCERAN, M. DALAUCHE, J. C. SCHWARTZ & P. VERROUST. 1988. Distribution of enkephalinase (membrane metalloendopeptidase, E.C. 3.4.24.11) in rat organs. Detection using a monoclonal antibody. Lab. Invest. **58:** 210–217.
55. HAY, D. W. P., S. G. FARMER, D. RAEBURN, R. M. MUCCITELLI, K. A. WILSON & J. S. FEDAN. 1987. Differential effects of epithelium removal on the responsiveness of guinea-pig tracheal smooth muscle to bronchoconstrictors. Br. J. Pharmacol. **92:** 381–388.
56. BARNES, P. J., F. M. CUSS & J. B. PALMER. 1985. The effect of airway epithelium on smooth muscle contractility in bovine trachea. Br. J. Pharmacol. **86:** 685–691.
57. FINE, J. M., T. GORDON & D. SHEPPARD. 1989. Epithelium removal alters responsiveness of guinea pig trachea to substance P. J. Appl. Physiol. **66:** 232–237.
58. JACOBY, D. B., J. TAMAOKI, D. B. BORSON & J. A. NADEL. 1988. Influenza infection causes airway hyperresponsiveness by decreasing enkephalinase. J. Appl. Physiol. **64:** 2653–2658.
59. BEASLEY, R., W. R. ROCHE, J. A. ROBERTS & S. T. HOLGATE. 1989. Cellular events in the bronchi in mild asthma and after bronchial provocation. Am. Rev. Respir. Dis. **139:** 806–817.
60. SPURZEM, J. R., A. B. THOMPSON, D. M. DAUGHTON, M. OEHLERKING, J. LINDER & S. I. RENNARD. 1989. Chronic inflammation is associated with an increased proportion of goblet cells recovered by bronchial lavage. Am. Rev. Respir. Dis. **139:** 336.
61. SPORN, M. B. & A. B. ROBERTS. 1989. Transforming growth factor-beta. Multiple actions and potential clinical applications. JAMA **262:** 938–941.
62. MASUI, T., L. M. WAKEFIELD, J. F. LECHNER, M. A. LAVECK, M. B. SPORN & C. C. HARRIS. 1986. Type bronchoalveolar lavage transforming growth factor is the primary differentiation-inducing serum factor for normal human bronchial epithelial cells. Proc. Natl. Acad. Sci. USA **83:** 2438–2442.
63. JETTEN, A. M., J. E. SHIRLEY & G. STONER. 1986. Regulation of proliferation and differentiation of respiratory tract epithelial cells by TGFβ. Exp. Cell Res. **167:** 539–549.
64. ROMBERGER, D., J. BECKMANN, L. CLAASSEN, R. ERTL & S. I. RENNARD. 1989. Transforming growth factor-beta 1 stimulates fibronectin release from bronchial epithelial cells. Am. Rev. Respir. Dis. **139:** 252.
65. VARTIO, T. 1983. Fibronectin: Multiple interactions assigned to structural domains. Med. Biol. **61:** 283–295.
66. RUOSLAHTI, E. 1988. Fibronectin and its receptors. Annu. Rev. Biochem. **57:** 375–413.
67. VARTIO, T., H. SEPPA & A. VAHERI. 1981. Susceptibility of soluble and matrix fi-

bronectins to degradation by tissue proteinases, mast cell chymase and cathepsin G. J. Biol. Chem. **256:** 471–477.

68. McDONALD, J. A., B. J. BAUM, D. M. ROSENBERG, J. A. KELMAN, S. C. BRIN & R. G. CRYSTAL. 1979. Destruction of a major extracellular adhesive glycoprotein (fibronectin) of human fibroblasts by neutral proteases from polymorphonuclear leukocyte granules. Lab. Invest. **40:** 350–357.

69. McDONALD, J. A. & D. G. KELLEY. 1980. Degradation of fibronectin by human leukocyte elastase. Release of biologically active fragments. J. Biol. Chem. **255:** 8848–8858.

Protease Actions on Airway Secretions

Relevance to Cystic Fibrosis[a]

JAY A. NADEL[b]

*Cardiovascular Research Institute and Departments of Medicine
and Physiology
University of California, San Francisco
San Francisco, California 94143*

Cystic fibrosis is a common genetic disorder associated with progressive airway obstruction and mucus hypersecretion. Inflammation and inflammatory cell infiltration are prominent in the airways of patients with cystic fibrosis, and this led to our "protease hypothesis" of mucus hypersecretion, which can be stated as follows:

1. Cells that play important roles in inflammatory responses similarly may play important roles in producing potent secretagogues.

2. Two "inflammatory" cells have been implicated in the secretory response: neutrophils and mast cells.

3. These cells produce potent secretagogue actions by releasing proteases in the airway.

4. Inhibition of these enzymes may provide an effective strategy for treating the hypersecretion in cystic fibrosis and in other diseases of airways associated with inflammation.

The rationale for the hypothesis and the strategies employed in investigating the problem will be outlined.

ROLE OF NEUTROPHIL PROTEASES IN HYPERSECRETION

Neutrophils are considered to play an important role in the pathogenesis of airway inflammation. They are greatly increased in the purulent respiratory tract secretions of patients with cystic fibrosis. They are the source of many important mediators of inflammation, including neutral proteases. The lysosomal enzymes cathepsin G and elastase are released from neutrophils during phagocytosis and cell death.[1,2] In contrast to healthy subjects in whom levels of catalytically active cathepsin G and elastase are negligible, patients with cystic fibrosis may have concentrations of neutrophil elastase in sputum exceeding 100 μg/ml (3.3 × 10^{-6} M).[3,4] Several lines of evidence suggest a connection between neutrophil elastase release and hypersecretion. Thus, neutrophil elastase causes discharge of goblet cells *in vivo*[5] and the release of surface glycoconjugates from airway epithelial cells in dogs[6] and hamsters.[7] However, most human airway secretions are believed to originate from submucosal gland secretory cells (their volume exceeds 40-fold that of goblet cells in normal airways[8]), and in cystic fibrosis, submucosal

[a] This work was supported in part by a Cystic Fibrosis Foundation RDP Center Grant.

[b] Address for correspondence: Jay A. Nadel, M.D., Cardiovascular Research Institute, Box 0130, University of California, San Francisco, CA 94130–0130 (415–476–1105)

gland secretory cells are even more conspicuous.[9] Nevertheless, the effects of proteases on gland secretion had not previously been studied. Therefore, the present study examined selectively the effect of neutrophil proteases on gland secretion. To accomplish this, we took advantage of a cultured line of tracheal gland serous cells that maintains characteristics of differentiated secretory cells developed by Finkbeiner and associates.[10] The results of these studies indicate that both neutrophil elastase and cathepsin G are potent noncytotoxic secretagogues. These studies are described in detail elsewhere.[11]

ROLE OF MAST CELL ENZYMES IN HYPERSECRETION

Mast cells are among the most potent inflammatory cells and are implicated in various inflammatory lung diseases. They are abundant in the airway submucosa,[12,13] where they can interact with submucosal glands. Because patients with cystic fibrosis have marked airway submucosal gland hypertrophy and hyperplasia,[9] we examined the effects of the mast cell proteases chymase and tryptase on secretion from cultured airway submucosal gland cells.[10] We discovered that chymase, but not tryptase, is a potent secretagogue for gland cells, and these findings, suggesting a possible role for chymase-containing mast cells in the pathogenesis of airway hypersecretion, are reported in detail elsewhere.[14]

METHODS

The methods are presented in detail in the original publications and will only be described briefly.

Culture of Bovine Tracheal Gland Serous Cells

Bovine tracheal gland serous cells were cultured as described previously[10] and maintained characteristics of differentiated serous cells. The cells incorporated radiolabeled precursors into macromolecules and secreted them in response to various mediators such as histamine, beta-adrenergic agonists, and prostaglandins. Cells were seeded onto tissue culture plastic coated with human placental collagen (15 μg/cm^2) at a density of 2×10^4 cells/cm^2 in medium containing 40% Dulbecco's modified Eagles' H21 medium, 40% Ham's F12 medium, 20% fetal calf serum, and 50 μg/ml gentamicin. Flasks were maintained at 37°C in 5% CO^2/95% air. The medium was changed every 3 days.

Release of ^{35}-Labeled Macromolecules

On day 9, confluent monolayers of serous cells cultured in human placental collagen-coated tissue culture flasks (surface area 75 cm^2) were incubated with 10 ml of medium containing 20 μCi/ml Na$_2$^{35}SO$_4$. After 24 hours, the medium containing the radiolabel was removed, and the cells were washed three times with Dulbecco's phosphate-buffered saline solution. Serum- and antibiotic-free medium was then added (10 ml/flask) and was changed every 30 minutes. At 210 minutes, the medium was collected and replaced with either fresh medium alone (baseline control) or medium containing the secretagogues. Antagonists were added concurrently. At 240 minutes, the medium was collected again. The harvested medium from the 180- to 210-minute and the 210- to 240-minute incubation periods was analyzed (Spectrapor tubing molecular mass cutoff, 12,000–14,000 D) against distilled water containing 10 mg/liter sodium azide to remove unincorporated $^{35}SO_4^{-2}$. Nondialyzable ^{35}S-labeled macromolecules were counted after the addition of scintillation fluid (Hydrofluor, National Diagnostics, Inc., Somerville, New Jersey) by scintillation spectroscopy to an accuracy of 2% (beta counter model LS7500, Beckman Instruments, Inc., Palo Alto, California). Secretion is expressed as a percentage increase of release of ^{35}S-labeled macromolecules during incubation with the agonists over the release during the immediately preceding time period (secretory index) and is corrected for the declining baseline (determined in controls incubated with medium alone).

Degranulation Supernatant from Mastocytoma Cells

As a source of mast cell degranulation supernatant and proteases, we used the BR line of canine mastocytoma cells established as a permanent line by serial passage in nude mice.[15] In response to the calcium ionophore A23187, BR mastocytoma cells release both chymase and tryptase together with histamine.[16]

Enzymatically disaggregated mastocytoma cells were activated with calcium ionophore A23187 in a two-stage reaction,[17,18] so no ionophore was detectable by high performance liquid chromatography (HPLC) in the supernatant (S. C. Lazarus, personal communication). Cells ($5 \times 10^8/3$ ml) were incubated with 3×10^{-6} M calcium ionophore in Ca^{2+}- and Mg^{2+}-free Tyrode's buffer, pH 7.4, containing 25 mM HEPES and 0.1% BSA at 4°C for 20 minutes; then they were washed three times with Ca^{2+}- and Mg^{2+}-free Tyrode's buffer and were resuspended in complete Tyrode's buffer at 37°C. After incubation for 30 minutes, the reaction was stopped at 4°C, and mediator-rich supernatant was separated from the cell pellet by centrifugation ($200 \times g$, 10 minutes, 4°C) and was stored at $-80°C$. In some experiments, mastocytoma cells were preincubated (15 minutes) with the combined lipoxygenase and cyclooxygenase inhibitor BW755c (10^{-4} M) before activation to prevent the generation of leukotrienes and prostaglandins.

Supernatant samples were analyzed for histamine using an o-phthalaldehyde spectrofluorometric method,[19] modified for autoanalyzer (Alpkem Corp., Clakamus, Oregon). Different batches of supernatant and supernatant obtained after preincubation with BW755c were adjusted to equal concentrations of histamine.

RESULTS

Neutrophil Proteases

Human neutrophil elastase and neutrophil cathepsin G both stimulated secretion of ^{35}S-labeled macromolecules markedly ($p < 0.0001$) in a concentration-dependent fashion (FIG. 1). For both elastase and cathepsin G, the threshold concentration causing a significant increase in secretion over baseline was $\geq 10^{-10}$ M, and the maximal secretory response of both proteases was similar. Elastase was more potent than cathepsin G ($p < 0.0001$), with responses at 10^{-8} M of $1,810 \pm 60\%$ and $970 \pm 20\%$ (mean $\pm$ SE) above baseline, respectively ($n = 5$). These responses were orders of magnitude greater than the responses to histamine, which caused secretion with a threshold of 10^{-6} M and a maximal response of only $185 \pm 8\%$ at 10^{-5} M ($p < 0.0001$, $n = 6$). The secretion induced by both enzymes caused no evidence of cytotoxicity as examined by lactic dehydrogenase release, vital dye exclusion, and morphologic examination.

The secretory responses to both enzymes were reduced or abolished by inhibitors (FIG. 2). Thus, the secretory response to 10^{-8} M cathepsin G was reduced by chymostain and by soybean inhibitor (FIG. 2A), and the secretory response to elastase (10^{-8} M) was reduced significantly by PMSF and was blocked by N-methoxysuccinyl-Ala-Ala-Pro-Val-chloromethylketone and by alpha$_1$-proteinase inhibitor (FIG. 2B). The inhibitory effect of the protease inhibitors on secretion was similar to their effect on the amidolytic activity of the proteases.

Having shown that purified neutrophil elastase and cathepsin G are potent

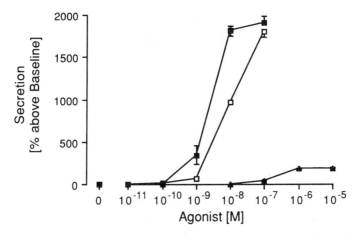

FIGURE 1. Effects of human neutrophil cathepsin G (*open squares*) and elastase (*solid squares*) and of histamine (*closed triangles*) on secretion of ^{35}S-labeled macromolecules from cultured tracheal gland serous cells. Secretion is stimulated markedly by cathepsin G and elastase ($p < 0.0001$, $n = 5$), whereas histamine has a comparatively small effect ($p < 0.0001$, $n = 6$). Values are mean $\pm$ SEM. (Reproduced with permission from Sommerhoff et al.[11])

secretagogues, we examined whether the results could be reproduced with products directly secreted from neutrophils. We purified human neutrophils from peripheral blood of healthy subjects ($\geq 95\%$ neutrophils), suspended them in complete HBSS (0.6×10^8 cells/ml), activated them with 5 μg/ml cytochalasin B and 10^{-6} M FMLP for 10 minutes, and prepared cell-free supernatant by centrifugation. When degranulation supernatant from 10^5 neutrophils was added per milliliter of incubation medium of serous cells, the secretory response averaged 2,095 $\pm$ 54% above baseline ($p < 0.001$, $n = 6$). The response was markedly diminished by preincubation with the human neutrophil elastase inhibitor N-methoxysuccinyl-Ala-Ala-Pro-Val-chloromethylketone (10^{-5} M). The remaining response (363 $\pm$ 22%) was further reduced by 50 μg/ml chymostatin (to 102 $\pm$ 19%).

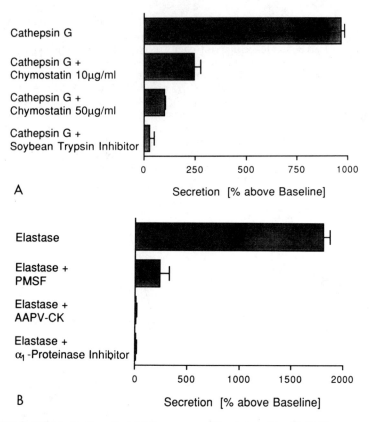

FIGURE 2. Effect of active site inhibitors on secretion induced by 10^{-8} M human neutrophil cathepsin G (**A**) and elastase (**B**). Soybean trypsin inhibitor, 100 μg/ml; PMSF, 10^{-4} M; N-methoxysuccinyl-Ala-Ala-Pro-Val-chloromethylketone (AAPV-CK), 10^{-5} M; α_1-proteinase inhibitor, 100 μg/ml. Values are mean ± SEM, $n = 5$. (Reproduced with permission from Sommerhoff et al.[11])

Mast Cell Proteases

Although purified tryptase had no significant effect on the release of ^{35}S-labeled macromolecules, purified chymase caused a profound secretory response ($p < 0.001$, $n = 5$; FIG. 3), with a threshold $\leq 10^{-9}$ M. At 10^{-8} M, the secretory index averaged 1,530 ± 80% (mean ± SE), and a maximum response was not reached in the concentrations tested (10^{-11} to 10^{-8} M). To test whether the active catalytic site of chymase was required to produce the secretory response, the inhibitory effects of the active site inhibitor, soybean trypsin inhibitor (100 μg/ml), or chymostatin (10 μg/ml) were examined in the presence of chymase. At these concentrations, the inhibitors reduced the amidolytic activity of chymase on peptide-4-nitroanilide substrates.[20] Both inhibitors reduced or abolished the secretory response to mast cell degranulation supernatant (FIG. 4).

The response to chymase was not associated with cytotoxicity as studied by lactic dehydrogenase release and vital dye exclusion. The secretory response was prevented by the metabolic inhibitors azide (100 mM, 30 minutes preincubation), dicumarol (1 mM), and 2,4-dinitrophenol (1 mM),[14] indicating that the secretory response to chymase required unimpaired cellular energy metabolism of the serous cell.

DISCUSSION

The present findings indicate that neutrophil elastase and cathepsin G and mast cell chymase are extremely potent secretagogues of cultured airway submucosal glands. Concentrations of other agonists, such as histamine, prostaglandins, and beta-adrenergic agonists, in excess of 10^{-8} M are required to induce secretory responses,[10,21,22] whereas the threshold concentration for proteases ($\geq 10^{-10}$ M) is lower by two orders of magnitude. Furthermore, only proteases induce maximal secretory responses ($\geq 80\%$ depletion of ^{35}S-labeled macromolecules from submucosal gland cells), and these responses are greater than 10-fold larger than those of other agonists.[14] In fact, in comparison with findings with other secretagogues described in the literature, neutrophil elastase apparently is the most potent secretagogue of airway submucosal glands ever described. Neutrophil cathepsin G also stimulates gland cell secretion, but its effect is less potent than is that of neutrophil elastase.

The potential importance of neutrophil enzymes in the pathogenesis of hypersecretion in cystic fibrosis is suggested by the immense neutrophil "load" present in the airways of these patients, and the presence of these neutrophils is the cause of the high concentration of neutrophil elastase in the sputum of patients with cystic fibrosis.[3,4]

Although the potential therapeutic efficacy of neutrophil elastase inhibitors is most obvious in the hypersecretion of cystic fibrosis, it is by no means limited to this disease. There are also compelling reasons to consider this therapeutic strat-

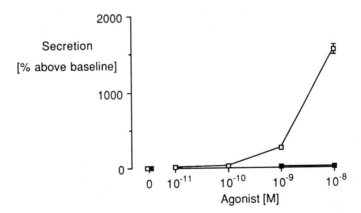

FIGURE 3. Effect of chymase (*open squares*) and tryptase (*closed squares*) on secretion of ^{35}S-labeled macromolecules from cultured tracheal gland serous cells. Secretion is greatly stimulated by chymase ($p < 0.0001$, $n = 5$), whereas tryptase has no significant effect. Values are mean $\pm$ SEM, $n = 5$. (Reproduced with permission from Sommerhoff et al.[14])

egy in selected patients with chronic bronchitis, an inflammatory disease characterized by chronic cough and excessive sputum production. Airway neutrophils occur in higher numbers in the airways and in bronchoalveolar lavage of cigarette smokers than nonsmokers,[23-25] and free neutrophil elastase activity may be present in the airways of cigarette smokers and in individuals with bronchitis.[26-30] Increased numbers of neutrophils are found in patients with chronic bronchitis.[31]

Exposure to air pollutants such as ozone and sulfur dioxide also causes an influx of neutrophils into the airways,[32-35] and this may be associated with bronchitis and hypersecretion.[34-36] Thus, there are multiple conditions in which neutrophil enzymes could play an important role in the pathogenesis of the airway hypersecretory state.

The concentration of neutrophil proteases required to produce significant secretagogue effects appears to be achievable during airway inflammation: neutro-

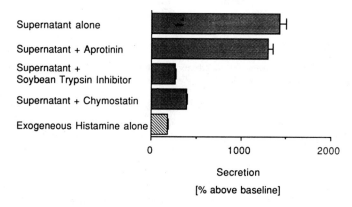

FIGURE 4. Effect of inhibitors of mast cell proteases on secretion induced by mastocytoma cell degranulation supernatant. Aprotinin (30 μg/ml), an inhibitor of tryptase, did not significantly change the response, whereas soybean trypsin inhibitor (100 μg/ml) and chymostatin (10 μg/ml) inhibitors of chymase decreased secretion to the level induced by exogenous histamine alone in an equivalent concentration (3 $\times$ 10^{-6} M). Values are mean $\pm$ SEM, $n = 6$. (Reproduced with permission from Sommerhoff et al.[14])

phils contain ~2-4 μg of cathepsin G[37,38] and 1-2 μg of elastase[39,40] per 10^6 cells. The release of 10% of the proteases from 10^5 neutrophils/cm^3 of submucosal tissue may be sufficient to cause a significant secretory response. In the presence of inflammation, the number of neutrophils in the submucosa is considerably higher (0.5 $\times$ 10^7/cm^3 before and 2 $\times$ 10^7/cm^3 after ozone in dogs[33]). Even if more than 99% of the proteases are inhibited by endogenous antiproteases, a secretory response could occur. In patients with cystic fibrosis and chronic bronchitis, sputum concentrations of free proteases may greatly exceed the concentrations used in this study.[3]

Chymase, an enzyme from mast cells, is another potent secretagogue in our preparation of cultured gland cells. This was the first study to demonstrate that a protease is capable of causing secretion from airway submucosal gland cells.[14] The threshold for this effect of chymase (10^{-10} M) is much lower than that for other nonprotease secretagogues ($\geq$10^{-8} M),[10,21,22] and the size of the maximal response is an order of magnitude greater than that to other agonists. The effect was noncytotoxic and was blocked by active site inhibitors of chymase.

The other major mast cell enzyme, tryptase, has no secretagogue effect. Supernatant obtained from degranulation of mastocytoma cells produced secretion of magnitude comparable to that caused by purified chymase. The threshold for secretion was reached when supernatant from 3×10^4 mastocytoma cells was added per milliliter of incubation medium of serous cells; histamine released from these cells did not play a significant role in the secretagogue effect. When supernatant from higher numbers of mast cells (10^7 cells) was added, the effect of histamine on secretion became measurable but was small (approximately 15% of the total effect; see FIG. 4).

In peripheral lung tissue of normal humans, chymase-containing mast cells are rare,[41] but the proportion of mast cells containing chymase is considerably higher in bronchi.[42] The proportion of chymase-containing mast cells in bronchoalveolar fluid is dramatically increased in patients with cystic fibrosis,[43] a disease characterized by profound hypertrophy and hyperplasia of the airway submucosal glands. This association of enlarged glands and chymase suggests two possibilities: (1) factor(s) in submucosal glands may influence the expression of chymase in mast cells; and (2) mast cell components (e.g., chymase) may be involved in the pathogenesis of gland enlargement in diseases such as cystic fibrosis.

The exact mechanism of gland secretion by these serine proteases is unknown. Most of our studies relating to mechanism were performed with chymase, so I will limit this part of the discussion to this enzyme. Our demonstration of secretion of granule-associated products by gland cells is not unique: Schick and Austen[44] have shown that chymotryptic proteases related to cathepsin G, chymotrypsin, and chymase cause degranulation of mast cells.

During mast cell degranulation with chymotrypsin, specific binding to mast cells did not occur, suggesting that simple ligand-receptor binding did not occur.[44] Low-affinity binding sites for cathepsin G and elastase have been described in the mediation of endocytosis into macrophages,[45,46] as well as high-affinity receptor binding sites for elastase on fibroblasts in responses not involving endocytosis.[47] The effect of chymase on gland secretion required an unimpaired cellular energy metabolism,[14] in contrast to the proteolytic cleavage of radiolabeled glycoconjugates from the airway epithelial cell surface.[6] Chymase has a molecular weight of 30,000 daltons and is therefore unlikely to penetrate cells, although pinocytosis of the enzyme cannot be ruled out. Recent information from our laboratory suggests that secretion by chymase involves an unusual stimulation-secretion coupling mechanism:[48] in contrast to mediators such as histamine and bradykinin, chymase-induced serous cell secretion did not increase cAMP, nor did the phosphodiesterase inhibitor 3-isobutyl-1-methylxanthine augment the secretory response. Furthermore, incorporation of ^{3}H-myoinositol into inositol phosphate was unaffected by chymase. Depletion of protein kinase C from cells by preincubation with phorbol 12-myristate had no effect on chymase-induced secretion. Finally, chymase-induced secretion caused no increase in cytoplasmic Ca^{2+} (measured with Fura-2 fluorescence), and the intracellular calcium antagonist TMB-8 (2×10^{-4} M) did not affect chymase-induced secretion.

Intracellular proteases have been shown to be required for exocytosis and other cellular functions in secretory cells (e.g., mast cells).[49,50] Therefore, it is possible that exogenous proteases may mimic these endogenous proteases if they are endocytosed. Alternatively, the catalytic activity of the proteases may be directed at targets on the cell surface. In this latter case, transduction of the signal from the cell surface occurs via a unique mechanism.

In summary, neutrophil elastase and cathepsin G and mast cell chymase cause profound stimulation of airway submucosal gland cells via an unknown mechanism involving the active site of the enzymes. The importance of these enzymes in the pathogenesis of increased and abnormal secretions associated with inflammatory airway diseases such as cystic fibrosis, chronic bronchitis, and asthma may be established using selective site-directed inhibitors of these enzymes.

ACKNOWLEDGMENT

The author thanks Patty Snell for assistance in manuscript preparation.

REFERENCES

1. WEISSMANN, G., R. B. ZURIER & S. HOFFSTEIN. 1972. Leukocytic proteases and the immunological release of lysosomal enzymes. Am. J. Pathol. **68:** 539–563.
2. OHLSSON, K. & I. OLSSON. 1977. The extracellular release of granulocyte collagenase and elastase during phagocytosis and inflammatory processes. Scand. J. Haematol. **19:** 145–152.
3. GOLDSTEIN, W. & G. DORING. 1986. Lysosomal enzymes from polymorphonuclear leukocytes and proteinase inhibitors in patients with cystic fibrosis. Am. Rev. Respir. Dis. **134:** 49–56.
4. SUTER, S., U. B. SCHAAD, H. TEGNER, K. OHLSSON, D. DESGRANDCHAMPS & F. A. WALDVOGEL. 1986. Levels of free granulocyte elastase in bronchial secretions from patients with cystic fibrosis: Effect of antimicrobiol treatment against *Pseudomonas aeruginosa*. J. Infect. Dis. **153:** 902–909.
5. BREUER, R., T. G. CHRISTENSEN, E. C. LUCEY, P. J. STONE & G. L. SNIDER. 1987. An ultrastructural morphometric analysis of elastase-treated hamster bronchi shows discharge followed by progressive accumulation of secretory granules. Am. Rev. Respir. Dis. **136:** 698–703.
6. VARSANO, S., C. B. BASBAUM, L. S. FORSBERG, D. B. BORSON, G. CAUGHEY & J. A. NADEL. 1987. Dog tracheal epithelial cells in culture synthesize sulfated macromolecular glycoconjugates and release them from the cell surface upon exposure to extracellular proteinases. Exp. Lung Res. **13:** 157–183.
7. BREUER, R., T. G. CHRISTENSEN, R. M. NILES, P. J. STONE & G. L. SNIDER. 1989. Human neutrophil elastase causes glycoconjugate release from the cell surface of hamster trachea in organ culture. Am. Rev. Respir. Dis. **139:** 779–782.
8. REID, L. 1960. Measurement of the bronchial mucous gland layer: A diagnostic yardstick in chronic bronchitis. Thorax **15:** 132–141.
9. SOBONYA, R. E. & L. M. TAUSSIG. 1986. Quantitative aspects of lung pathology in cystic fibrosis. Am. Rev. Respir. Dis. **134:** 290–295.
10. FINKBEINER, W. E., J. A. NADEL & C. B. BASBAUM. 1986. Establishment and characterization of a cell line derived from bovine tracheal glands. In Vitro Cell. Dev. Biol. **22:** 561–567.
11. SOMMERHOFF, C. P., J. A. NADEL, C. B. BASBAUM & G. H. CAUGHEY. 1990. Neutrophil elastase and cathepsin G stimulate secretion from cultured bovine airway gland serous cells. J. Clin. Invest. **85:** 682–689.
12. GUERZON, G. M., P. D. PARE, M.-C. MICHOUD & J. C. HOGG. 1979. The number and distribution of mast cells in monkey lungs. Am. Rev. Respir. Dis. **119:** 59–66.
13. SHANAHAN, F., I. MACNIVEN, N. DYCK, J. A. DENBURG, J. BIENENSTOCK & A. D. BEFUS. 1987. Human lung mast cells: Distribution and abundance of histochemically distinct populations. Int. Arch. Allergy Appl. Immunol. **83:** 329–331.
14. SOMMERHOFF, C. P., C. H. CAUGHEY, W. E. FINKBEINER, S. C. LAZARUS, C. B. BASBAUM & J. A. NADEL. 1989. Mast cell chymase: A potent secretagogue for airway gland serous cells. J. Immunol. **142:** 2450–2456.

15. LAZARUS, S. C., R. DEVINNEY, L. J. MCCABE, W. E. FINKBEINER, D. J. ELIAS & W. M. GOLD. 1986. Isolated canine mastocytoma cells: Propagation and characterization of two cell lines. Am. J. Physiol. 251: C935–C944.

16. CAUGHEY, G. H., S. C. LAZARUS, N. F. VIRO, W. M. GOLD & J. A. NADEL. 1988. Tryptase and chymase: Comparison of extraction and release in two dog mastocytoma lines. Immunology 63: 339–344.

17. LICHTENSTEIN, L. M. 1971. The immediate allergic response: In Vitro separation of antigen activation, decay and histamine release. J. Immunol. 107: 1122–1130.

18. KALINER, M. & K. F. AUSTEN. 1973. A sequence of biochemical events in the antigen-induced release of chemical mediators from sensitized human lung tissue. J. Exp. Med. 138: 1077–1094.

19. SIRAGANIAN, R. P. 1974. An automated continuous-flow system for the extraction and fluorometric analysis of histamine. Anal. Biochem. 57: 383–394.

20. CAUGHEY, G. H., N. F. VIRO, S. C. LAZARUS & J. A. NADEL. 1988. Purification and characterization of dog mastocytoma chymase. Identification of an octapeptide conserved in chymotryptic leukocyte proteases. Biochim. Biophys. Acta 952: 142–149.

21. SOMMERHOFF, C. P., W. E. FINKBEINER, J. A. NADEL & C. B. BASBAUM. 1987. Mediators of anaphylaxis stimulate ^{35}S-labeled macromolecule secretion from cultured bovine tracheal gland serous cells. Bull. Eur. Physiopathol. Respir. 23: 363s.

22. SOMMERHOFF, C. P., W. E. FINKBEINER, J. A. NADEL & C. B. BASBAUM. 1987. Prostaglandin D_2 and prostaglandin E_2 stimulate ^{35}S-labeled macromolecule secretion from cultured bovine tracheal gland serous cells. Am. Rev. Respir. Dis. 135: A363.

23. REYNOLDS, H. Y. & W. W. MERRILL. 1981. Airway changes in young smokers that may antedate chronic obstructive lung disease. Med. Clin. North Am. 65: 667–689.

24. HUNNINGHAKE, G. W., J. E. GADEK, O. KAWANAMI, V. J. FERRANS & R. G. CRYSTAL. 1979. Inflammatory and immune processes in the human lung in health and disease: Evaluation by bronchoalveolar lavage. Am. J. Pathol. 97: 149–198.

25. HUNNINGHAKE, G. W. & R. G. CRYSTAL. 1983. Cigarette smoking and lung destruction: Accumulation of neutrophils in the lungs of cigarette smokers. Am. Rev. Respir. Dis. 128: 833–838.

26. OHLSSON, K. & H. TEGNER. 1975. Granulocyte collagenase, elastase and plasma protease inhibitors in purulent sputum. Eur. J. Clin. Invest. 5: 221–227.

27. STOCKLEY, R. A. & D. BURNETT. 1979. Alpha 1-antitrypsin and leukocyte elastase in infected and non-infected sputum. Am. Rev. Respir. Dis. 120: 1081–1086.

28. WEITZ, J. I., K. A. CROWLEY, S. L. LANDMAN, B. I. LIPMAN & J. YU. 1987. Increased neutrophil elastase activity in cigarette smokers. Ann. Intern. Med. 107: 680–682.

29. JANOFF, A., L. RAJU & R. DEARING. 1983. Levels of elastase activity in bronchoalveolar lavage fluids of healthy smokers and non-smokers. Am. Rev. Respir. Dis. 127: 540–544.

30. TETLEY, T. D., S. F. SMITH, G. H. BURTON, A. J. WINNING, N. T. COOKE & A. GUZ. 1987. Effects of cigarette smoking and drugs on respiratory tract proteases and antiproteases. Eur. J. Respir. Dis. Suppl. 153: 93–102.

31. GIBSON, P. G., A. GIRGIS-GABARDO, M. M. MORRIS, S. MATTOLI, J. M. KAY, J. DOLOVICH, J. DENBURG & F. E. HARGREAVE. 1989. Cellular characteristics of sputum from patients with asthma and chronic bronchitis. Thorax 44: 693–699.

32. SELTZER, J., B. G. BIGBY, M. STULBARG, M. J. HOLTZMAN, J. A. NADEL, I. F. UEKI, G. D. LEIKAUF, E. J. GOETZL & H. A. BOUSHEY. 1986. O_3-induced change in bronchial reactivity to methacholine and airway inflammation in humans. J. Appl. Physiol. 60: 1321–1326.

33. HOLTZMAN, M. J., L. M. FABBRI, P. M. O'BYRNE, D. B. GOLD, H. AIZAWA, E. H. WALTERS, S. E. ALPERT & J. A. NADEL. 1983. Importance of airway inflammation for hyperresponsiveness induced by ozone in dogs. Am. Rev. Respir. Dis. 127: 686–690.

34. SELTZER, J., P. D. SCANION, J. M. DRAZEN, R. H. INGRAM, JR. & L. REID. 1984. Morphologic correlation of physiologic changes caused by SO_2-induced bronchitis in dogs: The role of inflammation. Am. Rev. Respir. Dis. 129: 790–797.

35. SHORE, S. A., S. T. KARIYA, K. ANDERSON, W. SKORNIK, H. A. FELDMAN, J. PEN-NINGTON, J. GODLESKI & J. M. DRAZEN. 1987. Sulfur-dioxide-induced bronchitis in dogs: Effects on airways responsiveness to inhaled and intravenously administered methacholine. Am. Rev. Respir. Dis. **135:** 840–847.

36. PHIPPS, R. J., S. M. DENAS, M. W. SIELCZAK & A. WANNER. 1986. Effects of 0.5 ppm ozone on glycoprotein secretion, ion and water fluxes in sheep trachea. J. Appl. Physiol. **60:** 918–927.

37. HECK, L. W., K. S. ROSTAND, F. A. HUNTER & A. BHOWN. 1986. Isolation, characterization, and amino-terminal amino acid sequence analysis of human neutrophil cathepsin G from normal donors. Anal. Biochem. **158:** 217–227.

38. SENIOR, R. M. & E. J. CAMPBELL. 1984. Cathepsin G in human mononuclear phagocytes: Comparison between monocytes and U927 monocyte-like cells. J. Immunol. **132:** 2547–2551.

39. BAUGH, R. J. & J. TRAVIS. 1976. Human leukocyte granule elastase: Rapid isolation and characterization. Biochemistry **15:** 836–841.

40. HECK, L. W., W. L. DARBY, F. A. HUNTER, A. BHOWN, E. J. MILLER & J. C. BENNETT. 1985. Isolation, characterization, and amino-terminal amino acid sequence analysis of human neutrophil elastase from normal donors. Anal. Biochem. **149:** 153–162.

41. SCHWART, L. B., A.-M. A. IRANI, K. ROLLER, M. C. CASTELLS & N. M. SCHECHTER. 1987. Quantitation of histamine, tryptase and chymase in dispersed human T and TC mast cells. J. Immunol. **138:** 2611–2615.

42. IRANI, A. A., N. M. SCHECHTER, S. S. CRAIG, G. DeBLOIS & L. B. SCHWARTZ. 1986. Two types of human mast cells that have distinct neutral protease compositions. Proc. Natl. Acad. Sci. USA **83:** 4464–4468.

43. SCHWARTZ, L. B., A. A. IRANI, S. MROUEH & A. SPOCK. 1987. Reversed ratio of T/TC types of human mast cells in cystic fibrosis. Clin. Res. **35:** 255A.

44. SCHICK, B. & K. F. AUSTEN. 1986. Rat serosal mast cell degranulation mediated by chymase, an endogenous secretory granule protease: Active site-dependent initiation at 1°C. J. Immunol. **136:** 3812–3818.

45. CAMPBELL, E. J., R. R. WHITE, R. J. RODRIGUEZ & C. KUHN. 1979. Receptor-mediated binding and internalization of leukocyte elastase by alveolar macrophages in vitro. J. Clin. Invest. **64:** 824–833.

46. CAMPBELL, E. J. 1982. Human leukocyte elastase, cathepsin G and lactoferrin: Family of neutrophil granule glycoproteins that bind to an alveolar macrophage receptor. Proc. Natl. Acad. Sci. USA **79:** 6941–6945.

47. CAMPBELL, C. H. & D. D. CUNNINGHAM. 1987. Binding sites for elastase on cultured human fibroblasts that do not mediate internalization. J. Cell. Physiol. **130:** 142–149.

48. SOMMERHOFF, C. P., G. H. CAUGHEY & J. A. NADEL. 1990. Classical second messengers are not involved in the stimulation-secretion coupling of chymase-induced degranulation of airway gland serous cells. FASEB J. **4:** A1940.

49. KIDO, H., N. FUKUSEN & N. KATANUMA. 1985. Antibody and inhibitor of chymase inhibit histamine release in immunoglobulin E-activated mast cells. Biochem. Int. **10:** 863–871.

50. MUNDY, D. I. & W. J. STRITTMATTER. 1985. Requirement for metalloendoprotease in exocytosis: Evidence in mast cells and adrenal chromaffin cells. Cell **40:** 645–656.

Potential Role of Proteases in Pulmonary Fibrosis

MOISÉS SELMAN[a,b] AND ANNIE PARDO[c]

[a]Instituto Nacional de Enfermedades Respiratorias
SSA
México DF, México

[c]Facultad de Ciencias
Universidad Nacional Autónoma de México
México DF, México

The pathogenesis of interstitial lung fibrosis includes a complex series of cellular and molecular mechanisms which, through a still unclear sequence of events, finally lead to an abnormal accumulation of collagens in the interstitium.

In the last few years we have been working on different possible mechanisms involved in the development of lung fibrosis. In this brief review we will discuss those related to the dynamics of the abnormal accumulation of collagen in the interstitium. This feature represents the final common pathway of the interstitial pulmonary diseases that evolve into fibrosis.[1]

Regardless of its etiology, pulmonary fibrosis is a general term used to describe interstitial lung diseases with histologic evidence of diffuse thickening of alveolar walls, in which the main protein stained is collagen. In this sense, it was surprising that in contrast to all other fibrotic tissues studied, the first reported biochemical data in lung fibrosis did not reveal an increase in the amount of collagen.[2]

This finding, accepted for years as a statement with scarce critical thought, was nevertheless an astonishing proposition. Could it be that the fibrotic reaction of the lung differed from that in other tissues?

Stimulated by this controversy, we determined collagen concentration in lung tissue samples obtained from 22 patients with diffuse lung fibrosis of different etiologies and 13 control subjects.[3-5] We measured hydroxyproline by two different methods which demonstrated that collagen concentration was significantly higher in patients than in controls (FIG. 1). A similar finding was later reported by other authors.[6]

These results agree with the morphologic observations in human fibrosis and the biochemical findings in animal models of pulmonary fibrosis, in which collagen content is usually increased.[1,7] Therefore, our subsequent studies were related to the possible mechanisms involved with the increase of collagen in pulmonary fibrosis.

In the last instance, the common feature of fibrosis must be an imbalance in collagen turnover rates so that synthesis exceeds degradation, resulting in excessive accumulation of this protein. Regarding collagen metabolism, attention has largely been focused on collagen synthesis, and there is widespread agreement that the rate of synthesis is increased in experimental lung fibrosis.[1] Unfortu-

[b] Address for correspondence: Dr. Moisés Selman L., Instituto Nacional de Enfermedades Respiratorias, Tlalpan 4502; CP 14080, México DF, México.

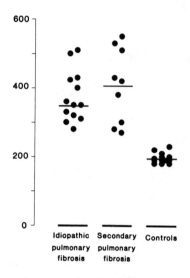

FIGURE 1. Collagen concentration (μg/mg dry weight) in lung parenchyma of patients with idiopathic and secondary pulmonary fibrosis and in control subjects. Mean values are indicated by horizontal bars ($p < 0.001$).

nately, studies of collagen metabolism in normal and fibrotic human lungs are scant and mainly are those related to collagen degradation.

Our first approach was employed in patients with idiopathic pulmonary fibrosis[5] (IPF) throughout the following protocol: lung samples obtained by biopsy were divided into four parts. One was used for morphologic studies, another to measure collagen concentration, another to analyze collagen synthesis, and the last one to study collagen degradation. Collagen synthesis was measured as described elsewhere.[8] Briefly, lung samples were incubated in Dulbecco's modified Eagle's medium containing 10% fetal calf serum, cofactors for proline hydroxylase, and antibiotics in a 95% air–5% carbon dioxide atmosphere at 37°C in a shaking water bath for 1 hour. Samples were placed in fresh medium with the same constituents also containing 30 μCi of (^{3}H)-proline and incubated for a further 4 hours, after which the tissue samples were homogenized and hydrolyzed, and (^{3}H)-hydroxyproline was separated and measured.

To analyze collagen degradation, Ryan and Woessner's method[9] with minor modifications was used. Briefly, lung tissue samples were homogenized in 50 mM Tris-HCl buffer with 5 mM $CaCl_2$ and were divided into six aliquots. Three of them were incubated alone and the remaining three in the presence of EDTA for 24 hours at 37°C. After incubation the homogenates were centrifuged and the supernatants filtered through a membrane with an exclusion limit of 100 kD. Collagen fragments were detected through the release of free hydroxyproline-containing material after acid hydrolysis.[10] In addition, undegraded collagen was quantified separately in each pellet, and collagenolytic activity was expressed as micrograms of collagen degraded per milligram of collagen incubated per hour.

Regarding collagen synthesis, the results did not show significant differences between normal lungs and those with IPF. At first this finding was unexpected, because in many animal models collagen synthesis is almost always increased.[1] However, this result, together with recent experiments performed in our laboratory, suggests that the increase in collagen production may be a transient stage in the development of lung fibrosis. Thus, for example, in our model induced in rats with paraquat and hyperoxia, collagen synthesis was significantly elevated from the fourth week on, coinciding with interstitial fibroblastic proliferation.[11] Several months later, however, collagen synthesis returned to almost normal levels in the majority of animals.[12]

On the other hand, our principal finding in the study on lungs with IPF, was a remarkable decrease in collagen degradation. As these patients were studied when their disorder was at an advanced stage, when there was a predominance of fibrosis over inflammation, we could not determine when in the evolution of the disease the decrease in collagenolytic activity had appeared and how relevant for fibrogenesis it really was. In this context, we analyzed lung collagen metabolism in patients with hypersensitivity pneumonitis,[13] an interstitial inflammatory disease in which almost most patients improve or heal, but in some the disease may progressively evolve to interstitial lung fibrosis. We studied 18 nonsmoking female patients with hypersensitivity pneumonitis and 7 control subjects using the same protocol as already mentioned. The patients were conventionally treated with prednisone and then followed-up for 2–3 years after which they were classified into three groups according to their clinical, radiologic, and functional responses: Group 1, healed (4 patients); Group 2, improved (8 patients); and Group 3, no improvement or condition worsened (6 patients). Concerning collagenolytic activity, noteworthy differences between the three groups were found. The patients who healed presented a several-fold increase in collagenolysis; the patients who improved exhibited normal or high values; and five of the six patients who did not show improvement or whose condition worsened displayed a decrease in collagenolytic activity of about half the normal values (FIG. 2). These findings supported our hypothesis that a decrease in local collagenolysis is a crucial event in the progression to fibrosis in some patients with interstitial lung inflammation.

To gain further insight into this hypothesis we also performed a sequential study in experimental lung fibrosis in rats, using a double aggression with paraquat plus high concentrations of oxygen.[11] This combined aggression resulted in

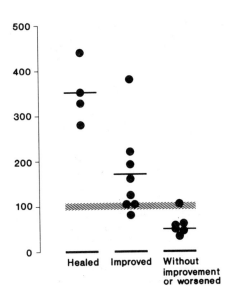

FIGURE 2. Collagen degradation in lung samples obtained from patients with hypersensitivity pneumonitis who followed different clinical courses. Results are expressed as a percentage of controls. *Hatched horizontal column* indicates normal range.

severe damage to the lower respiratory tract, with diffuse interstitial inflammation which evolves into fibrosis. Analysis of collagen degradation showed that in the first week, when only focal inflammation is present, no differences in collagenolytic activity were noted between experimental and control groups. In the second week, when diffuse and severe inflammation was observed, almost all experimental rats displayed a noteworthy increase in collagen degradation. From the third to the sixth week trimodal behavior of collagenolysis was noted, that is, some rats had high values, others normal, and another group decreased collagenolytic activity. Finally, after 8 weeks, when fibrogenesis was active, almost all rats displayed a statistically significant lower rate of collagenase activity than that in controls. Similar results were obtained in a silica-induced model[14] (FIG. 3). Therefore, these sequential studies, in which three different stages in lung collagen degradation are observed, suggest that low collagenolytic activity is associated with active fibrogenesis, and some animals may exhibit this abnormality in early stages of evolution of the disease.

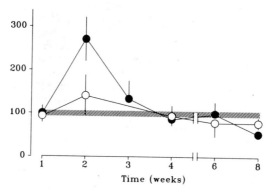

FIGURE 3. Collagenolytic activity in rat lung homogenates arising from two experimental models, expressed as a percentage of controls. *Open circles*: paraquat plus oxygen; *closed circles*: silica. *Hatched horizontal column* shows normal range.

Additionally, several studies in other fibrotic processes have demonstrated a decrease in collagenase or collagenolytic activity. This observation was well documented in human and animal cirrhotic livers[15,16] and in the skin of patients with progressive systemic sclerosis.[17] Moreover, it was recently reported that the reversibility of liver fibrosis after curative chemotherapy in murine schistosomiasis is closely related to an increase in both active and latent collagenase activity.[18]

Nevertheless, substantially different results in collagenolytic activity and the progression of interstitial lung diseases were reported when bronchoalveolar lavage fluid (BAL) was used. Gadek et al.[19] showed that 15 of 24 patients with IPF displayed BAL collagenase activity against type I collagen, whereas normal individuals and patients with sarcoidosis showed none. In sarcoidosis, however, O'Connor et al.[20] found BAL latent collagenase activity in 16 of 43 patients, most of them belonging to the group of patients with prolonged disease and need of treatment. On the other hand, Weiland et al.[21] evaluated the presence of BAL fluid collagenase activity in 10 patients with rheumatoid arthritis, five with and five without interstitial lung disease. Only lavage fluid from four of the five patients with pulmonary abnormalities contained active type I collagenase. In all these investigations an association between BAL collagenase activity and interstitial collagen abnormalities and/or the possible progression to fibrosis has been suggested. However, the association between increased collagenolytic activity and fibrosis, characterized by an exaggerated deposition of collagen, the substrate of this enzyme, is hard to understand. Moreover, it has been proved that excessive production of collagenase plays a key pathologic role in those diseases in which, by contrast, an exaggerated collagen breakdown occurs, such as proliferative rheumatoid synovitis,[22] alkali burns of the cornea,[23] periodontal disease,[24] and recessive dystrophic epidermolysis bullosa.[25]

In this context, we recently approached the question of the presence of BAL collagenase in the progression to interstitial fibrosis.[26] Therefore, we studied the presence of fibroblast type collagenase, which is the same as alveolar macrophage collagenase, and collagenolytic activity in BAL fluid from patients with chronic hypersensitivity pneumonitis, which, as already mentioned, may follow completely different clinical courses. Therefore, hypersensitivity pneumonitis represents a lung disorder in which the significance of the presence or absence of collagenase in bronchoalveolar lavage fluid may be analyzed in relation to the prognosis. We performed bronchoalveolar lavage in 25 nonsmoking patients and 4 healthy volunteers. The lavage fluid was filtered and centrifuged, and the supernatants were concentrated 20 times. We quantified collagenase by the ELISA method and collagenolytic activity using as substrate type I collagen labeled with (^{3}H)-acetic anhydride. The patients were treated with prednisone and followed-up from 2–3 years, after which they were classified as healed ($n = 7$), improved ($n = 10$), or worsened ($n = 8$) according to their clinical, radiologic, and functional evolution. Our results showed that 4 of 7 patients who healed, 6 of 10 patients who improved, and 4 of 8 patients whose condition worsened presented collagenase in BAL fluid. No relationship was noted between the presence or absence of the enzyme or its amount and the evolution of the disease. Similar results were obtained when collagenolytic activity was measured.

It therefore seems likely that the finding of increased BAL collagenase or collagenolytic activity might reflect the presence of activated inflammatory cells in the lavage fluid and/or the inflammatory stage of these diseases, but not necessarily the collagen turnover events occurring in the parenchymal extracellular matrix during active fibrogenesis. Thus, on the basis of our results on collagenolytic activity directly analyzed in lung tissue samples, we postulate that a progressive decrease in interstitial collagen degradation plays a critical role in the development of pulmonary fibrosis.

A hypothetical sequence of pathologic events, including collagen turnover regulation, that might occur during lung fibrogenesis is shown in FIGURE 4. After the initial lung damage in which any parenchymal cell can be injured or stimulated, an inflammatory process affecting the lower respiratory tract is noted. Afterwards a complex series of events link alveolitis with the beginning of fibrosis. Among them, the increased secretion of growth factors, responsible for fibroblast chemotaxis and proliferation, might play a relevant role. However, probably until this point, the pathologic events may still be resolved with partial or total restitution of the parenchyma. If the process continues, fibrosis expressed by the abnormal accumulation of collagens occurs; this phase is usually considered progressive and irreversible. Regarding collagen metabolism, collagen synthesis during inflammation might be in the normal range or even increased if we assume that cells from nonmesodermal origin are able to produce collagen as it occurs in other tissues and/or fibroblasts are stimulated to synthezise more collagen per cell before proliferation. In relation to collagen degradation, the inflammatory phase can be characterized by an increase in collagenolytic activity. During the fibro-blastic proliferative phase, an increase in collagen synthesis occurs, probably initially with type III collagen and afterwards with type V and type I collagen, which is the main genetic type of collagen found in advanced forms of lung fibrosis. In this phase a progressive decrease in collagenolytic activity can be observed. Finally, during the fibrotic phase, collagen synthesis gradually returns to normal, but as collagenolytic activity remains low, abnormal collagen turnover leads to continued collagen deposition. Nevertheless, this cellular and molecular sequence of events leading to pulmonary fibrosis, although divided in different and recognized phases, probably does not present clear-cut limits *in vivo*. Al-though obviously the role of increased collagen synthesis in the pathogenesis of

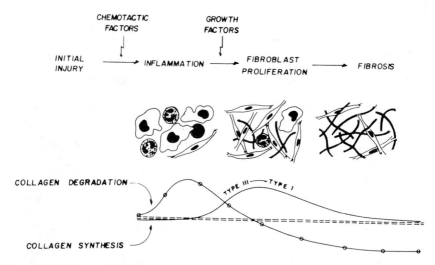

FIGURE 4. Hypothetical behavior of collagen metabolism in the sequence of events lead-ing to interstitial lung fibrosis. (See text.)

lung fibrosis cannot be dismissed, our attention has been focused mainly on abnor-malities related to collagen degradation.

Regulation of collagenase activity may occur at several levels; these include modulation of enzyme biosynthesis and secretion, control of latent enzyme acti-vation, and interaction with specific tissue inhibitors. (For review see ref. 27.) However, most of the mechanisms controlling *in vivo* collagenolysis remain ob-scure.

Human fibroblast collagenase is a Zn-metalloproteinase, indistinguishable from those synthesized by alveolar macrophages and endothelial cells, capable of catabolizing the breakdown of interstitial collagens types I, II, and III. With soluble substrates, type III collagen is degraded preferentially, but when fibrillar forms are used, type III and type I collagens are cleaved at the same rate. In contrast, neutrophil collagenases, which probably represents a different gene product, degrades soluble type I collagen faster than type III collagen; however, these differences also disappear when fibrillar collagen, the relevant physiologic substrate, is used. As *in vivo* collagen molecules spontaneously assemble into fibrils in the extracellular space, it can be assumed that in the local milieu both collagenases degrade genetically distinct interstitial collagens at similar rates.

Collagenase expression can be modulated by a series of different molecules including hormones, cytokines, arachidonic acid derivatives, endotoxins, and several growth factors that may stimulate or suppress, at least *in vitro*, collagenase synthesis in target cells.

After synthesis, fibroblast-type collagenase is promptly secreted as an inactive proenzyme that consists of a glycosylated form and an unmodified polypeptide of 57 and 52 kD, respectively. In contrast, neutrophil collagenase is stored in specific intracellular granules from where it is released. Activation of the proenzyme constitutes an important step in the regulation of collagenolytic activity, and some enzymatic activators have been proposed, including plasminogen activator, plasmin, kallikrein, stromelysin, tryptase, and cathepsin B. All modes of activation of latent human fibroblast collagenase are believed to involve the dissociation of Cys^{73} from the active-site zinc atom and its replacement by water with the subsequent exposure of the active site.[28] Another critical level in the control of extracellular-microenvironment proteolysis is constituted by collagenase inhibition. Thus, active collagenase can be "turned off" by several inhibitory molecules, including alpha$_2$-macroglobulin, a group of proteins known as tissue inhibitors of metalloproteinases (TIMP), and another group of not well-characterized small cationic proteins. Under normal conditions, the proenzyme, activator(s), and inhibitor(s) must be secreted in a concerted fashion and interact in dynamic equilibrium in order to obtain adequate collagenolysis. In addition, substrate must be "available" to the enzyme. Therefore, abnormal compartmentalization of collagen during fibrogenesis might also affect the appropriate action of collagenase.[29]

Because an important decrease in collagenolytic activity occurs during the development of lung fibrosis, we explored some of the collagenase regulatory mechanisms. In our first experimental approach, we analyzed the spontaneous and latent collagenolytic activity and the presence of collagenase inhibitory activity in pulmonary parenchyma obtained from five patients with IPF and six with chronic hypersensitivity pneumonitis.[30] Latent collagenase activity was revealed using 4-aminophenylmercuric acetate (APMA). After activation, control lungs displayed a two- or three-fold increase in collagenolytic activity, suggesting that in normal lungs about one-half to two-thirds of the total collagenolytic activity is in latent form, at least *in vitro*. By contrast, three of the lungs with IPF displayed low spontaneous collagenolysis, but more importantly, there was minimal or no activation at all, suggesting that total collagenolytic activity was already being expressed. Finally, in hypersensitivity pneumonitis, heterogeneous behavior in spontaneous collagenolytic activity was observed, and after APMA treatment in all but one patient, a severalfold increase in collagenase activity was obtained. However, only three reached normal values, two remained below controls, resembling IPF results, and the other presented unusually high enzyme activity.

On the other hand, we analyzed the presence of collagenase inhibitors in the supernatants of the same lung tissue homogenates. Our results showed that in both interstitial lung diseases, the inhibitory collagenase activity of the supernatants was severalfold higher than that in normal lungs. However, although both diseases displayed similar levels of enzyme inhibitory activity, we were unable in IPF to reveal normal levels of total collagenolytic activity. These experimental observations lead us to hypothesize that in lungs with IPF the production of collagenase might be decreased. To test this hypothesis, we quantified the constitutive level of collagenase secretion and its induction with phorbol myristate acetate (PMA) in six fibroblast cell lines derived from two normal lungs and four with IPF.[31] Three of them (one normal and two with IPF) were obtained from the American Type Culture Collection and the other three cell lines were obtained by lung biopsy. The fibroblasts were cultured under standard conditions, and the collagenase levels in fibroblast-conditioned media were measured by ELISA. Under basal conditions, the constitutive level of collagenase production presented noteworthy differences. In relation to ATCC cell lines, normal fibroblasts produced three- to fivefold more collagenase than did "fibrotic" cell lines. When PMA was added to the culture medium, a different increase in collagenase synthesis was observed, but although the rate of enzyme production increased sevenfold in one of the cell lines derived from fibrotic lung, collagenase levels remained below normal. Regarding the cell lines derived from our patients, normal fibroblasts under basal conditions produced at least fourfold more collagenase than did those from IPF fibroblasts. After PMA treatment one fibrotic cell line did not respond to the induction, whereas the other reached normal levels. These findings suggest that collagenase gene expression might be repressed in IPF cell lines, although it may have become activated, at least in some of them. Therefore, the possible mechanisms involved in the decrease of collagenolytic activity observed in human and experimental lung fibrosis should at least involve two possible mechanisms: (1) decreased production of the enzyme, and (2) a higher content of inhibitors relative to susceptible enzymes.

CONCLUSION

Several potential mechanisms can lead to excessive accumulation of collagens in the lung parenchyma during the development of interstitial fibrosis. Studies of both experimental and human pulmonary fibrosis performed in our laboratory have revealed that an important decrease in collagenolytic activity takes place, supporting the idea that an alteration in collagen catabolism may play a major role in lung fibrogenesis. The cellular and molecular mechanisms are currently under investigation.

REFERENCES

1. SELMAN, M. 1990. Pulmonary fibrosis: Human and experimental disease. *In* Connective Tissue in Health and Disease. M. Rojkind, ed. **1:** 123–188. CRC Press Inc. Boca Raton, Florida.
2. FULMER, J., R. S. BIENKOWSKY, M. J. COWAN, S. D. BREUEL, K. M. BRADLEY, V. J. FERRANS, W. C. ROBERTS & R. G. CRYSTAL. 1980. Collagen concentration and rates of synthesis in idiopathic pulmonary fibrosis. Am. Rev. Respir. Dis. **122:** 289–301.
3. SELMAN, M., R. CHAPELA, M. MONTAÑO, H. SOTO & L. DÍAZ. 1981. Increased lung collagen content in diffuse pulmonary fibrosis. Am. Rev. Respir. Dis. **123:** 63.
4. SELMAN, M., M. MONTAÑO, R. CHAPELA, H. SOTO & L. DÍAZ. 1982. Changes of collagen content in fibrotic lung disease. Arch. Invest. Med. (Méx.) **13:** 93–100.
5. SELMAN, M., M. MONTAÑO, C. RAMOS & R. CHAPELA. 1986. Concentration, biosynthesis and degradation of collagen in idiopathic pulmonary fibrosis. Thorax **41:** 355–359.

6. KIRK, J. M. E., P. E. DA COSTA, M. TURNER-WARWICK, R. J. LITTLETON & G. J. LAURENT. 1986. Biochemical evidence for an increased and progressive deposition of collagen in lungs of patients with pulmonary fibrosis. Clin. Sci. **70:** 39–45.

7. SELMAN, M., M. MONTANO, I. MONTFORT & R. PEREZ-TAMAYO. 1985. A new model of diffuse interstitial pulmonary fibrosis in the rat. Exp. Molec. Pathol. **43:** 375–387.

8. FIGUERAS, T. & A. PARDO. 1986. Collagen biosynthesis and degradation during deposit and resorptive phases of carrageenin granuloma. Collagen Rel. Res. **6:** 379–386.

9. RYAN, J. N. & J. F. WOESSNER. 1971. Mammalian collagenase: Direct demonstration in homogenates of rat uterus. Biochem. Biophys. Res. Commun. **44:** 144–149.

10. ROJKIND, M. & E. GONZALEZ. 1974. An improved method for determining specific radioactivities of proline-C^{14} and hydroxyproline-C^{14} in collagen and noncollagen proteins. Anal. Biochem. **57:** 1–7.

11. SELMAN, M., M. MONTANO, C. RAMOS, R. BARRIOS & R. PEREZ-TAMAYO. 1989. Experimental pulmonary fibrosis induced by paraquat plus oxygen in rats: A morphologic and biochemical sequential study. Exp. Molec. Pathol. **50:** 147–166.

12. SELMAN, M. UNPUBLISHED RESULTS.

13. SELMAN, M., M. MONTANO, C. RAMOS, R. CHAPELA, G. GONZALEZ & F. VADILLO. 1988. Lung collagen metabolism and the clinical course of hypersensitivity pneumonitis. Chest **94:** 347–353.

14. RAMOS, C., M. MONTAÑO, G. GONZÁLEZ, F. VADILLO & M. SELMAN. 1988. Collagen metabolism in experimental lung silicosis: A trimodal behavior of collagenolysis. Lung **166:** 347–354.

15. MONTFORT, I. & R. PEREZ-TAMAYO. 1978. Collagenase in experimental carbon tetrachloride cihrrosis of the liver. Am. J. Pathol. **92:** 411–420.

16. PEREZ-TAMAYO, R., I. MONTFORT, E. GONZALEZ & A. M. ALVIZOURI. 1982. Collagenolytic activity in cirrhosis of the liver. *In* Collagen Degradation and Mammalian Collagenase. M. Tsuchiya, R. Pérez-Tamayo, I. Okazaki & K. Maruyama, eds.: 135–149.

17. BRADY, A. H. 1975. Collagenase in scleroderma. J. Clin. Invest. **56:** 1175–1182.

18. EMONORD, H. & J. A. GRIMAUD. 1989. Active and latent collagenase activity during reversal of hepatic fibrosis in murine schistosomiasis. Hepatology **10:** 77–83.

19. GADEK, J. E., J. A. KELMAN, G. A. FELLS, S. E. WEINBERGER, A. L. HORWITZ, H. Y. REYNOLDS, J. D. FULMER & R. G. CRYSTAL. 1979. Collagenase in the lower respiratory tract of patients with idiopathic pulmonary fibrosis. N. Engl. J. Med. **301:** 737–742.

20. O'CONNOR, C., C. ODLUM, A. VAN BREDA, C. POWER & M. X. FITZGERALD. 1988. Collagenase and fibronectin in bronchoalveolar lavage fluid in patients with sarcoidosis. Thorax **43:** 393–400.

21. WEILAND, J. E., J. G. GARCIA, W. B. DAVIS & J. E. GADEK. 1987. Neutrophil collagenase in rheumatoid interstitial lung disease. J. Appl. Physiol. **62:** 628–633.

22. HARRIS, E. D., J. M. EVANSON, D. R. DiBONA & S. M. KRANE. 1970. Collagenase and rheumatoid arthritis. Arthritis Rheum. **13:** 83–94.

23. BROWN, S. I. & C. A. WELLER. 1971. The pathogenesis and treatment of collagenase-induced diseases of the cornea. Trans. Am. Acad. Ophthalmol. Otolaryngol. **74:** 375–383.

24. BIRKEDAL-HANSEN, H. 1980. Collagenase in periodontal disease. *In* Collagenase in Normal and Pathological Connective Tissues. D. E. Wooley & J. M. Evanson, eds.: 127–140. John Wiley & Sons. New York.

25. BAUER, E. A. & E. Z. EISEN. 1978. Recessive dystrophic epidermolysis bullosa: Evidence for increased collagenase as a genetic characteristic in cell culture. J. Exp. Med. **148:** 1378–1383.

26. SELMAN, M., A. PARDO, N. BARQUIN, R. SANSORES, R. RAMIREZ, C. RAMOS, M. MONTAÑO & G. STRICKLIN. 1991. The significance of collagenase inhibitors in bronchoalveolar lavage fluids. Chest. In press.

27. PARDO, A. & M. SELMAN. 1991. The collagenase gene family. Relationship between collagenolytic activity and fibrogenesis. *In* Interstitial Pulmonary Diseases. Selected Topics. M. Selman & R. Barrios, eds.: 306. CRC Press Inc. Boca Raton, Florida.

28. SPRINGMAN, E. B., E. L. ANGLETON, H. BIRKEDAL-HANSEN & H. E. VAN WART. 1990. Multiple modes of activation of latent human fibroblast collagenase: Evidence for the role of a Cys^{73} active-site zinc complex in latency and a "cysteine switch" mechanism for activation. Proc. Natl. Acad. Sci. USA **87:** 364–368.

29. PEREZ-TAMAYO, R., I. MONTFORT & A. PARDO. 1980. What controls collagen resorption in vivo? Med. Hypotheses **6:** 711–726.

30. MONTAÑO, M., C. RAMOS, G. GONZALEZ, F. VADILLO, A. PARDO & M. SELMAN. 1989. Lung collagenase inhibitors and spontaneous and latent collagenase activity in idiopathic pulmonary fibrosis and hypersensitivity pneumonitis. Chest **96:** 1115–1119.

31. PARDO, A. & M. SELMAN. 1989. Decreased collagenase production by fibroblasts derived from idiopathic pulmonary fibrosis. Matrix Metalloproteinase Conference. Destin, Florida.

Conference Summary

Pulmonary Emphysema: The Rationale for Therapeutic Intervention

ALLEN B. COHEN[a]

Department of Medicine and Biochemistry
University of Texas Health Center at Tyler
Tyler, Texas 75710

At the risk of being a pedantic, in this summary I will deal mostly with broad concepts, which will be oversimplified to make a few major points. If some of my criticisms hit too close to home for some, they will probably hit just as close to home for me. The protease inhibitor hypothesis, if it has done nothing else, has been the "NASA" of biologic science. It has had enormously productive spinoffs in many areas of biologic science. Alpha$_1$-protease inhibitor has been a bonanza for all kinds of research in areas such as protein chemistry, crystallography, genetics, and animal models of other genetic diseases. Proteases are now implicated in a wide variety of disorders and imbalances of proteases and inhibitors, and the effects of neutrophil oxidants are subjects of research in any disease in which neutrophils are present outside the circulation.

Other presentations in this volume discuss the natural history of chronic obstructive pulmonary disease (COPD), behaviors of neutrophil and macrophage proteinases and their inhibitors, elastin chemistry and biology, potential chemical markers of lung destruction, current and future treatments of emphysema, and the effect of proteases in other diseases.

I will begin by reminding everyone how important COPD really is. Chronic obstructive pulmonary disease is the fifth most common cause of death in this country.[1] It is the only one of the five most common causes of death that is continuing to increase in number (FIG. 1).[1] Starting with the Second World War, more women began to smoke, and they are now paying for that habit with an increase in COPD (FIG. 2).[1,2] During that same time the prevalence of COPD in the male population leveled off in this country. For those who are afraid that people will stop smoking and put us out of business, I have good news for you and bad news for the patients. Even though many people have quit smoking, and it has become politically and socially unacceptable to smoke in public, cigarette consumption in this country has just begun to level off (FIG. 3).[3] If you add in worldwide cigarette consumption, there has not even been a break in the curve (FIG. 4).[3] An enormous number of patients throughout the country now have COPD. However, our prevalence is low compared to that of Eastern Europeans (FIG. 5).[4] If someone is looking for a good reason to develop drugs for COPD, they certainly have it.

I will next look at the primary observation that initiated this field and some of

[a] Address for correspondence: Dr. Allen B. Cohen, Executive Associate Director, Professor of Medicine and Biochemistry, University of Texas Health Center at Tyler, P.O. Box 2003, Tyler, Texas 75710.

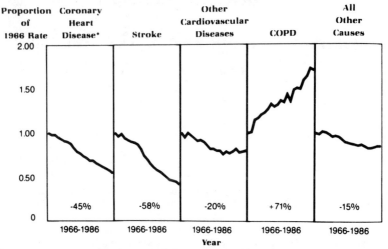

Twenty year trends in age-adjusted death rates, United States 1966-1986 (expressed as a proportion of the rate in 1966).

FIGURE 1. Taken from monograph entitled, *Clinical Epidemiology of Chronic Obstructive Pulmonary Disease,* 1989, p. 32; used with permission of publisher and author.

the hypotheses that this observation engendered. The primary observation made by Laurell and Eriksson was that patients with low blood concentrations of α_1-protease inhibitor tended to develop emphysema at a young age,[5] especially if they smoked cigarettes. Like other great observations or great hypotheses, this hypothesis to some extent has limited this field of study for years. Karlinsky and Snider's review[6] of causes of emphysema in animal models describes 15 or 20 different ways to cause emphysema. Almost all of us study one way. The observation of Laurell and Eriksson, which was an observation of fact, not an experiment, led to what I call the central hypothesis and three corollaries.

Primary Observation: Patients with low blood concentrations of α_1-protease inhibitor tend to develop emphysema at a young age, regardless of their smoking history.

Central Hypothesis: Proteases digest the lungs of these patients to produce emphysema.

Corollaries: (1) Elastases are the likely proteases. (2) Neutrophil elastase is the most likely protease. (3) Emphysema in cigarette smokers may also result from an imbalance between neutrophil elastase in the lungs and protease inhibitors available in the lungs to inhibit it.

How have these hypotheses served us? What is the purpose of stating a hypothesis? What is the gold standard for testing these hypotheses?

All of these hypotheses remain unproved. We have fallen into somewhat of a trap in that we have performed an enormous number of experiments to show what may happen and very few experiments to show what does happen.

As I was reviewing this field and thinking about these issues before writing this paper, a Japanese scientist, Dr. Shigeki Nagao, who is currently on sabbatical in my lab, showed me an article that has some very interesting implications for this field of study. The article, titled "Sushi Science and Hamburger Science," recently appeared in *Perspectives in Biology and Medicine*.[7] The author, Dr. T. Motokawa, was recently on sabbatical in North Carolina at a research institution. He draws the conclusion that Eastern (Japanese) science and Western science are divided by stark differences in philosophy that emanate from the dominant religious teachings. Basically, Dr. Motokawa indicates that Western scientists always attempt to see grand designs in nature. He notes that in Christianity, the noted character of God is that He is the one and only. He is the creator of the world and He created it with a purpose. The world is designed by Him. In Zen Buddhism, there is no one creator god. There is no purpose of creation. Concepts such as purpose are a creation of the mind of man and consequently are not reliable. Therefore, what man says about the facts hides the truth. The truth is the facts.

For the scientist then, the issue is what can he trust? According to Dr. Motokawa, in the West, the word may be mediator. The word is the creator. The word becomes the fact. However, in the oriental sciences, the fact is the truth itself and nothing else is. The word, therefore, conceals the fact. Have these hypotheses led us to carry out a lot of experiments that have concealed what is actually going on? Do we have a triumph of "sushi science" or "hamburger science" or a failure of one? The initial observation of Laurell and Eriksson was "sushi science." They

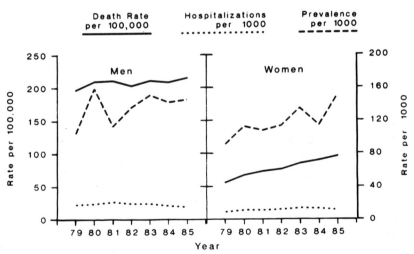

Prevalence, hospitalizations and mortality for COPD for ages 65-74 by sex; United States 1979-1985.

FIGURE 2. Taken from monograph entitled, *Clinical Epidemiology of Chronic Obstructive Pulmonary Disease*, 1989, p. 32; used with permission of publisher and author.

made an observation of a fact, and that has been a dominant force in biology ever since.

In this regard, it is interesting to note Dr. Travis's observation that the fragment of α_1-protease inhibitor that detaches from an inhibited enzyme is an inducer of synthesis of certain liver proteins. Could a totally different function of α_1-protease inhibitor lead to the lung destruction that we see in α_1-protease inhibitor-deficient patients?

Dr. Motokawa's point requires close scrutiny. I would like to build the case that although the protease inhibitor hypothesis of emphysema has engendered much good research, as a scientific tool it has failed. It has reduced creative development of new hypotheses in the area of emphysema for two and a half

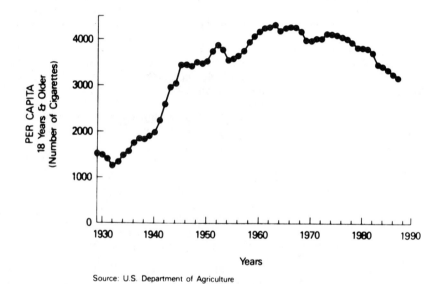

Source: U.S. Department of Agriculture

U.S. per capita cigarette consumption: 1955-1984

FIGURE 3. Taken from monograph entitled, *Clinical Epidemiology of Chronic Obstructive Pulmonary Disease,* 1989, p. 263; used with permission of publisher and author.

decades. I am making this statement to be provocative and I invite my colleagues to challenge my thesis. To date, none of the central hypotheses has been proven or refuted, and no treatment has been proven to help a patient.

Why have we failed? I don't think that we are dealing with a case of the grand unprovable hypothesis on which many professorial careers were dependent in past years. We are dealing with a relatively poor application of the scientific method as it has evolved in modern times. Too often we have designed experiments to provide descriptive data that are "compatible with" the Grand Hypothesis. Popper[8] describes the method of science as the formulation of bold conjectures to determine by deduction the consequences of these conjectures and then to collect data in a directed way to test each conjecture by attempting to falsify

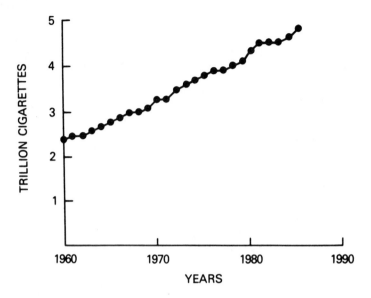

Cigarette consumption worldwide: 1960–1985

FIGURE 4. Taken from monograph entitled, *Clinical Epidemiology of Chronic Obstructive Pulmonary Disease,* 1989, p. 264; used with permission of publisher and author.

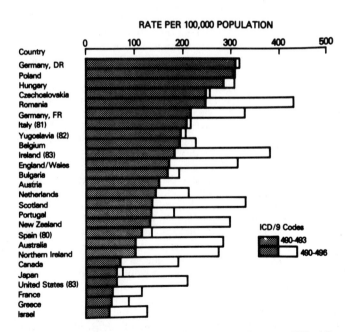

FIGURE 5. Taken from *American Review of Respiratory Disease,* 1989, **140:** number 3, part 2, p. S28; used with permission of publisher and author.

it.[8-10] Platt[10] went further to assert that there was a method of doing science that was learned by two branches of science and that these sciences are the only sciences that make true and rapid advances. These two fields are molecular genetics and particle physics. Platt first recognizes that the major scientific tool is the experiment that involves measurements. As noted from statistical theory, one cannot prove a hypothesis; one can only disprove it. Therefore, in the best case, the scientist designs a hypothesis and then designs the perfect null hypothesis. The investigator then defeats the null hypothesis. Therefore, scientific advancement is actually a destructive process: It advances by destroying what went before it. Platt further describes the Theory of Multiple Working Hypotheses. The concept is as follows: for every phenomenon, one puts forth all of the possible hypotheses. For each hypothesis, one sets up the perfect null hypothesis and defeats the null hypothesis if possible. In addition to moving science forward by the destructive mechanism, this method can discover phenomena with multiple causations.

I suggest that the hypotheses that I have listed have stimulated science but they have not been definitively proven or disproven. This is a failure of Western science. The closest we have come is another use of what I would call "sushi science," in which a scientist among us tried to determine "what is" and not "what might be," based on an artificial system. Damiano and colleagues[11] observed neutrophil elastase on lung elastin in areas of lungs with emphysema. Whether it turns out to be right or wrong, it was an elegant experiment. If their observation is correct, then there is very little doubt that neutrophil elastase causes *some* of the emphysema that occurs in cigarette smokers. Dr. Fox's group was unable to corroborate Damiano and colleagues' observation. One of the important results of this conference is that Dr. Fox's group agreed to trade samples and methods with Dr. Weinbaum's group to resolve the dispute.

Is there a way to pose a hypothesis that goes right to the heart of the matter? Is there an experiment that can determine the role that neutrophil elastase plays in lung destruction in patients with emphysema? I think there is such an experiment, and the experiment is called a clinical trial. On the basis of what we have heard in this conference, the new elastase inhibitors being developed by several companies may permit us to carry out such a trial. We have done about as much as we are going to be able to do in the laboratory to prove or disprove the neutrophil elastase hypothesis. Now is the time to administer a drug that eliminates elastase from the lungs to the extent possible and to determine what effect it has on emphysema. I do not think we will get any closer to proving or disproving the central hypotheses without that kind of an approach.

This brings us to Dr. Glass's presentation, which was a very fine and very elegant analysis of what it would take to do this kind of study. I suggest, however, that Dr. Glass has defined a "Cadillac" clinical trial. In the realm of both pulmonary research and cardiology research in this country we have become hooked on "Cadillac" research. In this financial milieu we ought to begin to think about how to build a "Ford," the vehicle that will get us there, but it will get us there cheaply and without as much elegance. Let me give you some examples. One of the things that Dr. Glass did to get his numbers for financial costs of this kind of trial was to call up clinical centers and ask how much it would cost to do various tests.

That is the wrong approach for a trial of a drug in smokers' emphysema. A center might be able to amass a cohort of 600 or 1,000 patients. Instead of getting the patient into the medical economic system, which is enormously overpriced, buy a spirometer, train the technicians in how to use it, and measure FEV_1. (I think that the scientists in this area will be able to document very thoroughly that

FEV_1 and mortality are the only measurements that are going to be useful and meaningful.) Do not determine diffusing capacities because they are not going to be a meaningful endpoint. One chest x-ray may be taken to assure that the patient does not have some other disease at the start. But here, too, there is a chance of achieving a savings. Much literature and a lot of practice support the idea that if you're doing something in the course of normal patient care, the research sponsor does not have to pay for it. We can certainly justify a chest x-ray in a patient with chronic pulmonary disease. We need to design a study that sharply focuses on the endpoints and works around rather than through the medical economic system.

Next, I would like to think about what the succeeding workshop should try to analyze. If, in fact, it is true that we should begin to focus on a clinical trial to see how much emphysema can be prevented by eliminating neutrophil elastase from the lungs, one of the issues that has to be reexamined is the biochemical markers. A good chemical marker of lung destruction would be very valuable because it may give the people in the drug companies the evidence they need to say, yes, it's worth going ahead with this expensive trial. We wish Drs. Weinbaum, Rosenbloom, and Kucich very good luck in their new approaches. I hope that the conflict between Drs. Mumford and Weitz on the value of elastase-generated fibrinopeptides as a predictor of lung destruction in patients with emphysema will soon be resolved. We have some possible indicators of lung breakdown, and we should pursue them with the best science available. For the second part of the next workshop we should get a group of people from academia, from the drug companies, and from the FDA together to design and agree upon a minimal study, that is, to design a "Ford." Another issue we would like to develop is to determine who within the medical scientific community should participate. Can the Federal Government play some kind of role? The Federal Government has sponsored clinical trials that have cost multiple millions of dollars. Maybe we could develop some kind of marriage between industry and the Federal Government. It is in the interest of the Federal Government and the mission of the National Institutes of Health to try to treat this pervasive disease. If some kind of "marriage" could then be developed, the workshop should develop a plan for choosing the very best drug for the trial. Therefore, the next meeting should develop the criteria for choosing the best drug and a mechanism for drug testing.

I will close with my favorite fortune cookie saying: The man who suggests that it cannot be done should not interrupt the man doing it.

NOTE ADDED IN PROOF: Dr. Damiano's important experiment has now been reproduced. YI-MIN, G., Z. YUAN-JUE & L. WEI-CI. 1990. Chinese Med. J. **103:** 588–594.

REFERENCES

1. HIGGINS, M. W. & T. THOM. 1989. Incidence, prevalence, and mortality: Intra- and intercountry differences. *In* Clinical Epidemiology of Chronic Obstructive Pulmonary Disease. M. J. Hensley & N. A. Saunders, eds.: 23–44. Marcel Dekker, Inc. New York.
2. FEINLEIB, M., G M. ROSENBERG, J. G. COLLINS, J. E. DELOZIER, R. POKRAS & F. M. CHEVARLEY. 1989. Am. Rev. Respir. Dis. **140:** s9–s18.
3. PARKER, S. R. & J. W. KUSEK. 1989. Smoking cessation. *In* Clinical Epidemiology of Chronic Obstructive Pulmonary Disease. M. J. Hensley & N. A. Saunders, eds.: 259–290. Marcel Dekker, Inc. New York.
4. THOM, J. T. 1989. Am. Rev. Respir. Dis. **140:** S27–S34.

5. ERIKSSON, S. 1965. Acta Med. Scand. **177**(Suppl. 432): 1–85.
6. KARLINSKY, J. B. & G. L. SNIDER. 1978. Ann. Int. Med. **117:** 1109–1134.
7. MOTOKAWA, T. 1989. Perspect. Biol. Med **32:** 489–504.
8. POPPER, K. R. 1968. The Logic of Scientific Discovery. 2nd ed. Harper & Row. New York.
9. ROTHMAN, J. 1986. Ann. Int. Med. **105:** 445–447.
10. PLATT, J. R. 1964. Science **146:** 347–353.
11. DAMIANO, V. V., A. TSANG, U. KUCICH, W. R. ABRAMS, J. ROSENBLOOM, P. KIMBEL, M. FALLAHNEJAD & G. WEINBAUM. 1986. J. Clin. Invest. **78:** 482–493.

Contribution of Additional Elastase Inhibitors Other Than Alpha₁-Protease Inhibitor and Bronchial Antileukoprotease to Neutrophil Elastase Inhibiting Capacity of Bronchoalveolar Lavage[a]

RAJA ABBOUD, ANGUS TSANG, AND MARTA TALLER

Respiratory Division
University of British Columbia
Vancouver General Hospital
Vancouver, B.C., Canada

We compared ELISA alpha$_1$-protease inhibitor (α_1-PI) and antileukoprotease (ALP) contents and neutrophil elastase inhibiting capacity (NEIC) and pancreatic elastase inhibiting capacity (PEIC) of bronchoalveolar lavage (BAL) in six healthy smokers (mean age 22 ± SD 2 years) and seven smokers (36 ± SD 6 years) who had smoked over 10 pack years (mean 18 ± SD 7 pack years). NEIC and PEIC to determine active α_1-PI were assayed on fresh unconcentrated BAL, using pure enzymes titrated against fully active pure α_1-PI. In the nonsmokers, we compared BAL done with 5 × 20 ml aliquots of saline solution from one lung with BAL done with 5 × 50 ml of saline solution from the other lung to determine the effect of the volume of lavage on the relative concentrations of ALP and α_1-PI in the recovered BAL.

Results in the 20-ml lavages were similar to those in the 50-ml lavages, but the relative concentration of ALP was greater in the 20-ml lavages and the total quantities of inhibitors recovered were less. ALP concentration in the first aliquot of lavage was highest and decreased in successive aliquots of lavage, especially in the 50-ml aliquots. In the later aliquots of 50-ml lavage, ALP molar concentration was about 20% of the α_1-PI concentration. In the nonsmokers, the fraction of total NEIC not due to active α_1-PI as determined from PEIC was 0.86 ± SE 0.13 nmol, slightly lower than the ALP content of 1.08 ± 0.15 nmol. In the smokers, the fraction of total NEIC not due to active α_1-PI was 1.43 ± 0.28 nmol, slightly greater than the ALP content of 1.17 ± 0.20 nmol. The difference may be due to an additional inhibitor, but it was less than 10% of the total NEIC in five of the seven smokers, which is within the variability in assays and was within 15–20% in two smokers, suggesting the possibility of an additional inhibitor in these two subjects. However, additional studies on concentrated BAL, including gel filtration column chromatography and FPLC using molecular sieve and ion exchange columns, did not separate an additional inhibitor from α_1-PI or ALP.

The data suggest that in most healthy nonsmokers and smokers, there is no evidence of significant amounts of an additional elastase inhibitor in BAL other than ALP and α_1-PI. However, small amounts of an additional inhibitor may be present in some smokers; confirmation of this would require separation and characterization.

[a] This work was supported by the Medical Research Council of Canada.

Neutrophil Sequestration in Rat Lungs[a]

G. M. BROWN,[b] E. DROST,[c] C. SELBY,[c] K. DONALDSON,[b]
AND W. MacNEE[c]

[b]Institute of Occupational Medicine
Edinburgh, Scotland

[c]Unit of Respiratory Medicine
City Hospital
Edinburgh, Scotland

The pulmonary microvasculature contains a large marginated pool of neutrophils (PMNs)[1] whose passage through the pulmonary circulation is delayed during cigarette smoking.[2] They may become activated by exposure to cigarette smoke and thus contribute to lung damage. We previously showed that exposure to cigarette smoke *in vitro* reduces the deformability of PMNs,[3] and we questioned whether this might influence the sequestration of PMNs in the lung. We therefore developed a rat model to investigate the factors that influence the passage of PMNs through the pulmonary circulation.

MATERIAL AND METHODS

Rats were transfused with a plasma expander (Hetastarch) to increase the blood yield. Peripheral blood PMNs (PBNs) were obtained by separation using Percoll and Sepracell gradients. Inflammatory PMNs (INs) were obtained by bronchoalveolar lavage of rat lungs inflamed with *C. parvum*. Red blood cells (RBCs) were obtained by heartstab from a control rat. PBNs and INs were then labeled with ^{51}Cr and RBCs with ^{99m}Technetium. Radiolabeled cells (6×10^6) were exposed to smoke for 1 minute in a tonometer system which agitated the cells at 37°C in a 1-ml volume of phosphate-buffered saline solution containing 1% bovine serum albumin. The radiolabeled PMNs and RBCs were mixed and injected via the iliac artery of anesthetized rats. After 10 minutes the chest cavity was opened, a small volume of blood was obtained by heartstab, and the heart and major blood vessels were immediately snared to prevent blood loss. The lungs were fixed overnight in gluteraldehyde, and then both blood and lungs were counted by gamma counter. PMN sequestration was expressed as the fold increase in the PMN : RBC ratio in lungs compared with peripheral blood. As a measure of cell deformability, the pressure developed by PMNs (1×10^5 in 2 ml) during filtration at constant flow (1.5 ml/m) through micropore membranes (pore dimensions 5 × 11 μ) was assessed over 6 minutes.

[a] This research was supported by the Chest, Heart and Stroke Association (England).

RESULTS

The sequestration of INs (90.5 [10.5], mean [SE]) in the lung was significantly greater than that of PBNs (8.1 [2.5]; p <0.001). A trend towards a greater-fold increase of PBN sequestration (16.2 [6.7]) in inflamed rat lungs compared with control lungs was also noted, but the difference was not statistically significant. Exposure to cigarette smoke caused significantly increased PMN sequestration in control rat lungs (27.2 [2.4]) compared with untreated PBNs (p <0.001) but not in comparison with PMNs agitated in the tonometer (34.5 [6.7]). The filtration pressure of INs (9.95 [1.6]), mean [SD]) was greater than that of smoke-exposed PBNs (6.42 [0.96]), agitated PBNs (1.47 [0.24]), and untreated PBNs (0.56 [0.3]).

DISCUSSION

We have shown that the increase in PMN sequestration is associated with alterations in PMNs caused by exposure to the inflammatory milieu of the lung (IN) or by treatment *in vitro*. The increased sequestration may be explained by decreased deformability of the PMN, reducing its ability to negotiate the pulmonary capillaries. Although alterations in the inflamed lung caused a modest increase in sequestration, the change was insignificant compared with increases due to changes in the PMN. These initial results suggest that changes in the neutrophil may contribute more to cell sequestration than do alterations in the pulmonary microvasculature.

REFERENCES

1. HOGG, J. C. *et al.* 1987. Physiol. Res. **67:** 1249–1295.
2. MACNEE, W., B. BIGGS, A. S. BELZBERG & J. C. HOGG. 1989. N. Engl. J. Med. **321:** 924–928.
3. MACNEE, W. & C. SELBY. 1990. Clin Sci. **79:** 97–107.

Synergistic Effect of Elastase and 60 Percent Oxygen on Pulmonary Airspace Enlargement

JEROME O. CANTOR,[a] STEPHEN KELLER,
JOSEPH M. CERRETA, AND GERARD M. TURINO

Columbia University College of Physicians and Surgeons
New York, New York 10019

St. John's University School of Pharmacy
Jamaica, New York 11439

It has been shown that exposure of normal lungs to 95% oxygen for several days can induce emphysema-like lesions.[1] The aim of this particular study was to assess the adverse effects of a more moderate concentration of oxygen on lungs that were already undergoing emphysematous changes. Syrian hamsters were given 30 units of porcine pancreatic elastase intratracheally, then exposed to 60% oxygen for 5 days, beginning 48 hours after instillment of the enzyme. Compared with elastase-treated animals exposed to room air, the hyperoxic group had significantly greater airspace enlargement 5 weeks after enzyme administration (FIG. 1). Additional controls instilled with saline solution and exposed to either 60% oxygen or room air showed no airspace enlargement (FIG. 1). With regard to elastin turnover, the elastase-hyperoxia group showed the highest level of ^{14}C-lysine incorporation into the desmosine and isodesmosine cross-links of elastin (FIG. 2). Furthermore, only those animals treated with both elastase and hyperoxia had a significant percentage of neutrophils (28%) in their lung lavage fluids following exposure to oxygen. These results provide additional evidence that oxidants may play an important role in the pathogenesis of emphysema.

The fact that exposure to 60% oxygen (a level that is only at the threshold of pulmonary toxicity) greatly amplified the effects of elastase injury suggests that the influence of oxidants may be greatest when underlying lung injury is already present. The mechanisms that may possibly be responsible for hyperoxia-enhanced airspace enlargement include: (1) elevated elastase activity in the lungs, (2) reduced amounts of functional alpha$_1$-antiproteinase due to oxidation of methionine,[2] (3) free radical-mediated injury to elastic fibers, and (4) impaired elastic fiber resynthesis.[3] Of these, an increase in elastase activity may be most likely, because a large number of neutrophils are present in the lungs of animals treated with both elastase and hyperoxia. Previous studies have shown that hyperoxia can cause neutrophils to rupture and release their contents.[4] Impaired elastic fiber resynthesis might also increase airspace enlargement, but labeling studies (FIG. 2) suggest that desmosine and isodesmosine formation is not adversely affected by hyperoxia. Similarly, deactivation of alpha$_1$-antiproteinase is probably not an important factor in these experiments, because most of the instilled elastase is cleared from the extracellular space well before 60% oxygen is

[a] Address for correspondence: Jerome Cantor, MD, 102 Antenucci Building, 428 West 59th St., New York, NY 10019.

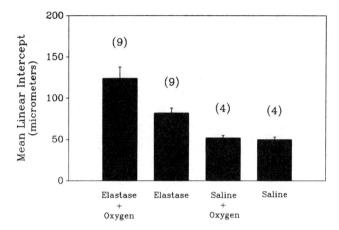

FIGURE 1. Graph showing changes in pulmonary airspace size, as determined by mean linear intercept measurements performed 5 weeks after elastase instillment. Exposure to hyperoxia caused a 51% increase in alveolar diameter among elastase-treated animals. (T bars indicate SE. Figures in parentheses refer to number of animals tested.)

administered. With regard to superoxide radicals, not enough evidence exists to confirm their role in the direct breakdown of elastin. Perhaps the most important question raised by this study is whether or not the use of oxygen in the treatment of emphysema accelerates the progression of the disease. Although alveolar damage induced by intratracheal instillment of elastase does not accurately mimic the human disorder, the results obtained with this model suggest the need to more closely examine the effects of oxygen on the lungs of persons who have emphysema.

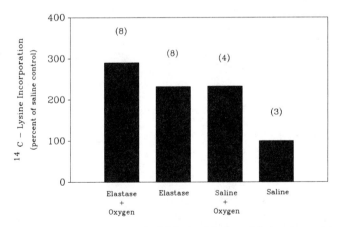

FIGURE 2. Graph showing differences in ^{14}C-lysine labeling of desmosine and isodesmosine cross-links of lung elastin, corrected for radioactive pool-size differences. Values are expressed as percentages relative to the saline-room air group. (Figures in parentheses refer to number of animals tested.)

REFERENCES

1. RILEY, D. J., M. J. KRAMER, J. S. KERR, C. U. CHAE, S. Y. YU & R. A. BERG. 1987. Damage and repair of lung connective tissue in rats exposed to toxic levels of oxygen. Am. Rev. Respir. Dis. **135:** 441–447.
2. MATHESON, N. R., P. S. WONG & J. TRAVIS. 1979. Enzymatic inactivation of human alpha-1-proteinase inhibitor by neutrophil myeloperoxidase. Biochem. Biophys. Res. Commun. **88:** 402–409.
3. OSMAN, M., J. O. CANTOR, S. ROFFMAN, S. KELLER, G. M. TURINO & I. MANDL. 1985. Cigarette smoke impairs elastin resynthesis in lungs of hamsters with elastase-induced emphysema. Am. Rev. Respir. Dis. **132:** 640–643.
4. STEINBERG, H., D. K. DAS, J. M. CERRETA & J. O. CANTOR. 1986. Neutrophil kinetics in O_2-exposed rabbits. J. Appl. Physiol. **61:** 775–779.

Eglin C and Heparin Inhibition of Platelet Activation Induced by Cathepsin G or Human Neutrophils

P. RENESTO, P. FERRER-LOPEZ, AND M. CHIGNARD

Unité de Pharmacologie Cellulaire
Unité associée IP/INSERM U285
Institut Pasteur
Paris, France

When purified human polymorphonuclear neutrophils (PMNs) and washed platelets are incubated together and challenged with N-formyl-Met-Leu-Phe (fMLP), platelet activation as measured by aggregation and serotonin release ensues. It has previously been demonstrated that such cell-to-cell cooperation is due to cathepsin G (cat G), a serine proteinase stored in the azurophilic granules of PMNs.[1-3] Experiments with purified cat G have determined that the range of concentrations for platelet activation is between 2 and 8 μg/ml.[3]

Eglin C, a proteinase inhibitor extracted from leeches, has the capability to interact with cat G.[4] We have indeed shown that 2 μg/ml (250 nM) of recombinant human eglin C prevented the biochemical activity of 180 nM purified cat G, as measured spectrophotometrically by hydrolysis of *N*-succinyl-Ala-Ala-Pro-Phe-paranitroanilide, a specific synthetic substrate. The very same eglin C concentration was able to inhibit platelet activation induced by 200 nM cat G, that is, a concentration inducing a submaximal platelet response. In fact, strictly comparable concentration-inhibition curves for biochemical and biologic cat G activities were obtained. Inhibition was specific (200 μg/ml eglin C failed to inhibit platelet activation induced by thrombin) and surmountable (1.2 μg/ml eglin C inhibited, by 100% and 50%, platelet activation triggered by 150 and 200 nM cat G, respectively).

Heparin is a glycosaminoglycan that, like eglin C, is able to inhibit cat G enzymatic activity.[5] However, the maximal inhibition that we observed was only 60% (heparin concentration up to 100 U/ml). As for eglin C, we investigated the capability of heparin to suppress cat G-induced platelet activation. A concentration-dependent response curve within the range of 10 to 70 mU/ml was obtained. Contrary to what was observed with the biochemical activity of cat G, the biologic activity was totally suppressed by high heparin concentrations. In fact, this inhibition resulted most probably from an electrostatic interaction between the anionic heparin and the cationic cat G, inasmuch as the addition of 180 mU/ml of protamine sulfate restored platelet activation.

The inhibitory effects of eglin C and heparin were then tested on the PMN-platelet cooperation model stimulated by 5×10^{-7} M fMLP. As expected, each substance displayed a concentration-dependent inhibition curve, with 25 μg/ml eglin C or 200 mU/ml heparin totally inhibiting platelet stimulation induced by activated PMNs (TABLES 1 and 2).

These experiments show that eglin C and heparin can prevent platelet activation induced by fMLP-activated PMNs and confirm the participation of cat G in the interaction between PMNs and platelets. In addition, as this cell-to-cell com-

TABLE 1. Inhibition by Eglin C of Biochemical and Biologic Cathepsin G Effects[a]

Eglin C (μg/ml)	Cat G Enzymatic Activity	Cat G-Induced Platelet Aggregation	Cat G-Induced Serotonin Release	fMLP-Induced Platelet Aggregation	fMLP-Induced Serotonin Release
0.2	37.0 ± 8.0	9.2 ± 8.7	2.6 ± 3.7	ND	ND
0.4	63.7 ± 8.3	30.6 ± 9.5	37.4 ± 15.4	ND	ND
0.8	86.3 ± 8.1	54.6 ± 13.1	62.3 ± 13.3	ND	ND
1.2	90.3 ± 1.1	74.0 ± 11.0	82.9 ± 14.4	1.9 ± 2.4	0.5 ± 0.9
1.6	95.1 ± 0.6	90.3 ± 12.2	91.6 ± 14.1	6.3 ± 7.3	8.8 ± 2.2
2.0	97.0 ± 3.2	100.0 ± 0.0	93.7 ± 3.7	27.3 ± 3.0	27.6 ± 5.1
2.4	98.5 ± 0.7	ND	ND	ND	ND
2.8	ND	ND	ND	40.5 ± 8.9	33.8 ± 9.5
3.6	ND	ND	ND	48.5 ± 12.1	56.4 ± 10.1
12.0	ND	ND	ND	96.3 ± 5.4	83.4 ± 23.3
24.0	ND	ND	ND	100.0 ± 0.0	94.8 ± 4.6

[a] Eglin C was added at different concentrations (column 1) to different experimental systems, and its inhibitory effect, expressed as a percentage of the control, was calculated.

- The enzymatic activity of 180 nM cat G was measured by hydrolysis of 1 mM N-succinyl-Ala-Ala-Pro-Phe-paranitroanilide (column 2).
- Cat G induced platelet aggregation (column 3) and serotonin release (column 4) were measured (2×10^8 platelets/ml) following challenge with 200 nM of the proteinase.
- fMLP-induced aggregation (column 5) and serotonin release (column 6) were measured on a mixture of PMNs (5×10^6 cells/ml) and platelets (2×10^8 cells/ml) preincubated with cytochalasin B (5 μg/ml) and challenged by 5×10^{-7} M fMLP.

Each value is the mean ± SD of three to six distinct experiments. ND = not determined.

TABLE 2. Inhibition by Heparin of Biochemical and Biologic Cathepsin G Effects[a]

Heparin (U/ml)	Cat G Enzymatic Activity	Cat G-Induced Platelet Aggregation	Cat G-Induced Serotonin Release	fMLP-Induced Platelet Aggregation	fMLP-Induced Serotonin Release
0.02	ND	18.6 ± 12.0	29.2 ± 8.9	11.6 ± 1.5	11.4 ± 9.0
0.03	17.5 ± 1.2	48.2 ± 8.0	42.4 ± 9.2	14.5 ± 5.6	22.2 ± 6.6
0.04	ND	73.3 ± 14.3	56.0 ± 14.0	18.4 ± 9.6	43.9 ± 26.7
0.05	38.1 ± 1.5	94.8 ± 5.3	79.4 ± 19.3	29.1 ± 9.3	36.1 ± 30.1
0.07	37.2 ± 1.5	98.2 ± 3.6	100.0 ± 0.0	59.5 ± 13.3	69.2 ± 16.3
0.1	56.1 ± 1.9	100.0 ± 0.0	100.0 ± 0.0	80.6 ± 12.7	63.2 ± 12.3
0.2	ND	ND	ND	87.8 ± 12.3	83.5 ± 16.5
0.3	ND	ND	ND	100.0 ± 0.	91.2 ± 8.8
1.0	60.6 ± 4.3	ND	ND	ND	ND
10	61.7 ± 2.0	ND	ND	ND	ND
100	57.0 ± 9.0	ND	ND	ND	ND

[a] Same legend as in TABLE 1. Each value is the mean ± SD of five to seven distinct experiments.

munication may be relevant to some pulmonary diseases such as adult respiratory distress syndrome, antiproteinases could have a therapeutic role in such pathology.

REFERENCES

1. CHIGNARD, M., M. A. SELAK & J. B. SMITH. 1986. Proc. Natl. Acad. Sci. USA **83:** 8609.
2. SELAK, M. A., M. CHIGNARD & J. B. SMITH. 1988. Biochem. J. **251:** 293.
3. FERRER-LOPEZ, P., P. RENESTO, M. SCHATTNER, S. BASSOT, P. LAURENT & M. CHIGNARD. 1990. Am. J. Physiol. **258:** C1100.
4. SEEMÜLLER, U., M. EULITZ, H. FRITZ & A. STROBL. 1980. Hoppe-Seyler's Z Physiol. Chem. **261:** 1841.
5. MAROSSY, K. 1981. Biochem. Biophys. Acta **659:** 351.

Does Cigarette Smoke Enhance the Proteolytic Activity of Neutrophils?

K. DONALDSON,[a] G. M. BROWN,[a] E. DROST,[b] C. SELBY,[b]
AND W. MacNEE[b]

[a]Institute of Occupational Medicine
Edinburgh, Scotland

[b]Unit of Respiratory Medicine
Rayne Laboratory
City Hospital
Edinburgh, Scotland

Neutrophils are found in increased numbers in the alveolar region of individuals who smoke cigarettes, and the release of proteinase by these cells in the alveolar region is considered to play an important role in the subsequent development of emphysema in a proportion of individuals who smoke.[1] We therefore undertook a series of studies to determine whether cigarette smoke, or its components, could directly stimulate human neutrophils to release elastase and degrade fibronectin.

MATERIALS AND METHODS

Human neutrophils were obtained from the peripheral blood of normal subjects by sedimentation through dextran and separation of the granulocyte fraction over plasma Percoll. Cigarette smoke exposure was accomplished in a tonometer system which allowed the neutrophils to be agitated in a 2-ml volume of phosphate-buffered saline solution with bovine serum albumin, in the presence of fresh cigarette smoke. Elastase levels in cells and supernatants were measured by radioimmunoassay using antibody specific for human neutrophil elastase. Fibronectin-degrading activity of the neutrophils was assessed by incubating the smoke-exposed cells on a matrix of radiolabeled fibronectin and measuring the release of radioactivity into the supernatant.[2]

RESULTS

Elastase Release

On exposure to vapor phase cigarette smoke in the tonometer system, neutrophils did not show an increased release of elastase, whereas intracellular elastase was reduced by smoke exposure; phorbol myristate acetate (PMA) was able to stimulate a modest release of elastase (TABLE 1). Cigarette smoke condensate, at a concentration of 1%, did not stimulate release of elastase in the tonometer system (TABLE 1). Exposure to a solution of 10% condensate was associated with a significant release of elastase into the supernatant, but this was a result of a

TABLE 1. Effect of Cigarette Smoke, PMA, and Condensate on Elastase Content and Secretion by Neutrophils[a]

Neutrophil Elastase Treatment	Neutrophils	Neutrophil Supernatant
Control	3,853 (456)	1.8 (0.7)
Smoke exposed	1,085 (590) p <0.01	2.2 (0.7) NSD
Control	ND	3.7 (4.0)
PMA exposed	ND	12.2 (8.0) p <0.01
Control	2,771 (1,370)	5.1 (3.3)
Condensate exposed	2,090 (1,188) NSD	3.6 (2.2) NSD

[a] *Upper panel*, mean (SD) of six experiments; *middle panel*, mean (SD) of six separate experiments; *lower panel*, mean (SD) of seven experiments. All data are ng/ml of human neutrophil elastase. NSD = no significant difference from control. PMA = phorbol myristate acetate 1 μg/ml.

direct toxic effect of the condensate because it was accompanied by a substantial decrease, from 98% to 43%, in the proportion of neutrophils able to exclude trypan blue.

Fibronectin Degradation

Neutrophils that had been exposed to particulate phase cigarette smoke in the tonometer system and subsequently incubated on a fibronectin matrix did not show an increased ability to degrade this extracellular matrix component but, in fact, displayed reduced activity (TABLE 2).

DISCUSSION

The release of increased amounts of neutrophil proteinase in smokers' lungs could be a result of (1) normal release of enzyme by a much expanded population, (2) stimulation of enzyme release, and (3) death of neutrophils with subsequent release of granule contents. In the present study we examined the likelihood of direct stimulation and toxicity of cigarette smoke and its components. The results suggest that cigarette smoke is not capable of directly stimulating elastase release or fibronectin-degrading activity in peripheral blood neutrophils. In experiments with condensate designed to repeat the conditions of Blue and Janoff,[3] the release

TABLE 2. Degradation of ^{125}I-Labeled Fibronectin by Control and Smoke-Exposed Neutrophils[a]

Treatment	Fibronectin Degradation
Control	9,262 (405)
Smoke exposed	7,032 (467) NSD

[a] Data given as mean (SD) counts per minute in eight separate experiments. NSD = no significant difference from control.

of elastase was accompanied by loss of ability to exclude trypan blue and so reflects a toxic effect of the smoke as opposed to a stimulatory one. The increased proteolytic activity seen in the lungs of smokers may, therefore, be a result of release from smoke-damaged cells or a result of the inflammatory response arising from it. Modification of neutrophil response by macrophage cytokines, however, cannot be ruled out and the role of alpha$_1$-protease inhibitor also needs to be considered.

REFERENCES

1. HOIDAL, J. & D. E. NIEWOEHNER. 1983. Chest **83:** 679–685.
2. BROWN, G. M. & K. DONALDSON. 1988. Thorax **43:** 132–139.
3. BLUE, M. L. & A. JANOFF. 1978. Am. Rev. Respir. Dis. **117:** 317–325.

Bronchial Inhibitor in Human Lung

B. FOX,[a] T. B. BULL,[b] A. GUZ,[c] AND T. D. TETLEY[c]

Departments of Histopathology,[a] Anatomy,[b] and Medicine,[c]
Charing Cross & Westminster Medical School
London, UK

Bronchial inhibitor (BI; antileukoprotease) can inhibit granulocyte elastase bound to elastin more readily than can alpha$_1$-proteinase inhibitor and may therefore be a significant controlling factor in the etiology of emphysema.

We have used two immunogold labeling electron microscopic techniques to localize BI in human lung tissue taken from lobectomy specimens. One method involved blocking nonspecific binding sites with preimmune goat serum, reacting sections with rabbit antibody specific for BI, then localizing the specific antibody with gold-labeled goat anti-rabbit antibody. This method resulted in staining of goblet cells and seromucous cells, and very occasional clumped label was found over elastin.

We also used gold-labeled protein A to locate bound BI antibody in adjacent serial sections. In this case, bovine serum albumin was used to block nonspecific binding sites (nonimmune serum contains Ig to which protein A binds). In addition to labeling of goblet cells and seromucous gland cells, there was patchy labeling of both elastic tissue and collagen. Using protein A without BI antibody resulted in similar labeling of elastin and collagen fibers, but there was no labeling of bronchial cells. Omission of BI antibody from the first method resulted in no labeling. Therefore, we do not believe that BI is associated with elastic tissue.

Different Evolution of Emphysema in Two Strains of Mice with Similar Serum Antielastase Deficit

C. GARDI, P. A. MARTORANA, M. M. DE SANTI,
P. VAN EVEN, P. CALZONI, AND G. LUNGARELLA[a]

Department of General Pathology
University of Siena
53100 Siena, Italy

and

Department of Pharmacology
Cassella AG
6000 Frankfurt/M. 61, FRG

The present study compares the development of emphysema in two strains of mice with a similar deficiency in serum elastase inhibitory capacity (EIC).

Against mouse leukocyte elastase (MLE), "tight-skin" (C57 B1/6J, *tsk+/+pa*) mice have significantly lower serum EIC values than do NMRI, BALB/C, or C57 B1/6J mice.[1] We show here that "pallid" mice (C57 B1/6J, *pa+/pa* also have a similar marked deficiency in serum EIC against MLE (FIG. 1).

In *tsk* mice pulmonary emphysema develops very early in life (3–4 weeks of age) and progresses thereafter.[2] The development of morphologic emphysema is accompanied by a marked decrease in lung elastin content[3] and a significant increase in alveolar elastolytic burden as detected by an immunologic technique.[4] In addition, in *tsk* mice, the emphysematous lesion is associated with a marked increase in lung collagen content coupled with ultrastructural defects of collagen fibers.[3]

By contrast, the development of emphysema in *pa* mice is a slow progressive process, occurring late in life. A few patchy areas of mild emphysema were first seen at 16 months of age.[2]

In *pa* mice, the development of a pulmonary lesion was preceded by an increase in the alveolar elastolytic burden, detected by an immunogold technique, 1–2 months after birth. Also, these mice showed a time-dependent decrease in lung insoluble elastin content. At 4 days of age the lung insoluble elastin content was normal. At 1 month after birth this parameter was not significantly decreased (-21%). At 8 (-45%) and 12 (-65%) months of age, however, lung elastin content was markedly and significantly decreased in *pa* mice (FIG. 2).

All these data show that although *tsk* and *pa* mice have a similar deficit in serum antielastase screen, the development of emphysema is completely different in these two strains. We suggest that the rapid and dramatic development of the emphysematous lesion in *tsk* mice is not dependent only on the EIC deficit. Other co-factors, that is, a defect in collagen metabolism, may play an important role.

[a] Address for correspondence: Dr. G. Lungarella, Department of Pathology, University of Siena, Via del Laterino n° 8, 53100 Siena, Italy.

329

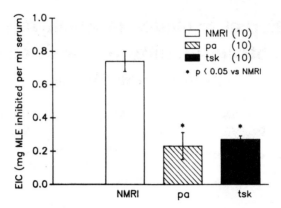

FIGURE 1. Comparison of serum elastase inhibitory capacity (EIC) in the various strains of mice. Values given are means ± SD. Pallid (*pa*) and tight-skin (*tsk*) mice show, against mouse leukocyte elastase (MLE), significantly lower values than do NMRI (about −64%).

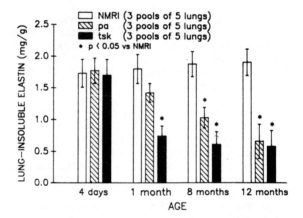

FIGURE 2. Lung insoluble elastin in the various strains of mice at different ages. Values represent means ± SD.

On the other hand, the slow developing, mild emphysema observed in *pa* mice may be a direct consequence of their EIC deficit.

REFERENCES

1. GARDI, C., P. A. MARTORANA, P. VAN EVEN, M. M. DE SANTI & G. LUNGARELLA. 1990. Exp. Mol. Pathol. **52:** 46–53.
2. MARTORANA, P. A., P. VAN EVEN, C. GARDI & G. LUNGARELLA. 1989. Am. Rev. Respir. Dis. **139:** 226–239.
3. GARDI, C., P. A. MARTORANA, M. M. DE SANTI, P. VAN EVEN & G. LUNGARELLA. 1989. Exp. Mol. Pathol. **50:** 398–410.
4. DE SANTI, M. M., C. GARDI, P. A. MARTORANA, P. VAN EVEN & G. LUNGARELLA. 1989. Exp. Mol. Pathol. **51:** 18–30.

Pulmonary Hemorrhage Induced by Intratracheal Administration of Human Leukocyte Elastase in Hamsters

ROBERT J. GORDON,[a] EDWARD FERGUSON,[a]
RICHARD DUNLAP,[b] CATHERINE FRANKE,[b] AND
PAUL J. SILVER [a]

[a]Department of Cardiovascular Pharmacology
Sterling Research Group
Rensselaer, New York 12144

[b]Life Sciences Research Laboratories
Eastman Kodak Company
Rochester, New York 14650

Among several animal models used for assessment of the effects of human leukocyte elastase (HLE),[1] the induction of pulmonary hemorrhage is commonly used for the *in vivo* evaluation of inhibitors of this enzyme.[2–4] We have characterized an adaptation of this model and now present data on the optimization of the model as it is used in our laboratory.

Male Golden Syrian hamsters (90–110 g) were pretreated systemically with test compound or vehicle and were anesthetized with enflurane (1.5% v/v in oxygen). With the aid of a laryngoscope, a 7-cm segment of PE100 polyethylene tubing was inserted through the mouth into the trachea of each animal. HLE (purified from purulent sputum, obtained from Elastin Products, Owensville, Missouri), 25 μg, or vehicle (0.5 ml of saline solution) was instilled into the catheter, followed by 0.3 ml of air. The catheter was then removed, and the animal was allowed to recover from anesthesia. Three hours later, the animal was euthanized by pentobarbital overdose, and the trachea was surgically exposed. A 3-ml saline lavage was instilled through a PE160 polyethylene tube that had been inserted via a small tracheal incision. After the animal's chest was briefly massaged, the lavage fluid was allowed to drain for collection. Absorbance (at 530 nm) of the lavage fluid, without further treatment, was used to quantify the degree of elastase-induced hemorrhage.

Absorbance was dependent on the intratracheal dose of HLE (FIG. 1, upper panel). Routinely, 25 μg of HLE per hamster was used because this dose was on the linear portion of the dose-response curve and it allowed detection of potent inhibition by methoxysuccinyl-alanyl-alanyl-prolyl-valine chloromethyl ketone (CMK), a known HLE inhibitor (FIG. 1, lower panel). Analogously, a 3-hour interval between HLE instillation and lavage was selected because it provided for maximum hemorrhage while still allowing inhibition by CMK (FIG. 2). Under these optimized conditions, the ED$_{50}$ values (mg/kg) of CMK were 2.5 (intraperitoneally, 15 minutes pre-HLE), 2.3 (subcutaneously, 15 minutes pre-HLE), and 0.079 (intravenously, 1 minute pre-HLE).

Absorbance was essentially linear versus lavage fluid dilution and therefore would also be linear versus inhibition. Absorbance correlated with the number of

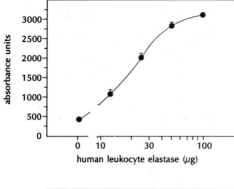

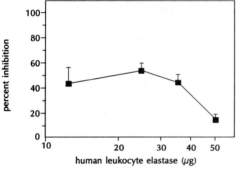

FIGURE 1. Production of pulmonary hemorrhage by elastase (*upper panel*) and its inhibition by CMK, 6 mg/kg ip (*lower panel*). Three hours after intratracheal instillation of elastase, bronchoalveolar lavage fluid was collected for measurement of absorbance at 530 nm. Hamsters receiving CMK were pretreated 15 minutes before elastase instillation. Each value is the mean ± SE of at least four animals.

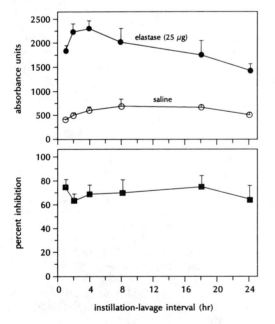

FIGURE 2. Time course of elastase-induced pulmonary hemorrhage (*upper panel*) and its inhibition by CMK, 6 mg/kg ip (*lower panel*). At various intervals after intratracheal instillation of saline solution or elastase (25 μg), bronchoalveolar lavage fluid was collected for measurement of absorbance at 530 nm. Hamsters receiving CMK were pretreated 15 minutes before elastase instillation. Each value is the mean ± SE of at least four animals.

erythrocytes in the lavage fluid ($r = 0.921$, $p < 0.001$, $n = 32$), indicating that absorbance reflected the degree of hemorrhage.

CMK also inhibited a nonprimate elastase, porcine pancreatic elastase (PPE), and prevented its hemorrhagic effect *in vivo*. However, CMK was slightly less potent as an inhibitor of PPE *in vitro* (inactivation rate = 830/M/sec versus 3150/M/sec for HLE) and against PPE-induced hemorrhage *in vivo* (ED_{50} = 6.1 mg/kg subcutaneously versus 2.3 mg/kg subcutaneously for HLE).

These results demonstrate the establishment of conditions that achieve an optimal balance between production of pulmonary hemorrhage and sensitivity to inhibition. The method represents an efficient means of evaluating HLE inhibition *in vivo*.

REFERENCES

1. SNIDER, G. L., E. C. LUCEY & P. J. STONE. 1986. Animal models of emphysema. Am. Rev. Respir. Dis. **133:** 149–169.
2. WILLIAMS, J. C., R. L. STEIN, C. KNEE, J. EGAN, R. FALCONE, D. TRAINOR, P. EDWARDS, D. WOLANIN, R. WILDONGER, J. SCHWARTZ, B. HESP, R. E. GILES & R. D. KRELL. 1988. Pharmacologic characterization of ICI 200,880: A novel, potent and selective inhibitor of human neutrophil elastase. FASEB J. **2:** A346.
3. BONNEY, R. J., B. ASHE, A. MAYCOCK, P. DELLEA, K. HAND, D. OSINGA, D. FLETCHER, R. MUMFORD, P. DAVIES, D. FRANKENFIELD, T. NOLAN, L. SCHAEFFER, W. HAGMANN, P. FINKE, S. SHAH, C. DORN & J. DOHERTY. 1989. Pharmacological profile of the substituted beta-lactam L-659,286: A member of a new class of human PMN elastase inhibitors. J. Cell. Biochem. **39:** 47–53.
4. HASSALL, C. H., W. H. JOHNSON, A. J. KENNEDY & N. A. ROBERTS. 1985. A new class of inhibitors of human leucocyte elastase. FEBS Lett. **183:** 201–205.

Purification of Elastin Antigen(s) from the Urine of Patients with Emphysema

U. KUCICH, J. HAMILTON, S. AKERS, P. KIMBEL,
S. METTE, J. ROSENBLOOM, AND G. WEINBAUM

*Research Division
Department of Medicine
The Graduate Hospital
Philadelphia, Pennsylvania 19146*

*Robert Wood Johnson Medical School
Camden, New Jersey 08103*

*School of Dental Medicine
University of Pennsylvania
Philadelphia, Pennsylvania 19104*

Destruction of lung elastin, with concomitant loss of elasticity, is a major cause in the development of chronic obstructive pulmonary disease (COPD). Elastin degradation is accelerated by excess elastolytic activity of neutrophilic origin in the lung interstitium,[1,2] resulting in the release of elastin-derived peptides (EDP) in the peripheral circulation. Earlier studies from our laboratory suggested that selected smokers and a high percentage of patients with COPD had an increased level of EDP in their plasma when compared to that of nonsmokers and most normal smokers.[3] In addition, the blood of patients with COPD contained elastin antigens of smaller molecular weight.[4] These findings prompted us to investigate EDP levels in urine, reasoning that more elastin antigen(s) of smaller molecular weight might be present in the urine of patients with COPD and allow us to characterize the *in vivo* elastin antigen(s) generated. Using an indirect enzyme-linked immunosorbent assay (ELISA), previously described,[3] we observed that patients with COPD had 262 ± 162 ng EDP/mg of creatinine ($n = 22$), whereas normal nonsmokers had 88 ± 60 ng/mg ($n = 24$, $p < 0.001$) (FIG. 1). Clearly, on the average, patients with COPD had three times the levels of EDP in their urine when compared to that of normal nonsmokers.

Chromatography of unconcentrated urine from a patient with COPD using a P-10 molecular sieving column revealed the presence of two immunoreactive peaks, 90% eluted in a defined peak with a molecular weight between 4 and 20,000 daltons and 10% eluted at approximately 1,000 daltons. Urine from a nonsmoker exhibited the high molecular weight antigen ($\frac{1}{3}$ to $\frac{1}{5}$ the amount of that in COPD urine), but no small molecular weight antigen. As a result of these preliminary experiments, we decided to purify the high molecular weight EDP, because it represented most of the urine immunoreactivity and because the small molecular weight EDP eluted together with most of the A_{280} absorbing material in urine. Therefore, COPD urine was concentrated $10\times$ by ultrafiltration with 60–70% EDP recovery and 10–30 μg EDP were purified in one step on a P-10 column with 90–100% recovery ($n = 3$). A representative chromatogram is shown in FIGURE 2, establishing that the EDP immunoreactivity eluted between 4 and 20,000 daltons with little contaminating protein, and its peak of immunoreactivity eluted at approximately 13,000 daltons. Concentration ($100\times$) of several 24-hour urine speci-

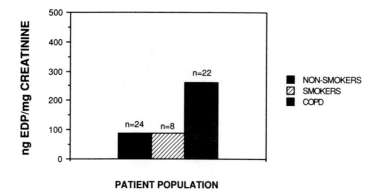

FIGURE 1. Urine elastin-derived peptide levels in control nonsmokers, smokers, and patients with COPD. Urine EDP levels were measured in duplicate at two different dilutions using an antibody raised against an elastase digest of human lung elastin (1 : 500 enzyme/elastin ratio).

mens from a single patient with COPD has allowed us to purify approximately 200 μg of EDP. The size distribution of human urine EDP was confirmed by fast protein liquid chromatography (FPLC) using a Superose 12 column. A single protein peak was obtained which coincided with the immunoreactive EDP peak and whose elution position was similar to that of our Cytochrome C standard. The size distribution of the elastin immunoreactivity in human urine agrees closely with the results reported by Goldstein and Starcher.[5] In those studies, intratracheal administration of elastase in hamsters resulted in the excretion of urinary

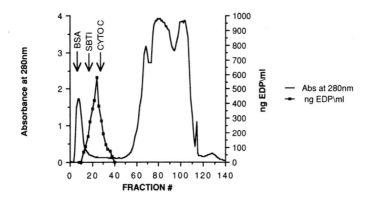

FIGURE 2. Fractionation of human urine by molecular sieving. 10× concentrated urine (10 ml) from a patient with COPD was chromatographed on a P-10 column (2.5 × 90 cm) in 0.02M phosphate buffer, 0.02% Na-azide, pH 7.2. Flow rate was 60 ml/h and fraction volume was 5 ml. Protein profile and immunoreactive EDP are shown as indicated in the figure. Antibody used was anti-human lung amorphous elastin. Molecular weight markers are: BSA (67,000), SBTI (21,500), and CYTO C (12,800).

desmosine-containing elastin peptides with sizes ranging between 9 and 27,000 daltons and a peak of immunoreactivity at 13,000 daltons. The increased amounts of EDP make the urine an excellent system for isolating elastin antigen(s) from patients with COPD. Antibodies generated against such *in vivo* EDP may provide the most discriminating probes of lung destruction, as well as aid in the development of an efficient system for monitoring the efficacy of a therapeutic regimen.

REFERENCES

1. GROSS, P. M., M. A. BABYAK, E. TOLKER & M. KASCHAK. 1964. Enzymatically induced pulmonary emphysema. J. Occup. Med. **6:** 481–484.
2. MARCO, V., B. MASS, D. MERANZE, G. WEINBAUM & P. KIMBEL. 1971. Induction of experimental emphysema. Am. Rev. Respir. Dis. **104:** 595–598.
3. KUCICH, U., P. CHRISTNER, M. LIPPMANN, P. KIMBEL, G. WILLIAMS, J. ROSENBLOOM & G. WEINBAUM. 1985. Utilization of a peroxidase-antiperoxidase complex in an enzyme-linked immunosorbent assay of elastin-derived peptides in human plasma. Am. Rev. Respir. Dis. **131:** 709–713.
4. KUCICH, U., W. R. ABRAMS, S. AKERS, P. KIMBEL, G. WEINBAUM & J. ROSENBLOOM. 1988. Determination of elastin peptide levels in human plasma using elastin domain-specific antibodies. Am. Rev. Respir. Dis. **137:** A372.
5. GOLDSTEIN, R. A. & B. C. STARCHER. 1978. Urinary excretion of elastin peptides containing desmosine after intratracheal injection of elastase in hamsters. J. Clin. Invest. **61:** 1286–1290.

Relationship between Intraalveolar Macrophage Population and Microscopic Emphysema

D. LAMB, A. McLEAN, AND W. WALLACE

Department of Pathology
University of Edinburgh
Edinburgh, UK

Histopathologists have long believed that smokers, including those with chronic obstructive airways disease, have an increased number of intraalveolar macrophages, particularly in the alveoli adjacent to respiratory bronchioles.[1] Bronchoalveolar lavage studies have consistently shown an apparent increase in intraalveolar macrophages (IAM) in patients with emphysema.[2] However, the relation between IAM number and early, microscopically recognized emphysema has not been assessed on tissue sections.

We examined tissue from 16 pulmonary resections for peripheral lesions; the subjects' ages ranged from 54–72 years; 14 were current smokers and 2 lifelong nonsmokers.

Microscopic emphysema was assessed using an automatic image analysis system, the IBAS II.[3] Lobes or lungs were fixed in a standard manner by inflation with formol-saline solution at 20 cm of water pressure, sagittally sliced, and randomly selected blocks were taken and embedded in glycol methacrylate. The amount of airspace wall was assessed by measuring the perimeter of airspace walls in 1 mm² microscope fields. The alveolar wall perimeter was then multiplied by 4/pi to give a value for alveolar surface area per unit of lung volume expressed as square millimeters per 1 mm³ lung tissue.

IAM were counted using an eyepiece graticule on the same material using 1 mm² fields from the 3-μ plastic sections. The IAM diameter was measured directly using a digitizing tablet according to the method of Lundberg and Vorwerk (quoted by Aherne and Dunnill[4]). From this the mean number of IAM per mm³ of lung tissue was calculated.

RESULTS

The number of IAM varied from 608–1485 per mm³ of fixed lung tissue. No correlation was noted between the value for IAM per mm³ and the mean alveolar wall surface area per unit volume (AWUV) (mm²/mm³) for the lung. However, when the macrophage number was expressed as IAM per mm² of alveolar surface area, there was a relation between increased macrophage number and a greater degree of microscopic emphysema, that is, low AWUV values ($r = 0.62$, $p < 0.01$) (see FIG. 1).

Macrophages/mm2

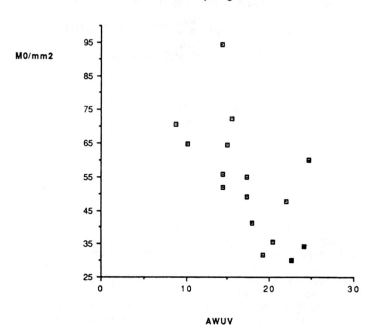

AWUV

FIGURE 1. Macrophage number/1 mm² surface area plotted against AWUV. A decrease in AWUV is associated with an increase in macrophage density for 1 mm² surface area ($r = -0.62$, p 0.05). Nonsmokers are shown as ■.

DISCUSSION

It is not surprising that the defensive or phagocytic function of macrophages is related to the surface area of the alveolar walls they are protecting. It is of particular interest that in smokers the macrophage number per unit alveolar surface area is elevated apparently in proportion to the degree of microscopic emphysema, that is, the number of macrophages increases as the AWUV decreases. Previous ideas on macrophage number and distribution within the acinar unit may have to be modified. It appears likely that the apparent marked increase in macrophage number in BAL in smokers may in part be related to the increased ability of macrophages to be washed out of an emphysematous lung.

REFERENCES

1. LAMB, D. 1975. Report of Tobacco Research Council.
2. FELLS, AOS & Z. A. COHN. 1986. J. Appl. Physiol. **60:** 353–369.
3. GOULD, G. A., W. MACNEE, A. MCLEAN, P. M. WARREN, A. REDPATH, J. J. K. BEST, D. LAMB & D. C. FLENLEY. 1988. CT measurements of lung density in life can quantitate distal airspace enlargement, an essential defining feature of human emphysema. Am. Rev. Respir. Dis. **137:** 380–392.
4. AHERNE, W. A. & M. S. DUNNILL. 1982. Morphometry.: 81–82. Edward Arnold. London.

Microscopic Emphysema and Its Variations with Age, Smoking, and Site within the Lungs

D. LAMB,[a] M. GILLOOLY,[a] AND A. S. J. FARROW[b]

[a]Department of Pathology
University of Edinburgh
Edinburgh, Scotland

[b]MRC Human Genetics Unit
Edinburgh, Scotland

Macroscopic emphysema can only be recognized by the naked eye when the airspace size reaches about 1 mm in diameter. Macroscopic emphysema, therefore, represents the end stage of a disease process inasmuch as 75% of the alveolar wall surface area must be lost before emphysema is visible to the naked eye.[1]

We have measured microscopic emphysema in terms of airspace wall surface area per unit volume (AWUV) using an objective automated image analysis technique.

MATERIAL AND METHODS

AWUV was measured on tissue sections from 62 cases, of which 49 were from lobar resections and 13 were autopsy specimens, all from subjects with known smoking histories (6 nonsmokers). Ages ranged from 23 to 85 years. All specimens were fixed by inflation with formol saline solution at a standard pressure of 20 cm of water, sliced sagittally, and 12 randomly selected blocks were embedded in glycol methacrylate, cut at 3 μ, and stained with hematoxylin and eosin. For autopsy cases, blocks were selected from upper, mid-, and lower portions of the upper and lower lobes.

AWUV was measured on 12 sections from each case using the Fast Interval Processor (FIP), a rapid scanning device that uses the optical density pattern on the tissue section to count intercepts of an electronic "test-line" with alveolar walls.[2] The mean linear intercept was calculated for each 1 mm^2 of histologic section, and from this figure a value for AWUV was derived.

For each case the AWUV value for at least 1,000 1-mm^2 fields was measured. The mean AWUV value was taken and compared with age and smoking history.

Using the autopsy lungs the distribution of AWUV values from the apex, middle, and base of each of the upper and lower lobes was assessed, and the mean AWUVs from these sites and from the upper and lower lobes were compared.

All statistics were performed using the Minitab package.

RESULTS

The mean AWUV range was 9.25–21.98 mm^2/mm^3. A significant negative correlation was found between AWUV and age ($r = -0.506$, $p < 0.01$, $n = 62$).

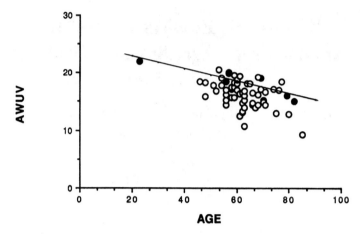

FIGURE 1. AWUV values for all cases plotted against age. Nonsmokers are shown as black circles. The line shown is the regression line for nonsmokers.

When the smoking and nonsmoking subgroups were considered separately, the two subgroups showed different relationships between AWUV and age (FIG. 1). The nonsmokers appeared to show a linear decrease in AWUV values with age ($r = -0.901$, $p < 0.05$, $n = 6$). The smokers showed a more complex relationship, some having a decrease in AWUV values with age similar to that of the nonsmokers and others showing a marked increased loss in AWUV values.

When AWUV values for the different zones within lobes or between the mean AWUVs of upper and lower lobes were compared, an analysis of variance showed no significant differences. We believe, therefore, that one lobe may be taken as representative of the whole lung in this population of cases, and in lungs with microscopic emphysema there is no apex to base gradient of abnormality.

CONCLUSIONS

Microscopic emphysema, as identified by a decrease in AWUV values, apparently has a homogeneous distribution throughout the lung and does not show variation between upper and lower lobes.

A linear decrease in AWUV values appears with age in nonsmokers, whereas many smokers showed an increased fall in AWUV values when compared to nonsmokers of similar age.

REFERENCES

1. LAMB, D. 1990. Chronic obstructive airways disease. *In* Respiratory Medicine. R. A. L. Brewis, G. J. Gibson & D. M. Geddes, eds. Baillière Tindall. London.
2. SHIPPEY, G., R. BAYLEY, S. FARROW, R. LUTZ & D. RUTOVITZ. 1981. A fast interval processor (FIP) for cervical prescreening. Anal. Quant. Cytol. **3:** 9–316.

A Polymer-Bound Elastase Inhibitor Is Effective in Preventing Human Neutrophil Elastase-Induced Emphysema

E. C. LUCEY,[a,b] P. J. STONE,[c] G. A. DIGENIS,[d] AND
G. L. SNIDER[a,c]

[a]Pulmonary Section
Boston Veterans Administration Medical Center and
[c]Department of Biochemistry
Boston University School of Medicine
Boston, Massachusetts 02130

[d]Division of Medicinal Chemistry and Pharmacognosy
University of Kentucky
Lexington, Kentucky

A small molecular weight, reversible elastase inhibitor is capable of potentiating the emphysema induced by human neutrophil elastase (HNE).[1] We postulate that the HNE inhibitor complex is transported into the alveolar interstitium, followed by dissociation. Because of its small size, the free inhibitor is rapidly cleared, and the reactivated HNE attacks elastic fibers, giving rise to emphysema. In the present study we tested whether the binding of a small molecular weight, reversible elastase inhibitor to a large water soluble polymer produces an agent capable of inhibiting HNE-induced emphysema in hamsters.

The agent tested was a polymer-bound peptidyl carbamate inhibitor (PPCI). The carbamate moiety, methoxysuccinyl-L-alanyl-L-alanyl-L-prolyl-CH_2N(i-Pr)CO_2-p-nitrophenol, was synthesized as previously described.[2] PPCI was synthesized by covalently linking, via an aminoethyl spacer, the carbamate hemisuccinyl-L-alanyl-L-alanyl-L-prolyl-CH_2N(I-Pr)CO_2-p-nitrophenol to a linear, water soluble polymer of N(2-hydroxyethyl)-D,L-aspartamide. The average molecular weight of PPCI is 22 kD. The dissociation constant is $8 \times 10^{-3} \times s^{-1}$.

IN VITRO STUDY OF ELASTASE INHIBITION

Inhibition of HNE by PPCI was assessed using ^{3}H-elastin substrate suspended in 4 ml of buffer. PPCI in buffer was added to the elastin suspension followed by 10 μg of HNE and incubated at 37°C for 4 hours. By interpolation, 50% of the elastolytic activity of 10 μg of HNE was inhibited by the addition of 1.8 μg of PPCI.

[b] Address for correspondence: Edgar C. Lucey, Ph.D., Research, Pulmonary, #151, Boston V.A. Medical Center, 150 South Huntington Ave., Boston, MA 02130.

CLEARANCE OF PPCI FROM HAMSTER LUNGS

Retention in the lungs of functionally active PPCI was studied in six hamsters instilled with 0.5 ml of saline solution containing 1 mg of freshly dissolved PPCI. Three hamsters received saline solution instillation only. Groups of three hamsters were studied 1 or 8 hours after PPCI instillation. The saline-treated control group was studied 1 hour after instillation. PPCI exhibited a functional half-life of 7.3 hours in the lavageable compartment of the hamster lung when given intratracheally.

PREVENTION OF HNE-INDUCED LUNG INJURY BY PRETREATMENT WITH PPCI

Hamsters were given two transoral intratracheal instillations 1 or 8 hours apart. There were six groups with eight animals in each. Hamsters were anesthetized by CO_2 inhalation before each administration of either 0.5 ml of saline solution or saline solution containing PPCI or 300 μg HNE. All animals survived the treatment and were studied 56 days later.

Doses of 100, 300, and 900 μg PPCI given intratracheally to hamsters all produced significant protection against the emphysema induced by 300 μg of HNE given 1 hour after the PPCI. Pretreatment of hamsters with 900 μg of PPCI 8 hours before the instillation of 300 μg HNE resulted in significant protection against induction of emphysema.

We conclude that the binding of low molecular weight inhibitors to large water soluble polymers is an effective means of slowing the rate of clearance and increasing the effectiveness of synthetic elastase inhibitors.

REFERENCES

1. STONE, P. J., E. C. LUCEY & G. L. SNIDER. 1990. Induction and exacerbation of emphysema in hamsters with human neutrophil elastase inactivated reversibly by a peptide boronic acid. Am. Rev. Respir. Dis. 141: 47–52.
2. DIGENIS, G. A., B. J. AGHA, K. TSUJI & M. SHINOGI. 1986. Peptidyl carbamates incorporating amino acid isoteres as novel elastase inhibitors. J. Med. Chem. 29: 1468–1476.

On the Activity of MR 889, a New Synthetic Proteinase Inhibitor

M. LUISETTI,[a] P. D. PICCIONI, M. DONNINI,
R. ALESINA, A. M. DONNETTA, AND V. PEONA

Istituto di Tisiologia e Malattie Respiratorie
IRCCS Policlinico San Matteo
Università di Pavia
Pavia, Italy

Synthetic proteinase inhibitors are promising tools for the treatment of pathologic conditions characterized by proteinase/proteinase inhibitor imbalance, such as pulmonary emphysema, chronic bronchitis, adult respiratory distress syndrome (ARDS), and cystic fibrosis.[1] We investigated the inhibitory activity of a new compound, 2-[3-thiophencarboxythio]-N-[dihydro-2(3H)-thiophenone-3-il]-propionamide, a thiolactic acid derivative (MR 889 kindly supplied by Medea Res. Ltd, Milan, Italy; MW 315.419). MR 889 was preincubated for 30 minutes with porcine pancreatic elastase (PPE, 1 μg), human neutrophil elastase (HNE, 1 μg), and bovine α-chymotrypsin (CHY, 50 μg). At low concentrations ranging from 10^{-5}–10^{-6} M), MR 889 showed a mean 50% inhibition of PPE, HNE, and CHY activity on specific low molecular weight nitroanilide substrates. The efficiency of MR 889 seemed to overcome that of PMSF, a compound currently used in laboratory inhibition assays.

In pathologic conditions such as ARDS, alpha$_2$-macroglobulin (α_2-MG) may be responsible for proteinase residual activity deriving from α_2-MG-trapped HNE.[2] That residual activity is scarcely inhibitable by high molecular weight inhibitors such as α_1-PI. To investigate the inhibitory activity of MR 889 on α_2-MG/HNE complex, α_2-MG and HNE at different molar ratios were preincubated for 1 hour. The formation of α_2-MG/HNE complex was confirmed by SDS-PAGE (FIG. 1) and by the activation of HNE towards specific nitroanilide substrate.[3] The addition of MR 889 10^{-3} M resulted in an inhibition ranging from 89%–98% of α_2-MG-trapped HNE residual activity, whereas PMSF 10^{-3} m inhibition was 76%–90%.

Finally, MR 889 was tested *in vitro* towards free elastase-like activity (ELA) and chymotrypsin-like activity (CLA) determined by specific nitroanilide substrates (Me(0)SAAPVpNa and SAAPPhepNa, respectively) in sputum sol-phase[4] obtained from six patients with chronic bronchitis. Results (summarized in TABLE 1) show that MR 889 displayed satisfactory inhibition, particularly towards CLA.

We conclude that MR 889, a reversible, slow binding, fully competitive proteinase inhibitor,[5] has potential for therapeutic use in pathologic conditions characterized by an excess of proteinase in extracellular fluids.

[a] Address for correspondence: Dr. Maurizio Luisetti, Istituto di Tisiologia e Malattie Respiratorie dell' Università, IRCCS Policlinico San Matteo, via Taramelli 5, 27100 Pavia, Italy.

343

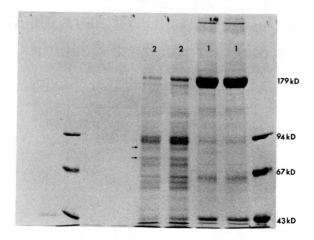

FIGURE 1. *Lane 1*: SDS-PAGE of α_2-MG under reducing conditions (2-mercaptoethanol), with the 180-kD subunits. *Lane 2*: α_2-MG and HNE were preincubated for 1 hour at 37°C (molar ratio 1:2) and then run on gel. The two additional bands between 94 kD and 67 kD (*arrows*) are likely to be ascribable to the bait region cleavage.

TABLE 1. Inhibition of Free ELA and CLA in Chronic Bronchitis Sputum Sol-phase

			Inhibitory profile (% inhibition)				
	ELA	CLA	MR 889		PMSF	CMK	STI
Sample	(μg/ml)	(ng/ml)	(10^{-3} M)	(10^{-4} M)	(10^{-2} M)	(0.2 mM)	(20 μg)
1	17	—	29.4	0	91.9	99.1	—
2	20	—	21.1	0	95.4	96.7	—
3	20	—	84.3	0	97.2	100	—
4	—	1,580	100	71.7	92.3	—	100
5	—	420	100	100	97.2	—	100
6	—	200	87.8	81.8	86.5	—	100

[a] Abbreviations: ELA = elastase-like activity; CLA = chymotrypsin-like activity; CMK = Me(0)SAAPVchloromethyl ketone; STI = soybean trypsin inhibitor.

REFERENCES

1. POWERS, J. C. & Z. H. BENGALI. 1986. Am. Rev. Respir. Dis. **134:** 1097–1100.
2. WEWERS, M. D., D. J. HERZYK & J. E. GADEK. 1988. J. Clin. Invest. **82:** 1260–1267.
3. TWUMASI, D. Y., I. E. LIENER, M. GALDSTON & V. LEVYTSKA. 1977. Nature **267:** 61–63.
4. JACKSON, A. H., S. L. HILL, S. C. AFFORD & R. A. STOCKLEY. 1984. Eur. J. Respir. Dis. **65:** 114–124.
5. BAICI, A., R. PELLOSO & R. HORLER. 1990. Biochem. Pharmacol. **39:** 919–924.

Development of Cor Pulmonale in Tight-Skin Mice with Genetic Emphysema[a]

P. A. MARTORANA,[b,c] M. WILKINSON,[d] M. M. DE SANTI,[e]
P. VAN EVEN,[b] C. GARDI,[e] AND G. LUNGARELLA[e]

[b]Department of Pharmacology
Cassella AG
6000 Frankfurt/M. 61, FRG

[d]Department of Pathology
University Hospital
Queen's Medical Centre
Nottingham NG7 2 UH, UK

[e]Department of General Pathology
University of Siena
53100 Siena, Italy

In man, one of the worst prognostic features in emphysema is the development of pulmonary hypertension and right ventricular hypertrophy (RVH). The investigation of RVH in emphysema is made difficult by the inadequacy of the available animal models. It was therefore of interest to investigate the development of RVH in the tight-skin mouse, a new model of genetic emphysema.[1] In these mice, emphysema first occurs at 3–4 weeks of life and progresses thereafter.[2]

Male tight-skin (C57 BL/6J-$tsk+/+pa$) mice and female pallid (C57 BL/6J $pa+/+pa$) mice were obtained from Jackson Laboratory (Bar Harbor, Maine) and allowed to mate. The colony was maintained, whenever possible, by sibling mating, using $tsk+/+pa$ males and homozygous $pa+/+pa$ females. In all studies the pallid $pa+/+pa$ mice served as controls for the $tsk+/+pa$ mice. Male and female pa and tsk mice 2, 4, 8, 16, and 24 months old were used in this study.

At all ages body weight of tsk mice was lighter than that of pa mice. This was significant at 8 (-11%), 16 (-21%), and 24 (-22%) months.[3] In tsk mice right ventricular (RV) weight was significantly lighter at 2 months, held similar values as in the pa mice at 4 and 8 months, and was heavier at 16 ($+9\%$, ns) and 24 ($+52\%$, $p <0.001$) months.[3] In the mouse, as in a wide range of mammalian species, heart mass is directly proportional to adult body mass.[4] Thus, it would be expected that the RV mass of the tsk mice from the eighth month of age onward should have been significantly lighter than that of the pa mice. As this was not the case, it could mean that in the tsk mice RVH first started to develop between 8 and 16 months of age.

[a] This work was supported in part by the Special Trustees for Nottingham University Hospitals and by Collaborative Research Grant No. 0692/88 from NATO.

[c] Address for correspondence: Dr. P. A. Martorana, Cassella, AG, Department of Pharmacology, Hanauer Landstr. 526, 6000 Frankfurt/M. 61, FRG.

The ratio of RV/body weight was significantly greater in the *tsk* mice at 8 (+ 15%), 16 (+ 39%), and 24 (+ 96%) months. Similarly, the ratio of right ventricle/ left ventricle plus septum increased in the *tsk* mice from the eighth month onward, reaching levels of statistical significance at 16 (+ 37%) and 24 (+ 60%) months of age.[3] Additionally, in the *tsk* mice RV myocyte cross-sectional area, assessed on micrographs, was greater than that in the *pa* mice at 16 (+ 8%, ns) and 24 (+ 36%, $p < 0.005$) months of age (FIG. 1).

At 24 months of age, that is, at the time of the maximal increase in myocyte cross-sectional area in the *tsk* mice, capillary density was similar in both groups (capillaries/mm²; *pa*: 2602 ± 142; *tsk*: 2777 ± 82; [M ± SEM]). In hypertrophy the increase in the lateral dimension of the myocytes results in a lateral spreading of the adjacent capillaries and consequently in a decrease in their numerical density per unit area (capillary density). Consequently, the finding of unchanged capillary density in hypertrophic hearts indicates that capillary proliferation had

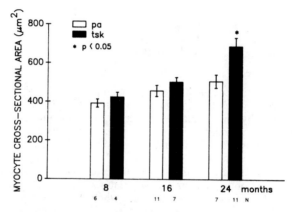

FIGURE 1. Myocyte cross-sectional area in the right ventricle of 24-month-old pallid (*pa*) and tight-skin (*tsk*) mice.

occurred. Similarly, capillary proliferation was recently reported in adult rats with long-term pressure-induced RVH.[5] Alternatively, the myocytes, in the process of hypertrophy, could have grown around the capillaries without altering the geometry of the capillary network and consequently the numerical capillary density. In fact, the appearance of capillaries internalized in a deep cleft of the membrane of myocytes was seen relatively often in the right ventricle of *tsk* mice at 24 months (FIG. 2). Similar "tunnel capillaries," which have previously been reported,[6,7] are considered to be one of the adaptations in cardiac hypertrophy and to contribute to the maintenance of the oxygen diffusion distance. In fact, in the right ventricle of the *tsk* mice, neither areas of myocyte loss nor replacement fibrosis were seen.

These results show the development of RVH in adult to senescent *tsk* mice with genetic emphysema. Because RVH in human emphysema is a slowly evolving process that develops late in life, the *tsk* mouse can be considered a unique model of the human condition.

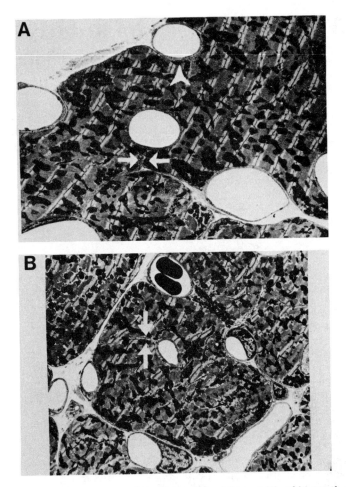

FIGURE 2. Electron micrographs of myocytes of the right ventricle of 24-month-old tight-skin mice. **(A)** A capillary in a deep cleft (*between the two arrows*) of a myocyte; a second capillary (*arrowhead*) appears to be in the process of being invaginated into the same myocyte. **(B)** A capillary located in the center of a myocyte; the *two arrows* indicate the membrane cleavage.

REFERENCES

1. SZAPIEL, R. H., J. D. FULMER, G. W. HUNNIGHAKE, N. A. ELSON, O. KAWANAMI, V. J. FERRANS & R. G. CRYSTAL. 1981. Am. Rev. Respir. Dis. **123:** 680–685.
2. MARTORANA, P. A., P. VAN EVEN, C. GARDI & G. LUNGARELLA. 1989. Am. Rev. Respir. Dis. **139:** 226–232.
3. MARTORANA, P. A., M. WILKINSON, P. VAN EVEN & G. LUNGARELLA. 1990. Am. Rev. Respir. Dis. **142:** 333–337.
4. HOU, P.-C. L. & W. W. BURGGREN. 1989. Respir. Physiol. **78:** 265–280.
5. OLIVETTI, G., C. LAGRASTA, R. RICCI, E. H. SONNENBLICK, J. M. CAPASSO, & P. ANVERSA. 1989. Am. J. Physiol. (Heart Circ. Physiol. 26) **257:** H1766–H1772.
6. IMAMURA, K. 1978. Jap. Circ. J. **42:** 979–1002.
7. IZUMI, T., M. YAMAZOE & A. SHIBATA. 1984. J. Mol. Cell. Cardiol. **16:** 449–457.

The *in Vitro* and *in Vivo* Recovery of PMN Elastase:Beta-Lactam Inhibitor Complexes by Cation Exchange Chromatography

E. E. OPAS, P. S. DELLEA, D. S. FLETCHER, P. DAVIES, AND J. L. HUMES

Department of Inflammation Research
Merck Sharp & Dohme Research Labs
Rahway, New Jersey 07065

The recovery from biologic fluids of synthetic inhibitor:polymorphonuclear (PMN) elastase complexes would be a useful marker of the biochemical efficacy of the drug in man. The cephalosporin-based L-658,758 (3-acetoxymethyl-7α-methoxy-8-oxo-5-thia-1-azabicyclo-[4.2.0]-oct-2-ene-2-(2-(S)-carboxypyrrolidine-carboxamide)5,5-dioxide) is a mechanism-based, acylating inhibitor of human PMN elastase. With cation-exchange chromatography, L-658,758:PMN elastase complexes can be concentrated and isolated from biologic fluids. Preformed synthetic inhibitor:PMN elastase complexes were purified on CM-52 (Whatman) cation exchange columns after *in vitro* incubation in human bronchoalveolar lavage fluid. Free drug and greater than 90% of endogenous protein (including α_1-proteinase inhibitor:elastase and α_2-macroglobulin:elastase complexes) were eluted from the cationic exchange resin with low salt buffer (50 mM potassium phosphate, 0.075% Tween-20, pH 7.4). Elastase and elastase complexed with synthetic inhibitor L-658,758 were subsequently eluted with high salt (1 M NaCl, 50 mM potassium phosphate, 0.075% Tween-20, pH 7.4). When radiolabeled drug was used, quantitation of synthetic complexes was performed by the determination of the radioactivity in the high salt fractions from the column. The intratracheal administration of human PMN elastase into the hamster results in lung hemorrhage[1] which can be inhibited by the prior instillation of ^{14}C-L-658,758. Radiolabeled complexes can be recovered after their formation *in vivo* in the hamster lung under these conditions.

The quantitation of nonradiolabeled synthetic inhibitor:PMN elastase complexes would enable the clinical assessment of biochemical efficacy. The key to these analyses are threefold: (1) the isolation and concentration of drug:enzyme complexes from biologic fluids by ion-exchange chromatography; (2) the time and temperature-dependent dissociation of synthetic inhibitor from elastase yielding active enzyme (FIG. 1); and (3) the quantitation of reactivation of elastase dissociated from nonradiolabeled drug compared to the reactivation of enzyme from preformed radiolabeled complexes. The sensitivity of the assay is such that with 5 hours of reactivation at 47°C, as little as 0.1 pmol per sample can be detected (TABLE 1). Complexes of nonradiolabeled synthetic inhibitor and PMN elastase formed *in vivo* in the hamster hemorrhage model were isolated and quantitated successfully in this manner.

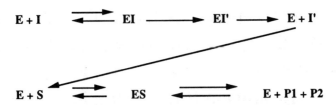

FIGURE 1. Quantitation of nonradiolabeled synthetic inhibitor:PMN elastase complexes. E (PMN elastase); I (synthetic inhibitor); I' (altered inhibitor); S (MeOSuc-Ala-Ala-Pro-Val-para-nitroaniline); P1 (MeOSuc-AAPV); and P2 (p-NA). The synthetic inhibitor is altered upon binding to the enzyme and cannot form complexes after its dissociation. The activity of the dissociated, viable enzyme can be measured by substrate utilization as monitored by absorbance at 405 nm.

TABLE 1. Subnanomolar Quantitation of L-658,758:PMN Elastase Complexes by Measuring Enzyme Reactivation

Time (min)	Complex[a]		
	4.1 nM	1.4 nM	0.4 nM
60	.083 ± .009	.016 ± .007	—
131	.230 ± .011	.055 ± .010	.008 ± .012
203	.384 ± .014	.100 ± .011	.021 ± .016
271	.547 ± .013	.154 ± .015	.033 ± .012
312	.644 ± .005	.180 ± .012	.038 ± .010

[a] Preformed complexes of ^{14}C-L-658,758:PMN elastase were purified by CM-52 chromatography as described, diluted serially into a 96-well plate, and incubated with substrate under the following conditions: 1 M NaCl, 50 mM potassium phosphate, pH 7.4, MeOSuc-AAPV-pNA [1mM], 20% DMSO, 47°C, final volume 0.25 ml. Absorbance at 405 nm was measured in a Flow Instrument Titertek plate reader. Data are expressed as mean determinations of eight replicates ± deviation.

REFERENCE

1. Fletcher, D. S., D. G. Osinga, K. M. Hand, P. S. Dellea, B. M. Ashe, R. A. Mumford, P. Davies, W. Hagmann, P. E. Finke, J. B. Doherty & R. J. Bonney. 1990. A comparison of α_1-proteinase inhibitor, methoxysuccinyl-ala-ala-pro-val-chloromethylketone, and specific β-lactam inhibitors in an acute model of human polymorphonuclear leukocyte elastase-induced lung hemorrhage in the hamster. Am. Rev. Respir. Dis. **141:** 672–677.

Lipopolysaccharide-Induced Alveolar Wall Destruction in the Hamster Is Inhibited by Intratracheal Treatment with r-Secretory Leukocyte Protease Inhibitor

J. STOLK,[a] A. RUDOLPHUS, AND J. A. KRAMPS

Department of Pulmonology (C3-P)
University Hospital
Leiden, The Netherlands

Pulmonary emphysema in man is characterized by destruction of alveolar lung septa. The destruction is most likely caused by elastase from polymorphonuclear leukocytes (PMN). Pulmonary emphysema in animals can be induced by a single intratracheal instillation of a large quantity of elastase.[1] This exogenous high proteinase load causes alveolar destruction within days. This animal model does not relate to the natural development of emphysema in man, in whom tissue destruction by pertubation of the endogenous elastase/elastase inhibitor balance takes many years. It was observed that repeated intratracheal instillations of lipopolysaccharide (LPS) in the hamster cause pulmonary emphysema.[2] We investigated whether LPS instillations cause PMN recruitment in the airway lumen and studied the effect of instilled recombinant secretory leukocyte protease inhibitor (r-SLPI) on LPS-induced emphysema.

METHODS

Syrian hamsters received a single instillation of 0.5 mg *Escherichia coli* LPS and were sacrificed at different time points. The lungs were lavaged with 2.5 ml of saline solution. BAL fluid and cells were separated by centrifugation and assayed for myeloperoxidase (MPO), elastase, and LDH activity. Isolated hamster PMNs were incubated in Hanks medium containing different concentrations of LPS (0.5–500 μg/ml). Samples were assayed for elastase activity with synthetic substrate. Two different groups of hamsters were treated for 4 weeks, twice a week, with intratracheal instillations of 0.5 mg LPS in 200 μl saline solution ($n = 8$) or 200 μl saline solution ($n = 4$). A third group ($n = 8$) received 0.5 mg LPS mixed with 0.5 mg r-SLPI (gift from Synergen Inc., Boulder Colorado) in 200 μl of saline solution, followed by a dose of 0.5 mg r-SLPI in 200 μl of saline solution 7 hours later. This treatment was given twice a week for 4 weeks. Mean linear intercept (L_m) in lung tissue sections was measured 4 weeks after the last treatment.

---

[a] Address for correspondence: Jan Stolk, Department of Pulmonology (C3-P), University Hospital, NL 9300 RC Leiden, The Netherlands.

RESULTS

MPO activity in cell pellets of BAL was measured as a marker of PMN recruitment by LPS. A single dose of LPS resulted in two peaks, one around 10 hours, another after 50 hours. Elastase activity in BAL supernatants after a single LPS dose showed a biphasic increase in activity, with peaks at 8 and 48 hours (FIG. 1). In the same samples, LDH activity in BAL supernatants, expressed as a percentage of activity in cell pellets, did not exceed 6% (FIG. 1), indicating that released elastase activity was not caused by cell lysis. Stimulation of isolated hamster PMNs with LPS concentrations up to 500 µg did not result in elastase release or cell lysis. Measurements of emphysema (L_m) in hamsters treated with

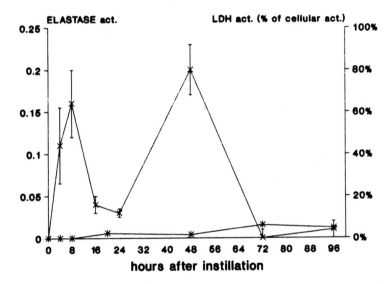

FIGURE 1. Elastase activity (x—x) and LDH activity (*—*) in BAL supernatant after a single instillation of 0.5 mg *E. coli* LPS. For details see text.

repeated instillations of LPS or saline solution revealed values of 89.3 ± 1.5 µm and 69.2 ± 1.0 µm, respectively (mean ± SEM). In the LPS/r-SLPI group the L_m was 82.6 ± 1.1 µm (FIG. 2). The effect of r-SLPI resulted in 33% inhibition of LPS-induced alveolar wall destruction ($p = 0.015$).

DISCUSSION

Intratracheal instillation of LPS is chemotactic for PMNs, leading to a biphasic influx of these cells in lung alveoli. An endogenous factor is most likely responsible for the release of elastase from hamster PMNs. The observation that a potent elastase inhibitor, like r-SLPI, is able to inhibit partly the development of

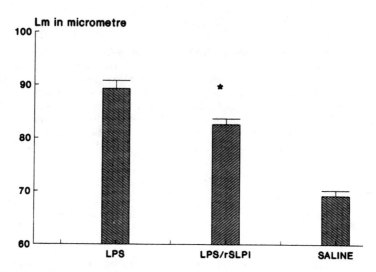

FIGURE 2. Inhibition of *E. coli* LPS-induced emphysema, as measured by mean linear intercept (L_m), by recombinant secretory leukocyte protease inhibitor (r-SLPI). For details see text. *r-SLPI significantly inhibited LPS-induced emphysema ($p = 0.01$).

emphysema at the concentration we used suggests that elastase is involved in alveolar destruction. We conclude that LPS-induced emphysema is probably caused, at least partly, by endogenous elastase released by recruited PMNs.

REFERENCES

1. SNIDER, G. L., E. C. LUCEY & P. J. STONE. 1986. Animal models of emphysema. Am. Rev. Respir. Dis. **133:** 149–169.
2. RUDOLPHUS, A., D OSINGA, J. STOLK, D. FLETCHER, K. KEENAN & J. A. KRAMPS. 1990. Induction of emphysema and secretory cell metaplasia (SCM) by intratracheal instillations of *E. coli* lipopolysaccharide (LPS) in hamsters. A new animal model. Am. Rev. Respir. Dis. **141:** A736.

Neutrophil Traffic Through the Lungs in Man[a]

C. SELBY,[b] E. DROST, P. K. WRAITH, AND W. MacNEE

Unit of Respiratory Medicine
Rayne Laboratory
City Hospital
Edinburgh EH10 5SB, Scotland, UK

Neutrophils (PMNs) are clearly implicated in the pathogenesis of pulmonary emphysema[1] because they are the major source within the lungs of the proteolytic enzyme elastase. Clearly, knowledge of how these cells pass through the lungs is important in our understanding not only of this condition, but also of others in which PMN-mediated lung injury occurs.

Using a τ camera-computer, we studied the passage of reinjected ^{111}Indium PMNs across the central circulation compared to that of simultaneously injected ^{99m}Technetium erythrocytes (RBCs). In normal elderly individuals ($n = 7$) and patients with chronic obstructive pulmonary disease (COPD, $n = 14$), 12 ± 3% PMNs are retained in their first passage through the lungs (FPR) *relative to the transit of RBCs*. In patients with COPD studied early during an acute exacerbation ($n = 7$) of their disease, FPR was higher (22 ± 14%, $p = 0.01$).

We believe FPR is influenced by the geometric constraints imposed upon PMNs[2] (mean diameter 7 μm) when passing through the pulmonary microcirculation[3] (mean capillary segment diameter 5 μm). Hence we measured *cell deformability* as the *in vitro* "cell filterability": the plateau pressure generated when 10^5 cells · ml^{-1} pass at constant flow through a micropore membrane whose pore dimensions mimic those of the pulmonary capillary segments. Deformability measured in this way correlates significantly with FPR ($r = 0.83$, $p < 0.01$).

Over the subsequent 30 minutes, PMNs wash out from the lungs exponentially. Regional sequestration of PMNs in the lungs at 5 and 10 minutes after reinjection in normal elderly individuals is related to local RBC transit time as a measure of blood velocity ($r = 0.87$, $p < 0.001$), a relation previously described in young healthy subjects.[4] This inverse relation to blood velocity suggests cell *margination* within small blood vessels.[5] In patients with COPD this relation is lost. Moreover, PMNs wash out slower from the lungs of patients with acute COPD than of patients with stable COPD despite shorter RBC transit times and therefore faster blood velocity which might be expected to propel more PMNs through the pulmonary vasculature. Such enhanced PMN retention in acute COPD may be associated with further lung damage. Preliminary studies suggest that this increased PMN burden within the lungs during acute COPD may be due to a temporary change in PMN rheology.

Further studies of the factors that influence PMN transit through and retention within the pulmonary vasculature will help to determine the role of PMNs in acute and chronic lung injury.

[a] This work was supported by the Norman Salvesen Emphysema Research Trust.
[b] To whom correspondence should be addressed.

353

REFERENCES

1. NIEWOEHNER, D. E., 1988. J. Lab. Clin. Med. **111:** 15–27.
2. SCHMID-SCHONBEIN, G. W., Y. SHIH & S. CHEIN. 1980. Blood **56:** 566–875.
3. WEIBEL, E. R. 1963. Morphometry of the Human Lung. Academic Press. New York.
4. MACNEE, W., B. WIGGS, A. S. BELZBERG & J. C. HOGG. 1989. N. Engl. J. Med. **321:** 924–928.
5. SCHMID-SCHONBEIN, G. W., S. USAMI, R. SKALAK & S. CHEIN. 1980. Microvasc. Res. **19:** 45–70.

Isolation of Urinary Desmosine by HPLC, Amino Acid Analysis, and Quantification by Isotope Dilution

P. J. STONE, E. C. LUCEY, J. BRYAN-RHADFI,
G. L. SNIDER, AND C. FRANZBLAU

Department of Biochemistry
Boston University School of Medicine
Pulmonary Section
Boston V.A. Medical Center
Boston, Massachusetts 02118–2394

Urinary desmosine (DES) has usually been measured by radioimmunoassay or enzyme-linked immunosorbent assay. Reports in the literature suggest that components in the urine may erroneously elevate values of urinary DES measured using these methods.[1,2] We have employed prefractionation by Sephadex G-15 chromatography of the hydrolyzed urine sample to remove contaminants that interfere with the measurement of DES. Desmosine and isodesmosine (IDES) are separately measured using high performance liquid chromatography. An isotope dilution technique is used to overcome the difficulty of variable losses during prefractionation; before processing, urine samples are spiked with ^{14}C-DES.

Groups of six hamsters were anesthetized and received intratracheal instillations of 0.5 ml of saline solution containing 300 µg porcine pancreatic elastase (PPE) or 300 µg human neutrophil elastase (HNE). Another group of six hamsters were untreated. Collection of urine was initiated immediately after treatment and pooled for each treatment group. At the end of the 3-day collection period, hamsters were anesthetized and studied by lavaging the lungs three times with heparin-saline solution to remove exudate. Lungs were injected with 5 ml of fixative, excised, and degased. Lung volume displacement and mean linear intercept (MLI), a measurement of airspace size that increases in emphysema, were measured as previously described.[3]

Before analysis the urine was centrifuged at 30,000 × g for 15 minutes to remove food material and other particulates. Six hamster-days of urine were pooled, a known amount of ^{14}C-DES was added, and the sample was stored at −20°C. For analysis the spiked urine was combined with an equal volume of 12 N HCl and the sample refluxed at 110°C under nitrogen for 24 hours.

Human urine was collected for 24 hours in the presence of 0.02% sodium azide at 2°C from male laboratory workers aged 25–50 years who had never smoked. Informed consent was obtained. Creatinine was measured. Aliquots representing 10% by volume of the 24-hour pool were stored at −20°C after the addition of a known amount of ^{14}C-DES. For analysis the sample was reduced in volume with a rotary evaporator under reduced pressure, brought up to 2 ml with water, combined with 20 ml of 12 N HCl, and hydrolyzed as just described.

Hydrolyzed urine samples were dried under a stream of nitrogen gas, brought up in 10 ml of 1% acetic acid, and loaded on disposable columns (Biorad) packed with Sephadex G-15 in 1% acetic acid. The early eluting fractions containing ^{14}C radioactivity were collected, reduced to a volume of 1–2 ml with a stream of nitrogen, and loaded on a 2.6 × 100 cm column (Pharmacia) packed with Sephadex G-15 in 1% acetic acid and run at room temperature. Eluted fractions were assessed for absorbance at 280 nm and for radioactivity. Desmosine eluted with a K_{av} of around 0.26 ahead of most of the 280 nm of absorbing material. The column was flushed with 1% acetic acid until the effluent had no measurable absorbance at 280 nm as compared with 1% acetic acid. In early experiments, column fractions containing ^{14}C radioactivity were pooled; 50% of the pool was analyzed by amino acid analysis and the remainder by HPLC.

The paired-ion C_{18} reversed phase HPLC procedure used by Black and co-workers[4] for pyridinoline and deoxypyridinoline, lysine-derived collagen cross-links, was modified; the gradient was initiated at 0% solvent B. Isodesmosine eluted at 23.1% B, 0.7 minutes before DES, which eluted at 23.8% B. Overall recovery of ^{14}C-DES was approximately 35%. When samples were spiked with both ^{14}C-DES and ^{14}C-IDES, recoveries of ^{14}C-DES and ^{14}C-IDES were not different. After these initial studies had validated the HPLC method for quantification of DES and IDES, individual column fractions were assessed by HPLC.

The DES values for pooled urine from hamsters receiving HNE (300 μg) or PPE (300 μg) were 3 and 11 times those of untreated hamsters, respectively (TABLE 1). Mean linear intercept values increased proportionate to the elevation of urinary DES, giving a correlation between increase in airspace size and elevation of urinary DES.

TABLE I. Daily DES and IDES Excretion Values for Treated Hamsters (mean ± SEM, no. of determinations on pooled urine)

Treatment	DES (μg)	IDES (μg)	MLI (μm)
None	0.074 ± 0.008 (8)	0.087 ± 0.005 (8)	54 ± 2 (9)
HNE	0.212 ± 0.012 (2)	0.245 ± 0.019 (2)	64 ± 2 (6)
PPE	0.816 ± 0.005 (2)	0.826 ± 0.072 (2)	86 ± 5 (6)

For four normal men aged 25–49 years who had never smoked, mean ± SEM 24-hour urinary values were 8.8 ± 0.8 μg DES per day and 8.1 ± 0.4 μg IDES per day. Values for individuals did not change when measured on different occasions. These measured values are much lower than those previously reported[5,6] for both human and hamster urine specimens. The values we have obtained for hamsters seem more compatible with literature turnover estimates of elastin in aorta and lung elastin from small laboratory animals.[7,8] This assay should prove useful in studying conditions in which DES excretion is elevated.

REFERENCES

1. GUNGA-SMITH, Z. 1985. Anal. Biochem. **147:** 258–264.
2. LAURENT, P., L. MAGNE, J. DE PALMAS, J. BIGNON & M.-C. JAURAND. 1988. J. Immunol. Methods **107:** 1–11.
3. LUCEY, E. C., P. J. STONE, D. E. CICCOLELLA, R. BREUER, T. G. CHRISTENSEN, R. C. THOMPSON & G. L. SNIDER. 1990. J. Lab. Clin. Med. **115:** 224–232.
4. BLACK, D., A. DUNCAN & S. P. ROBINS. 1988. Anal. Biochem. **169:** 197–203.

5. HAREL, S., A. JANOFF, S. Y. YU, A. HUREWITZ & E. H. BERGOFSKY. 1980. Am. Rev. Respir. Dis. **122:** 769–773.
6. KUHN, C., W. ENGLEMAN, M. CHRAPLYVY & B. C. STARCHER. 1983. Exp. Lung Res. **5:** 115–123.
7. LEFEVRE, M. & R. B. RUCKER. 1980. Biochim. Biophys. Acta **630:** 519–529.
8. RUCKER, R. B. & M. A. DUBICK. 1984. Environ. Health Prespect. **55:** 179–191.

The Blotchy Mouse, Lung Desmosine, and Emphysema

T. D. TETLEY, G. J. PHILLIPS, A. GUZ, AND B. FOX

Departments of Medicine and Histopathology
Charing Cross & Westminster Medical School
London, UK

Lysyl oxidase cross-links tropoelastin to form mature insoluble elastin. In animals, inhibition of lysyl oxidase results in pulmonary emphysema. In the blotchy (Blo) mouse model of emphysema, inheritance of the blotchy allele at the mottled locus of the X chromosome results in a deficiency of this enzyme and pulmonary emphysema. Low desmosine levels might be expected in affected tissues. However, although Starcher et al.[1] showed low lung desmosine in 11-day-old Blo compared to wild-type (WT) mice, Rowe et al.[2] showed normal or slightly elevated lung desmosine in adult Blo compared to WT mice.

We measured total lung desmosine (radioimmunoassay) and hydroxyproline (Hypro, i.e., collagen) levels in 3–12-week-old mice (TABLE 1); inheritance of the Blo allele resulted in enlarged airspaces.

TABLE 1. Results of Lung Desmosine and Hydroxyproline Measurements

	3-Week		6-Week		12-Week	
	WT	Blo	WT	Blo	WT	Blo
Des (nmol)	6.4 ± 0.8	4.9 ± 0.6[a]	7.5 ± 1.2	7.5 ± 1.7	7.8 ± 1.2	8.7 ± 1.4
Hypro (mg × 10)	1.7 ± 0.4	1.7 ± 0.5	3.8 ± 0.3	3.0 ± 0.8	3.1 ± 0.6	3.0 ± 1.0

[a] < 3-week WT, $p < 0.05$; mean ± SD.

Thus, insufficient lysyl oxidase activity at times of rapid lung growth or repair may lead to abnormal tissue structure and emphysema. In man, reduced lysyl oxidase activity may contribute to cigarette smoke-induced emphysema as a result of destruction of connective tissue followed by rapid repair.

REFERENCES

1. STARCHER, B. C., J. A. MADRAS & A. S. TEPPER. 1977. BBRC **78:** 706–712.
2. ROWE, D. W., E. B. McGOODWIN, G. R. MARTIN, M. D. SUSSMAN, D. GRAHN, B. FARIS & C. FRANZBLAU. 1974. J. Exp. Med. **139:** 180–192.

Ten-Year Changes of Protease Inhibitors in the Sons of Patients with COPD

A. YOSHIOKA, H. NAGATA, M. YAMAMOTO,
Y. AKIYAMA, M. NISHIMURA, K. MIYAMOTO, F. KISHI,
AND Y. KAWAKAMI

First Department of Medicine
Hokkaido University School of Medicine
Sapporo, Japan

Familial aggregation has been reported in the development of chronic obstructive pulmonary disease (COPD, pulmonary emphysema, and/or chronic bronchitis) even without α_1-antitrypsin deficiency.[1-3] Thus, a search has been made for the endogenous factors causing and modifying COPD. However, evidence is accumulating, suggesting that protease inhibitors other than α_1-antitrypsin may be involved in the pathogenesis of COPD.[4] We therefore measured four kinds of plasma protease inhibitors including α_1-antitrypsin along with pulmonary function variables in the sons of patients with COPD and controls with an interval of 10 years. We investigated if the aging process and smoking habits have any effect on plasma concentration in those protease inhibitors.

MATERIAL AND METHODS

Subjects included 16 sons of patients with COPD and 24 controls. They were all healthy, with no signs of respiratory disease. The diagnosis of COPD was made on the basis of history, physical examination, pulmonary function tests, and roentgenographic findings. We conducted the initial study in 1978–1981 and the present follow-up study in 1988–1990. The subjects were all free from acute inflammatory disease as proven by clinical findings and blood tests at the time of the studies. On both occasions, using a single radial immunodiffusion method, we measured plasma concentrations of protease inhibitors, including α_1-antitrypsin, α_2-plasmin inhibitor, antithrombin III, and α_2-macroglobulin. At the same time, we performed respiratory function tests in all subjects.

RESULTS

The results are presented in TABLES 1 and 2. The plasma concentration of α_1-antitrypsin decreased from 192.3 ± 9.4 (SE) mg/dl in the initial study to 149.4 ± 7.2 mg/dl in the present study in control subjects (p <0.01) and from 217.5 ± 10.0 mg/dl to 146.7 ± 5.9 mg/dl in the sons of patients with COPD (p <0.01). The annual decrease in plasma concentration of α_1-antitrypsin was larger in the sons of patients with COPD (-6.9 ± 0.6 mg/dl per year) than in the controls (-3.7 ± 1.0 mg/dl per year, p <0.05). The plasma concentration in α_2-plasmin inhibitor significantly increased from 5.8 ± 0.3 mg/dl initially to 6.6 ± 0.2 mg/dl finally in controls (p <0.01), whereas it was significantly decreased in the sons of patients with

359

TABLE 1. Plasma Protease Inhibitors at Initial and Present Studies in Two Groups (Means ± SE)

	Initial Study	Present Study	t Test
(a) Control Subjects ($n = 24$)			
α_1-Antitrypsin (mg/dl)	192.3 ± 9.4	149.4 ± 7.2	p <0.01
α_2-Plasmin inhibitor (mg/dl)	5.8 ± 0.3	6.6 ± 0.2	p <0.01
Antithrombin III (mg/dl)	29.5 ± 0.8	27.3 ± 0.4	p <0.05
α_2-Macroglobulin (mg/dl)	146.0 ± 6.1	180.0 ± 10.9	p <0.05
(b) Sons of Patients with COPD ($n = 16$)			
α_1-Antitrypsin (mg/dl)	217.5 ± 10.0	146.7 ± 5.9	p <0.01
α_2-Plasmin inhibitor (mg/dl)	6.4 ± 0.2	5.5 ± 0.3	p <0.01
Antithrombin III (mg/dl)	33.1 ± 0.9	26.4 ± 0.7	p <0.01
a_2-Macroglobulin (mg/dl)	172.4 ± 12.1	176.6 ± 10.2	NS

COPD from 6.4 ± 0.2 mg/dl to 5.5 ± 0.3 mg/dl (p <0.01). The plasma antithrombin III concentration decreased from 29.5 ± 0.8 mg/dl to 27.3 ± 0.4 mg/dl in controls (p <0.05), and from 33.1 ± 0.9 mg/dl to 26.4 ± 0.7 mg/dl in the sons of patients with COPD (p <0.01). The annual decrease in antithrombin III was greater in the sons of patients with COPD ($-0.7 ± 0.1$ mg/dl per year) than in controls ($-0.2 ± 0.1$ mg/dl per year, p <0.01). The plasma concentration of α_2-macroglobulin increased from 146.0 ± 6.1 mg/dl to 180.0 ± 10.9 mg/dl in controls, but no changes were noted in the sons of patients with COPD. The annual change in α_2-macroglobulin was not different in the two groups. Differences in smoking habits in two groups were not significant. The annual changes in protease inhibitors had no correlation with any changes in smoking index, body weight, plasma total protein, plasma albumin, and respiratory function variables.

DISCUSSION

In this study, we intended to search for endogenous factors causing or modifying COPD. We found that plasma concentrations of α_1-antitrypsin, α_2-plasmin inhibitor, and antithrombin III decreased more during a 10-year period of aging in the sons of patients with COPD than in controls. Inasmuch as the sons of patients with COPD are more likely than controls to have the disease in the future, the

TABLE 2. Annual Changes in Plasma Protease Inhibitors in Two Groups (Means ± SE)

	Control Subjects ($n = 24$)	Sons of Patients with COPD ($n = 16$)	t Test
α_1-Antitrypsin (mg/dl/yr)	−3.7 ± 1.0	−6.9 ± 0.6	p <0.05
α_2-Plasmin inhibitor (mg/dl/yr)	0.09 ± 0.02	−0.10 ± 0.03	p <0.01
Antithrombin III (mg/dl/yr)	−0.2 ± 0.1	−0.7 ± 0.1	p <0.01
α_2-Macroglobulin (mg/dl/yr)	3.4 ± 1.3	0.5 ± 0.7	NS

characteristics we found in the sons of patients with COPD may have some relevance in the pathogenesis of COPD.

Although there is no direct evidence that a deficiency in protease inhibitors other than α_1-antitrypsin causes COPD, α_2-plasmin inhibitor or antithrombin III is known to inactivate not only its specific enzyme but also other proteases such as trypsin and chymotrypsin,[5] thus potentially inactivating human neutrophil elastase in the lung. Therefore, accelerating decreases in plasma concentrations of α_1-antitrypsin, α_2-plasmin inhibitor, and antithrombin III may jointly contribute to the future development of COPD.

REFERENCES

1. WIMPFHEIMER, F. & L. SCHNEIDER. 1961. Am. Rev. Respir. Dis. **83:** 697–703.
2. HOLE, B. V. & K. WASSERMAN. 1965. Ann. Intern. Med. **63:** 1009–1017.
3. READ, J. & T. SELBY. 1961. Br. Med. J. **2:** 1104–1108.
4. MOOREN, H. W. D., J. A. KRAMPS., C. FRANKEN., C. J. L. M. MEIJER & J. H. DIJKMAN. 1983. Thorax. **38:** 180–183.
5. TRAVIS, J. & G. S. SALVESEN. 1983. Ann. Rev. Biochem. **52:** 655–709.

Early Diagnosis of Pulmonary Emphysema in Smokers

JAIME W. FIDALGO-GARRIDO[a] AND
JOSÉ L. MARTINEZ-CARRASCO

Hospital La Paz
Universidad Autónoma Madrid
Centro Salud Fuencarral
Madrid, Spain

Pulmonary emphysema is considered one of the most prevalent diseases associated with smoking. Furthermore, its chronic course, irreversibility, and severe impairment that it causes are significant. The evolution of emphysema is outstandingly silent, slow, and progressive, underlining the importance of susceptibility in smokers; only a relatively low proportion of smokers (10–15%) develop incapacitating emphysema; usually a long asymptomatic period precedes the onset of symptoms. We must try to diagnose the disease by screening asymptomatic smokers before the condition becomes fatally irreversible. Little is known about when and in whom this screening should be performed; we cannot study all smokers all the years. The principal aim of our experimental study is to identify the groups at risk in which this screening would prove efficacious.

Patients were selected from a primary care setting of an urban area in Madrid. All 115 adults, which included 6 women, were heavy smokers with a total consumption greater than 15 pack-years. Patients were not randomized. Evaluation included medical history, physical examination, routine blood tests, chest radiograph, and EKG. We classified the smokers by their basal spirometry, using FEV_1 as a measure of pulmonary obstruction. The EKG and chest radiograph criteria for pulmonary emphysema used were those previously described by Sylvester and Thurlbeck, respectively. Basal spirometry was performed using a Vitalograph spirometer, with $FEV_1 \geq 80\%$ defined as normal pulmonary function. All parameters were carefully evaluated by two examiners trying to unify the different scales and criteria used.

The population studied ranged in age from 33 to 85 (mean 55.8 ± 10.5) with a peak between 50 and 70 years. The distribution by cigarette consumption, ranging from 15 to 84 pack-years (mean 37.0 ± 15.0), was quite homogeneous. Seventy-four smokers were found to have normal pulmonary function (mean 98.5 ± 12.3). The other 41 revealed airway obstruction ($FEV_1 < 80\%$) with a mean of 54.9 ± 18.1.

When we studied the relation between age and pulmonary obstruction, we found that the disease is usually diagnosed in patients older than 50 ($p < 0.05$) or 55 ($p < 0.001$), both years being statistically significant (TABLE 1). Basal spirometry (FEV_1 index) apparently is not useful in the diagnosis of the disease in patients younger than 40 years of age. Moreover, we found a positive relation be-

---

[a] Deceased.

TABLE 1. Relationship between Age and Pulmonary Obstruction[a]

Age (yr)	Obstruction	No Obstruction	Total
≤49	5 (17.2%)	24	29
50–54	3 (15%)	17	20
55–59	9 (47.4%)	10	19
60–64	10 (43.5%)	13	23
≥65	14 (58.3%)	10	24

[a] Number of smokers with obstruction, <55 years of age, 8 (16.3%); number of smokers with obstruction, ≥55 years, 33 (50%); $p < 0.001$. Number of smokers with obstruction, <50 years of age, 5 (17.2%); number of smokers with obstruction, ≥50 years, 36 (41.9%); $p < 0.05$.

tween cigarette consumption of over 35 pack-years and pulmonary obstruction ($p < 0.05$).

TABLE 2 demonstrates the differences in some of the variables studied in both groups: those with and those without obstruction. We found statistical significance in all variables, except hemoglobin. It is important to stress that both the EKG and the chest radiogram are useful in predicting pulmonary obstruction and emphysema, but not in diagnosing.

Early diagnosis of pulmonary emphysema should be considered one of the priorities in preventive (primary and secondary prevention–early diagnosis) programs in primary care. The physician must have adequate knowledge of the natural history of the disease and access to reliable pulmonary function testing. Spirometry permits airway obstruction to be detected in the initial stages, before severe and irreversible functional impairment has developed. Basal spirometry testing has almost no risk, can be easily performed in the physician's office, is economical and reliable, and can be repeated and contrasted over time, forecasting the changes in pulmonary function. Despite these features, it is still used little in generalized primary care where in actuality it ought to be almost essential.

In conclusion, we have studied in our Center the problem of early diagnosis of pulmonary emphysema in smokers, and especially the factors, such as age and cigarette consumption, that most determine the groups at risk. Age seems to be the most important factor: the group between 50 and 60 years old shows the higher

TABLE 2. COPD Protocol: Obstruction Analysis

	Obstruction: FEV$_1$ ≤79	No Obstruction: FEV$_1$ ≥80	p
Number	41	74	
FEV$_1$	54.9 ± 18.1	98.5 ± 12.3	
Age (yr)	60.4 ± 8.9	53.3 ± 10.5	<0.01
Smoking consumption	41.9 ± 16.1	34.3 ± 13.6	<0.01
Chronic bronchitis	33	38	<0.01
Dyspnea	13	2	<0.001
Hemoglobin (♂)	16.1 ± 1.6	16.5 ± 1.2	NS
Obstruction on EKG	19	7	<0.001
Radiologic hyperinflation	11	5	<0.01

likelihood of positive results from mass screening. Our results permit screening (by basal spirometry) of all smokers included in any of the following groups at risk: (1) Smokers from 50–60 years old; (2) heavy smokers of more than 40 pack-years; (3) smokers with a history or chest radiograph suggestive of pulmonary emphysema.

When we think about the huge economical and human cost of pulmonary emphysema, early diagnosis of the disease is clearly the only effective way for changing its fatal progressive course.

REFERENCES

1. FLETCHER, C., R. PETO, C. TINKER & F. E. SPEIZER. 1976. The Natural History of Chronic Bronchitis and Emphysema.: 272. Oxford University Press. Oxford, England.
2. BLACK, L. F. 1982. Early diagnosis of chronic obstructive pulmonary disease. Mayo Clin. Proc. **57:** 765–772.
3. Medical Section of the American Lung Association. 1985. Cigarette smoking and health. Am. Rev. Respir. Dis. **132:** 1113–1138.

Rapid Screening for Human Alpha$_1$-Proteinase Inhibitor Deficiency[a]

C. A. LLOYD AND K. D. ARBTAN

Athens Research and Technology, Inc.
Athens, Georgia 30604

We have developed a test to detect individuals who have extreme genetic α_1-proteinase inhibitor (α_1-PI) deficiency, such as the ZZ and null genotypes.[1] The test measures functional α_1-PI in whole blood collected onto Guthrie filter papers, the sample collection material typically used when screening neonates for metabolic abnormalities.

Reagents were purified and standardized as previously described.[2–6] The steps involved in the screening are as follows:

1. Apply the blood specimen to the Guthrie filter paper and allow to air dry for 2 hours at 25°C. Cut ⅛-inch discs from the paper and place them in the wells of a 96-well microtiter plate.

2. Add 50 μl of solution containing 0.05 mol/l of Hepes buffer, pH 7.4, 0.5 mol/l NaCl, 1.0 g/l Brij, and 0.2 mol/l methylamine. Incubate the discs at 25°C for 2 hours to extract the α_1-PI from the filter paper and to inactivate the α_2M.

3. Add 1.44 μg active PPE to each well and incubate for 5 minutes. Add 2 μl Suc-L-Ala-L-Ala-L-Ala-4-trifluoromethylcoumarin-7-amide (33.4 mg/ml in dimethylsulfoxide), and after 30 minutes spot 2 μl of the reaction mixture onto Whatman #1 filter paper, allow to air dry thoroughly, and then examine under long ultraviolet light. α_1-PI-deficient samples will fluoresce.

One hundred-microliter volumes of whole blood collected from one PiMM and two PiZZ individuals (donated by Dr. Jack Pierce, Washington University School of Medicine, St. Louis, Missouri) were transferred randomly into 48 numbered Eppendorf tubes. A technician who was not involved in aliquoting the samples or preparing the sample key performed the assay for functional α_1-PI using discs punched from blood dried onto Guthrie filter paper. The samples deficient in α_1-PI, as indicated in FIGURE 1 (bottom), corresponded to the PiZZ sample placement shown in the key at the top of the figure.

[a] This work is supported by National Institutes of Health Grants 1 R43 HL 40770–01 and 2 R44 40770–02.

MM	MM	MM	MM	MM	MM	MM	ZZ	MM	MM
MM	MM	ZZ	MM	MM	MM	MM	MM	MM	MM
MM	MM	MM	MM	MM	MM	ZZ	MM	MM	MM
MM	MM	MM	MM	MM	MM	MM	MM	MM	MM
MM	MM	MM	MM	MM	MM	MM	ZZ	No Blood	No Blood
								No PPE	No PPE

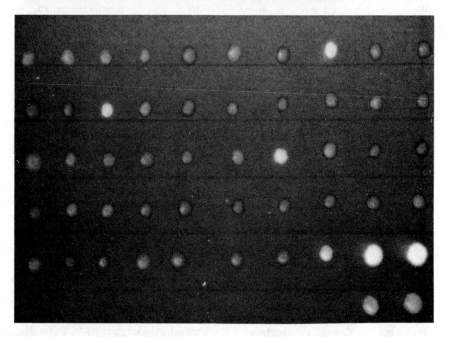

FIGURE 1. (*Top*) Diagram of placement of two blood sample types (denoted MM and ZZ) on Whatman #1 filter paper, and (*bottom*) results of the rapid assay of α_1-PI deficiency. The filter paper was photographed under ultraviolet light to detect fluorescing spots. These spots indicate samples with low concentrations of functional α_1-PI, indicative of the ZZ genotype.

REFERENCES

1. LLOYD, C. & J. TRAVIS. 1989. Clin. Chem. **35:** 1971–1975.
2. PANNELL, R., D. JOHNSON & J. TRAVIS. 1974. Biochem. **13:** 5439–5445.
3. VIRCA, G. D., J. TRAVIS, P. K. HALL & R. C. ROBERTS. 1978. Anal. Biochem. **89:** 274–278.
4. KURECKI, T., L. F. KRESS & M. S. LASKOWSKI. 1979. Anal. Biochem. **99:** 415–420.
5. CHASE, T. & E. SHAW. 1967. Biochem. Biophys. Res. Commun. **29:** 508–514.
6. BIETH, J., B. SPIESS & C. G. WERMUTH. 1974. Biochem. Med. **11:** 350–357.

Index of Contributors

On Wiley, but should it be? 06/05